AA001117

2006 International Conference on Advanced Semiconductor Devices and Microsystems

Smolenice, Slovakia
October 16-18, 2006

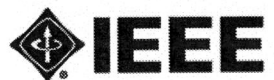

IEEE Catalog Number: 06EX1383
ISBN: 1-4244-0396-0

Copyright © 2006 by The Institute of Electrical and Electronics Engineers, Inc.
All Rights Reserved

Copyright and Reprint Permissions: Abstracting is permitted with credit to the source. Libraries are permitted to photocopy beyond the limit of U.S. copyright law for private use of patrons those articles in this volume that carry a code at the bottom of the first page, provided the per-copy fee indicated in the code is paid through Copyright Clearance Center, 222 Rosewood Drive, Danvers, MA 01923.

For other copying, reprint or republications permission, write to IEEE Copyrights Manager, IEEE Operations Center, 445 Hoes Lane, Piscataway, New Jersey USA 08854. All rights reserved.

IEEE Catalog Number:	06EX1383
ISBN:	1-4244-0396-0
LOC:	2006925541

Additional Copies of This Publication Are Available from:

IEEE Service Center
445 Hoes Lane
Piscataway, NJ 08854
IEEE Service Center
445 Hoes Lane
Piscataway, NJ 08854
Phone: (800) 678-IEEE
          (732) 981-1393
Fax: (732) 981-9667
E-mail: customer-service@ieee.org

2006 International Conference on Advanced Semiconductor Devices and Microsystems

Smolenice, Slovakia
16-18 October 2006

IEEE Catalog Number: CFP06469-POD
ISBN: 978-1-42440-396-7

Table of Contents

Carbon Nanotubes in Electronics..1
Siegmar Roth, Jiangling Wang, Viera Skakalova

Superior nMOSFET scalability using Fluorineine co-implantation and spike annealing..........................7
S. Kubicek, T. Hoffmann, E. Augendre, B. Pawlak, T. Chiarella, C. Kerner, S. Severi, A. Falepin, A. De Keersgieter, T.Noda, M. Jurczak, P. Absil, S. Biesemans

Formation of sharp-apex pyramids for active tips used in scanning probe microscopy.....................11
J. Soltys, D. Gregusova, R. Kudela, A. Satka, I. Kostic, V. Cambel

RF characterization and modeling of AlGaN/GaN based HFETs and MOSHFETs...............................15
A. Fox, M. Marso, G. Heidelberger, P. Kordoi

Electronic transport in carbon nanotubes: From individual nanotubes to thin and thick networks..........19
V. Skákalová, A. B. Kaiser, Y.-S. Woo, S. Roth

Electron-electron Interaction Induced Parabolic Negative Magnetoresistance in Two-dimensional Electron Gas in InGaAs/InP..22
B. Podör, G. Reményi

Terahertz-Radiation Photomixers on Nitrogen-Implanted GaAs...26
M. Mikulics, M. Marso, S. Stancek, E. A. Michael, P. Kordoi

Simulation of Advanced Tunneling Devices..30
A. Heigl, G. Wachutka

Simulation of the p-type RTD..34
J. Voves

Investigation of Si delta-doped InGaAs/GaAs QW MSM photodetectors......................................38
M. Florovic, J. Kovác, R. Srnánek, J. Jakabovic, J. Chovan, B. Sciana, D. Radziewicz, M. Tlaczala

Micro-Raman scattering: A versatile tool for characterization of semiconductor structures..............42
G. Irmer

Measurement of the Germanium fraction in strained and relaxed SiGe by Spectroscopic Ellipsometry........48
J. Moers, D.M. Buca, M. Goryll, R. Loo, M. Caymax, S. Mantl

X-ray diffraction characterization of Low Temperature grown GaAs/InP epilayers.........................52
C. Ferraria, F. Dubeckyb, R. Kudelab, J. Johnc, R. Srnanekd

Investigation of Nickel Silicide Contact Layers for Power Diodes.......................................56
A. -atka, R. Srnánek, A. Vincze, D. Donoval, G. Irmer, J. Ková

Reliability issues in advanced High k/metal gate stacks for 45 nm CMOS applications....................60
G. Groeseneken, M. Aoulaiche, S. De Gendt, R. Degraeve, M. Houssa, T. Kauerauf, L. Pantisano

Nb-Ti/Al/Ni/Au Ohmic Metallic System to AlGaN/GaN..65
T. Lalinsky, G. Vanko, Z. Mozolova, J. Liday, P. Vogrincic, A. Vincze, F. Uherek, Hascik, I. Kostic

Electrical and optical behaviour of nanocrystalline CdS/InP heterojunction p-n diodes..................69
V. Rakovics, Zs. J. Horváth, B. Podör

High Purity p-type InP Grown by LPE with Rare-Earth Admixtures...70
J. Grym, O. Procházková, J. Zavadil, K. Zdánský

Performance study of bulk semi-insulating InP radiation detectors with different electrode metallizations..74
B. Zatkoa, F. Dubeckýa, O. Procházkováb, V. Necasc

GaAs and GaN based SAW chemical sensors: acoustic part design and technology...........................78
L. Rufer, T. Lalinsky, D. Grobelny, S. Mir, G. Vanko, Zs. Oszi, Z. Mozolova, J. Gregu

Physical properties of transparent conductive oxides prepared by RF reactive sputtering................82
V. Tvarozek, I. Novotny, P. Sutta, S. Flickyngerova, L. Harmatha, E. Vavrinsky, M. Nigrovicova, J. Mullerova

Table of Contents

Translinear Subthreshold MOS Filter for the Wireless Sensors Applications..................86
Adam Boura, Miroslav Husak

Direction Sensitivity Matrix with PLL Temperature Sensor..................90
M. Husak, A. Boura, J. Jakovenko

Differential blood analysis by thin film interdigitated arrays of electrodes 94
R. Víglaský, M. Nigrovicová, V. Tvarožek, M. Weis

Influence of the doping material on the benzene detection 98
P.Ivanov, F. Blanco, I. Gràcia, N. Sabaté, X. Vilanova, X. Correig, L. Fonseca, E. Figueras, J. Santander, R. Rubio, C. Cané

Design and simulation of humidity micro-sensors structure based on Polymers..................102
Pavel Suchánek, Miroslav Husák

4H-SiC Diode with a RuOx and a RuWOx Schottky Contact Irradiated by Fast Electrons..................106
L. Stuchlíková, L. Harmatha, D. Búc, J. Benkovská, B. Hlinka, G. G. Siu

Rapid thermal annealing and performance of Al2O3/GaN metaloxide-semiconductor structures..................110
K. Cico, J. Kuzmik, D. Gregusova, T. Lalinsky, A. Georgakilas, D. Pogany, K. Frohlich

Helium Irradiation for Advanced Lifetime Control in Silicon: New Recombination Centers and Their Interaction Stimulated by Isochronal Annealing..................114
Volodymyr Komarnitskyy, Pavel Hazdra

Electrical and Memory Properties of Non-volatile Memory Structures with Embedded Si Nanocrystals..................118
Zs. J. Horváth, P. Basa, T. Jászi, A. E. Pap, P. Szöllosi, K. Nagy, V. Hardy

Leakage current and Physical properties of Tantalum oxide thin films for Micro capacitor integration..................122
Insung- Kim, Jaesung-Song, Bokki-Min

Leakage characteristics of advanced MOS capacitors with hafnium silicate dielectric and Ru electrode..................126
M. Tapajna, K. Hlieková, K. Fröhlich, E. Dobrocka, F. Roozeboom

First results observed with test X-CT system using GaAs radiation detector working in single photon counting regime..................130
F. Dubecky, B. Zatko, I. Frollo, J. Juras, J. Pribil, J. Jakubek, J. Mudron

Changes of GaAs neutron detectors properties after fast neutron irradiation..................134
Milan Ladziansky, Andrea Sagatova-Perdochova, Bohumir Zatko, Vladimir Necas, Frantisek Dubecky

Control of defects and impurities at GaN and AlGaN surfaces for FET and sensor applications 138
Junji Kotani, Masamitsu Kaneko, Kazushi Matsuo, Tamotsu Hashizume

Comparison of AlGaN/GaN MSM varactor diodes based on HFET and MOSHFET layer structures..................146
M. Marso, A. Fox, G. Heidelberger, P. Kordoi, H. Lüth

Dark current of AlGaAs/GaAs n-QWIP prepared on patterned (001) GaAs substrate by MOVPE..................150
Strichovanec P., Kúdela R., Vávra I., R. Srnánek, Novák J.

Study of electrically active defects in GaAs/InAs/GaAs QDs structures by DLTS and TEM..................154
M. Prezioso, E. Gombia, R. Mosca, L. Nasi, A. Motta, P. Frigeri, G. Trevisi, L. Seravalli, S. Franchi

Technology related issues regarding fabrication of AlGaN/GaN-based MOSHFETs with GdScO3 as dielectric 158
G. Heidelberger, M. Roeckerath, R. Steins, M. Stefaniak, A. Fox, J. Schubert, N. Kaluza, M. Marso, H. Lüth, P. Kordoi

Investigation of GaN/ ZnO heterostructures properties 162
J. Kovac, J. Skriniarova, P. Kudela, I. Novotny, J. Bruncko, D. Donoval, J. Jakabovic, M. Michalka, L. Janos, A.sVincze, D. Hasko

Table of Contents

Preparation and properties of AlGaN/GaN MOSHFETs with MOCVD Al2O3 as gate oxide.........166
R. Stoklas, K. Cico, D. Greguiová, J. Novái, P. Kordoi

Post-metallization H2 annealing of electrically active defects in Ta2O5/nitrided Si stacks.........170
A. Paskaleva, E. Atanassova

2D electron transport through potential barrier prepared by LAO on shallow GaAs/AlxGa1-xAs/InGaP heterostructure.........174
J. Martaus, V. Cambel, R. Kudela, D. Gregusova, J. Soltys

Recent Developments in Microsystem- and Nanodevice-TCAD.........178
Gerhard Wachutka

Analysis of device geometry on the ruggedness of power DMOS transistor supported by 3-D modeling and simulation.........180
Andrej Vrbicky, Daniel Donoval, Juraj Marek, Ales Chvala, Peter Beno

New 600V Lateral Superjunction Power MOSFETs Based on Embedded Non-Uniform Column Structure.........184
K. Permthammasin, G. Wachutka, M. Schmitt, H. Kapels

FEM Simulation and Characterization of Microcantilevers Resonators.........188
Margarita S. Narducci, Eduard Figueras, Isabel García, Luís Fonseca, Carles Cané

Electric energy harvesting inside self powered microsystem.........192
Janicek Vladimir, Husak Miroslav

Monitoring of Psychosomatic Properties of Human Body by Skin Conductivity Measurements using Thin Film Microelectrode Arrays.........196
E. Vavrinský, V. Stopjaková, L. Majer, V. Tvarožek, M. Weis, P. Marman

PREPARATION AND CHARACTERIZATION OF MICROHOTPLATE FOR GAS SENSORS.........200
A. Reháková, D. Tengeri, I. Hotový, T. Lalinský, V. Veháek, L. Spiess, H. Romanus, Š. Hašík

Reduced-order modeling of capacitive MEMS microphones using mixed-level simulation.........204
M. Niessner, W. Bedyk, G. Schrag, G. Wachutka, B. Margesin, A. Faes

Recent Advances in Organic Electronic.........208
J. Kovác, J. Jakabovic, L. Peternai, O. Lengyel, M. Kytka

Energy Band Diagram of the Ru/Hf0.75Si0.25Oy/Si Gate Stack.........216
K. Fröhlich, J. P. Espinos, M. Tapajna, K. Huieková, R. Lupták

Towards a Microtechnology based 4-channel infrared detector unit for a miniaturised NDIR system.........220
L. Fonseca, J. Santander, R. Rubio, N. Sabaté, P. Ivanov, E. Figueras, I. Gràcia, C. Cané

Nano-structures for light management in optoelectronic devices.........224
M. Zeman, J. Krc

Preparation of p-type ZnO thin films by RF diode sputtering.........228
K. Shtereva, I. Novotny, V. Tvarozek, R. Srnanek, J. Kovac, P. Sutta

Coupling Capacitances of Connecting-lead Systems in Integrated Circuits.........234
J. Novak, J. Foit, V. Janicek

The Effect of Rapid Thermal Annealing on Oxygen Precipitation in Nitrogen Doped Silicon Substrate.........238
L. Stuchlíková, L. Harmatha, M. Tapajna, P. Ballo, P. Písecný, M. Benkovic, J. Jakabovic

Microstructure of HfO2 and HfxSi1-xOy Dielectric Films Prepared on Si for Advanced CMOS Application.........242
M. Franta, A. Rosová, M. Tapajna, E. Dobrocka, K. Fröhlich

Deep Defects in MOVPE Grown SiC/AlGaN/GaN Heterostructures.........246
D. Kindl, P. Hubik, J. Kri.tofik, J. J. Mare, Z. Vyborny, M. R. Leys, S. Boeykens

Diagnostics of LT GaAs/ InP structures by micro-Raman spectroscopy.........250
R. Srnánek, G. Irmer, R. Záluský, F. Dubecký, R. Kúdela, A. Vincze, I. Novotný, J. John

Table of Contents

RF plasma deposition of thin amorphous silicon carbide films using a combination of silan and methane 254
J. Huran, I. Hotový, J. Pezoltd, N. I. Balalykin, A.P. Kobzev

Photoluminescence and electrical characterization of transparent Eu and Pd-doped TiO2 thin films 258
J. Domaradzki, A. Borkowska, D. Kaczmarek

Structural, optical and electrical properties of nanocrystalline TiO2 – HfO2 thin films 262
J. Domaradzki, D. Kaczmarek

Microelectromagnetic matrix for local assembling of magnetic nanoparticles 266
S. Luby, L. Chitu, E. Majkova, R. Senderak, I. Kostic, P. Hrkut, L. Matay, S. Hascik, T. Lalinsky, I. Capek, A. Satka

Nickel ohmic contact on silicon carbide 270
Machac P., Barda B., Sajdl P.

Two-dimensional electron gas as the THz radiation detector 274
Michal Horák

Influence of mechanical strain on essential characteristics of GMR structures 278
V. Ác, B. Anwarzai, S. Luby, E. Majkova

Energy band diagram and charge distribution in AlGaN/GaN heterostructure studied by classical approach 282
J. Osvald

Spectroscopic ellipsometric study of LPCVD-deposited Si nanocrystals in SiNx and Si3N4 286
P. Basa, P. Petrik, M. Fried

Annealing behaviour of low temperature grown GaAs investigated by SIMS 289
A. Vincze, J. Kovác, R. Srnánek

Evaluation of parameters of Schottky junctions with large excess currents 293
Zs. J. Horváth

Deposition of AZ5214-E Layers on Non-planar Substrates with a "Draping" Technique 294
P. Eliái, D. Greguiová, P. Strichovanec, I. Kostic, J. Novák

Carbon Nanotubes in Electronics

Siegmar Roth [1,2], Jiangling Wang [3], and Viera Skakalova [4]

[1] Max Planck Institut für Festkörperforschung, Stuttgart, Germany
[2] Sineurop Nanotech GmbH, Stuttgart, Germany
[3] Shanghai Yangtze Nanomaterials Co. Ltd., Shanghai, China
[4] Danubia Nanotech s.r.o., Bratislava, Slovakia

Some people claim that carbon nanotubes will be the most important material of the 21st century. The future will tell us whether this is true, but the past teaches that most predictions are wrong. In any case, carbon nanotubes certainly are the most popular material of the present: Since their discovery in 1991 [1] some 20 000 publications have appeared and some 1000 patents have been filed. Fig. 1 shows the computer model of a carbon nanotube.

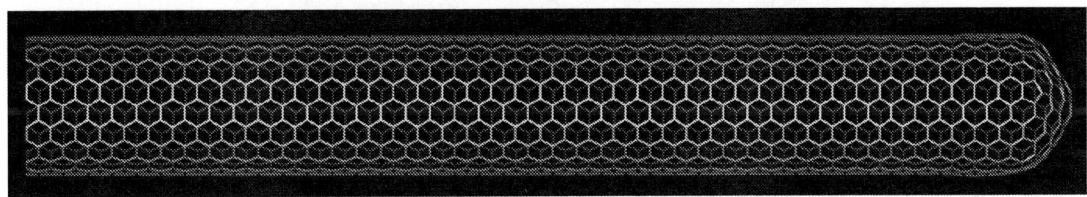

Fig. 1: Computer model of a single-walled carbon nanotube.

Carbon nanotubes are seamless tubes of graphitic monolayers, about 1 or 2 nanometers in diameter and up to several micrometers or even millimeters long. Depending on the diameter and on the details of seamless joining, the nanotubes are metallic or semiconducting. Because of the strong carbon-carbon bond and because of phase space arguments in narrow constrictions, there are only few scattering events in nanotubes and the conductivity is mainly ballistic, which for practical purposes means, for metallic nanotubes it is very high. The energy gap of semiconducting nanotubes can range from a few meV up to 1 eV. There are single-walled and multi-walled carbon nanotubes. Multi-walled nanotubes consist of up to 70 concentric tubes. For monographs on carbon nanotubes see Refs. [2-5].

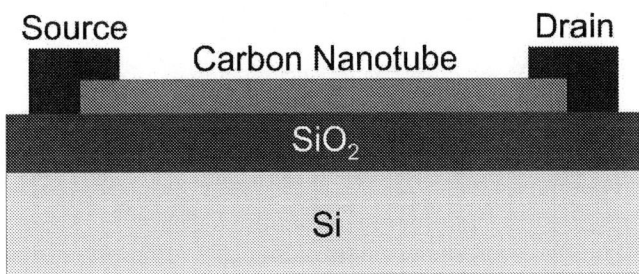

Fig. 2: Schematic diagram of nanotube field-effect transistor

If there are semiconducting nanotubes, it is tempting to make nanotube transistors. Several groups have prepared and investigated such transistors [6-11]. These transistors are expected to be smaller than silicon transistors, faster, and less energy-consuming. Fig. 2 shows a schematic diagram of a nanotube field-effect transistor (TUBE-FET): A semiconducting single-walled carbon nanotube is placed on the oxide layer of a doped silicon chip. Gold leads are applied by electron beam lithography as source and as drain contacts. Usually the nanotube is already p-doped from the purification process (which generally involves an oxidation step) or it gets p-doped by interaction with the gold leads (different work function between carbon and gold). Consequently the nanotube is conducting and the transistor is in the "on" state. To switch the transistor off, a voltage is applied between the nanotube and the doped silicon chip (which serves as a gate contact). The on-off ratio of such a transistor can be as large as 5 or 6 orders of magnitude and it has been shown that these nanotube transistors outperform the best silicon transistors also with respect to other parameter. Fig. 3 shows an AFM image of a single-walled carbon nanotubes placed over 4 gold leads prepared by electron beam lithography [12]. Any two of the gold leads can be used as the source and the drain contact of the field-effect transistor.

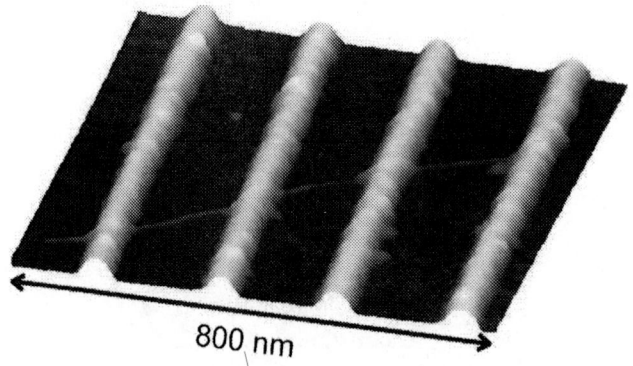

Fig. 3: AFM image of a carbon nanotube laying over 4 gold leads prepared by electron beam lithography [12].

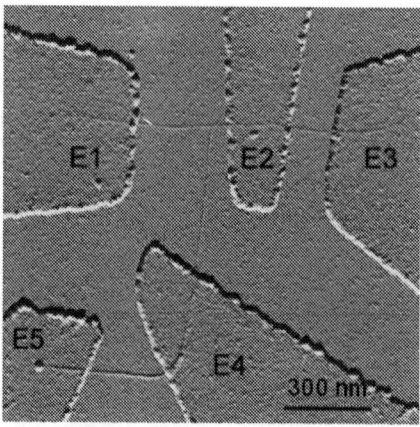

Fig. 4: "All-carbon" transistor with nanotube not only as conducting channel but also with nanotube gate. The dielectric is a linker molecule at the T-junction. The device has been "synthesized by wet-chemical methods" and adsorbed on a silicon chip prior to the application of the lithographic gold leads [11].

Fig. 4 presents the "all-carbon" transistor, where the gate is also a carbon nanotube [11]. This transistor has been "synthesized by wet-chemical methods", and in principle it would allow for really nanometer-sized devices and extremely high integration density. Phaedon Avouris' group at IBM has prepared a complimentary field-effect device consisting of a nanotube over tree gold electrodes and of which one part is p-doped and the other n-doped [13]. n-doping has been achieved by covering one part of the nanotube by a polymer film and exposing the other to ammonia, so that there the original p-doping is overcompensated (Fig. 5). Such a device can serve as a voltage inverter (or as a logic NOT gate). An other way of saving space and going to higher integration densities is to put the transistors vertical and to use a wrap-around gate. Such transistors have been patented by Infineon and by Samsung. An artist's view is shown in Fig. 6 [14]. Today we have quite a good understanding on how transistors can be made from individual nanotubes and on how individual nanotubes behave compared to silicon, but we lack a technique of selecting the wanted type of nanotubes and then positioning it in the right place. Developing a technology which could one day replace large scale integrated silicon circuits is an endeavour of decades, not of years.

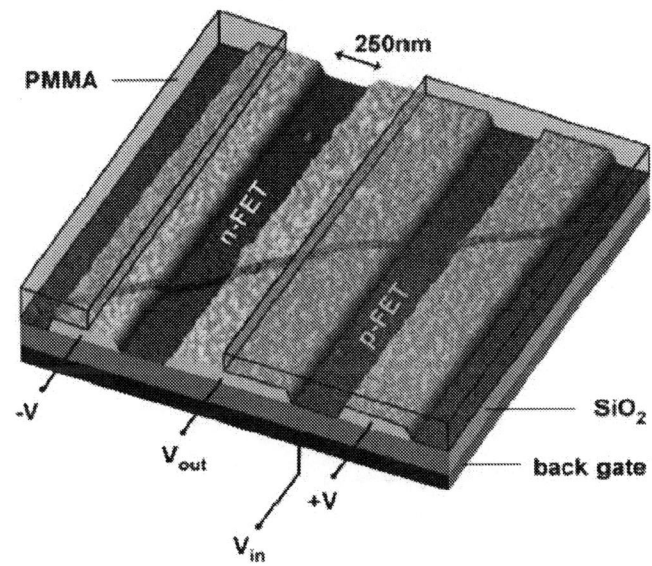

Fig. 5: Voltage inverter consisting of a p-type and an n-type carbon nanotube field-effect transistor [13].

Fig. 6: Vertical nanotube field-effect transistor with wrap-around gate electrode [14].

The limiting problem of today's electronics is not the active elements (transistors), it is wiring. Wiring at all levels, from the cables between computer and peripherals down to interconnects between the transistors on a chip. A special challenge are VIAs (vertical interconnect access), connecting different layers on a chip. It is very difficult to pass high current densities through very thin wires of conventional metals because of surface and grain boundary scattering. In addition, such thin wires tend to decompose, releasing small metal particles which migrate around within the circuitry (electromigration). Because of the strong covalent bonds between carbon atoms, as compared to the rather week metallic bonds, there is much less electromigration from carbon nanotubes, and the maximum current density in carbon nanotubes is by three orders of magnitude higher than in copper wires. Therefore several companies are working on VIAs based on carbon nanotubes. Fig. 7 shows the electronmicrograph of a multi-walled nanotube growing out of a hole etched into a silicon chip [15]. Experts believe that a hybrid technology - with active elements based on silicon and with interconnects based on carbon - could be ready within a few years. An "all-carbon" technology with metallic nanotubes connecting transistors from semiconducting nanotubes is a dream for a very far future. Yet, a disordered version of this dream has already been realized today: the transparent nanotube transistor.

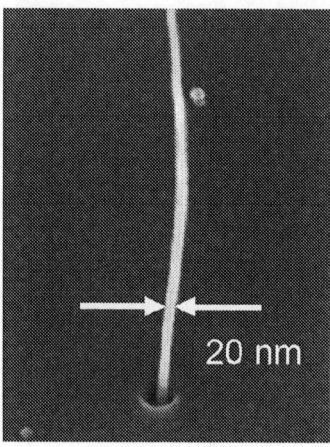

Fig. 7: Multi-walled carbon nanotube growing out of a hole etched into a silicon chip, first step towards nanotube-based VIAs [15].

Individual nanotubes are difficult to handle. But there are many applications in which large ensembles of nanotubes are used. This is particularly true for nanotube-polymer composites, i.e. for polymers filled with carbon nanotubes. Such filling might improve the electrical and mechanical properties of the polymers. Of course, polymers filled with larger particles do already exist (e.g. aluminium flakes, silver beats, or stainless steel fibres as conductive fillers, carbon fibres for mechanical reinforcement), and non-tubular nanoparticles have also been known for a long time as polymer fillers (carbon black for chemical purposes in car tires, carbon black as pigments, carbon black as conductive fillers). Conductive polymers can be used for dissipation of electrostatic charges, for electromagnetic shielding, and for electrical heating (of car seats or of outside mirrors). There are certain advantages if the fillers are thin and long (lower percolation threshold, so that less filler material is needed) and there are certain advantages if the filler particles are nanometer size (homogeneity down to nanometer scale as needed in ultrathin adhesive layers).

Because of the low percolation threshold that can be obtained with nanotube filling, combined with nanoscale homogeneity, transparent conductive films can be made. These films can compete with ITO layers (indium-tin oxide), which is the standard material for transparent electrodes (needed in solar cells and in light emitting devices). A very interesting example of the application of transparent nanotube networks is the flexible transparent nanotube transistor [16]. A photograph of such a transistor is shown in Fig. 8. and a schematic drawing in Fig. 9. The essential feature of this transistor is that there are two transparent nanotube layers, separated by a thin insulating layer of parylene. Both layers consist of a mixture of metallic and semiconducting nanotubes (Today only mixtures of nanotubes can be synthesized and there is no efficient way of separating them). For the lower layer the nanotube concentration is chosen such that the metallic nanotubes form a continuous network. This layer behaves like a metal and serves as the gate electrode of the transistor. In the upper nanotube layer the concentration of the tubes is only half as large. Now most metallic nanotubes do not touch and the network is only continuous if both metallic and semiconducting nanotubes are taken into account. This network behaves semiconductor-like and forms the conducting channel of the transistor. Actually, this layer can be considered as a random ensemble of individual semiconducting nanotubes, each connected by metallic nanotubes as local source and drain contacts.

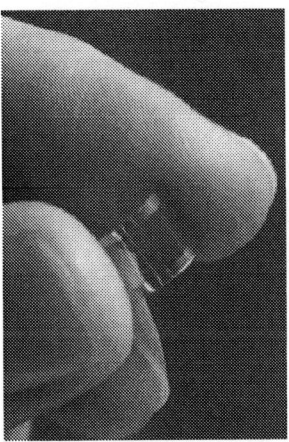

Fig. 8: Photograph of flexible transparent nanotube transistor [16].

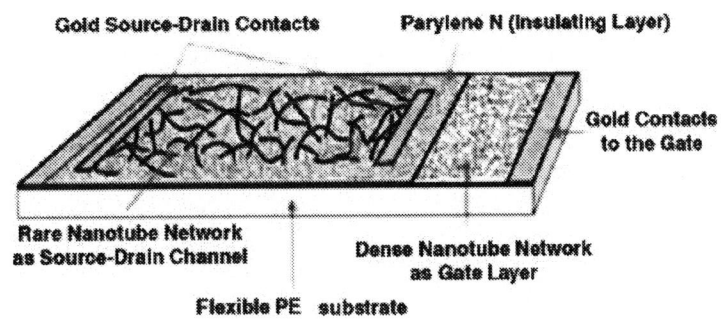

Fig. 9: Schematic drawing of flexible transparent nanotube transistor [16].

References

[1]   S. Iijima: Helical microtubulus of graphitic carbon. Nature **354**, 56-58 (1991).

[2]   T.W. Ebbesen (Ed.): "Carbon Nanotubes - Preparation and Properties" CRC Press (1997).

[3]   M.S. Dresselhaus, G. Dresselhaus, Ph. Avouris (Eds.) "Carbon Nanotubes" Springer (2000).

[4]   P.J.F. Harris: "Carbon Nanotubes and Related Structures" Cambridge University Press (1999).

[5]   R. Saito, G. Dresselhaus, and M.S. Dresselhaus: "Physical Properties of Carbon Nanotubes" Imperial College Press, London (1998).

[6]   S.J. Tans, A.R.M. Verschueren, and C. Dekker. Room-temperature transistor based on a single carbon nanotube. Nature **393**, 49–52 (1998).

[7] H.W.C. Postma, T. Teepen, Z. Yao, M. Grifoni, and C. Dekker. Carbon nanotbe single-electron transistors at room temperature. Science **293**, 76–79 (2001).

[8] R. Martel, T. Schmidt, H.R. Shea, T. Hertel, and Ph. Avouris. Single- and multi-wall carbon nanotube field-effect transistors. Applied Physics Letters **73**, 2447-2449 (1998).

[9] S. Heinze, J. Tersoff, R. Martel, V. Derycke, J. Appenzeller, Ph. Avouris. Carbon nanotubes as Schottky barrier transistors. Physical Review Letters **89**, 106801-1 – 106801-4 (2002).

[10] A. Javey, J. Guo, Q. Wang, M. Lundstrom, and H. Dai. Ballistic carbon nanotube field-effect transistors. Nature **424**, 654–657 (2003).

[11] P.W. Chiu, M. Kaempgen, and S. Roth. Band-structure modulation in carbon nanotube T-junctions. Physical Review Letters **92**, 246802-1 – 246802-4 (2004).

[12] J. Muster, M. Burghard, S. Roth, G.S. Duesberg, E. Hernández, and A. Rubio. Scanning force microscopy characterization of individual carbon nanotubes on electrode arrays. Journal of Vacuum Science and Technology B **16**, 2796 (1998).

[13] Ph. Avouris, R. Martel, V.Derycke, J.Appenzeller, Carbon nanotube transistors and logic circuits, d 10598, USA, Physica B **323**,6-14 (2002).

[14] F. Kreupl, G.S. Duesberg, A.P. Graham, M. Liebau, E. Unger, R. Seidel, W. Pamler, and W. Hönlein. "Carbon nanotubes in microelectronic applications". In: Physics, Chemistry and Application of Nanostructures : Reviews and Short Notes to Nanomeeting 2003 Minsk, Belarus 20-23 May 2003. S.V. Gaponenko, V.S. Gurin (Eds.). World Scientific Pub Co Inc (2003/04)

[15] F. Kreupl, A.P. Graham, M. Liebau, G.S. Duesberg, R. Seidel, E. Unger. IEDM Tech. Dig., pp. 683 - 686, December 2004, cond-mat/0412537.

[16] E. Artukovic, M. Kaempgen, D.S. Hecht, S. Roth, and G. Grüner. Transparent and flexible carbon nanotube transistors. Nano Letters **5**, 757-760 (2005).

Superior nMOSFET scalability using Fluorineine co-implantation and spike annealing

S. Kubicek, T. Hoffmann, E. Augendre, B. Pawlak*, T. Chiarella, C. Kerner, S. Severi, A. Falepin,
A. De Keersgieter, T.Noda[+], M. Jurczak, P. Absil and S. Biesemans

IMEC, Leuven, Belgium
kubicek@imec.be
*Philips Research, Leuven, Belgium
[+]Matsushita Electric assignee in IMEC

We report the simultaneous improvement of both on- and off-properties for nMOSFETs by means of Fluorineine co-implantation at extension level, using conventional spike annealing. For the first time, spike-annealed NFETs with Fluorineine co-implanted source/drain extensions (SDE) are shown to outperform conventional As-implanted and C co-implanted devices in the deca-nanometric range. Parameters such as on-current, drain-induced barrier lowering (DIBL), external resistance (R_{EXT}) vs. effective channel length (Leff) trade-off are examined.

I. INTRODUCTION

Device scaling imposes that SDE (Source and Drain extensions) become shallower to support gate length reduction and less resistive to preserve current drivability. Single dopant implantation and annealing cannot meet these requirements anymore and various approaches are being investigated in order to improve the trade-off between junction depth (x_j) and resistivity (ρ_{sheet}). All approaches (epitaxial regrowth [1], Flash anneal [2], Laser anneal [3], F/C co-implantation [4,5]) rely on a pre-amorphization implant (PAI) and minimization of dopant interaction with end-of range (EOR) Si interstitials (SiI). While other techniques reduce SiI diffusion through thermal budget reduction, co-implantation is based on SiI trapping by co-implanted C or F atoms that are inserted between EOR defects and the dopant profile (as illustrated in Fig. 1.

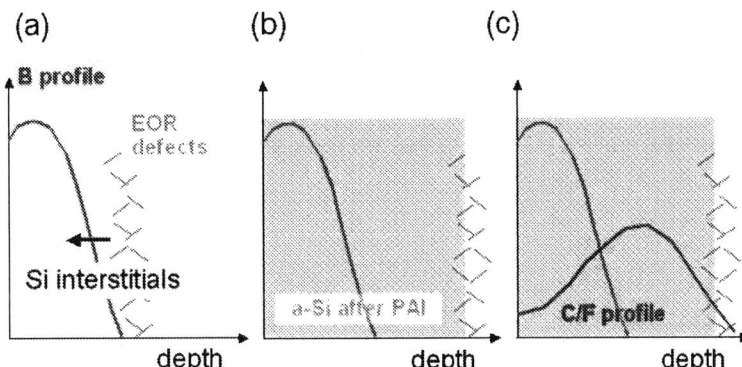

Figure 1.   Principle of the of the co-implantation approach: PAI sets EOR defects away from the junction, a C/F peak traps EOR-emitted SiI and prevents them from impacting the dopant profile

A little device data [5] has been presented so far on spike-annealed Fluorineine coimplantation and the demonstration of concurrent on- and off-state improvement in deca-nanometric devices and its comparison with conventional and C co-implanted devices is still missing. This is the purpose of this paper.

II. EXPERIMENTAL: DEVICE FABRICATION

N-and pFETs were fabricated on 300 mm wafers using SiON as gate dielectric (1.4 nm equivalent oxide thickness), polysilicon as gate material. After gate patterning, all devices receive the same pocket implants but different SDE conditions.

Spacers are formed at low-temperature (<500°C) and As and P n-HDD is implanted. After spike annealing at 1050°C with high ramp-up rate, partial thin Ni mono-salicidation of all electrodes completes the fabrication.

III. RESULTS AND DICUSSION

Fig. 2 compares nMOS drive currents at constant off-state current (60nA/um) as function of gate leakage density (J_G) for different process conditions.

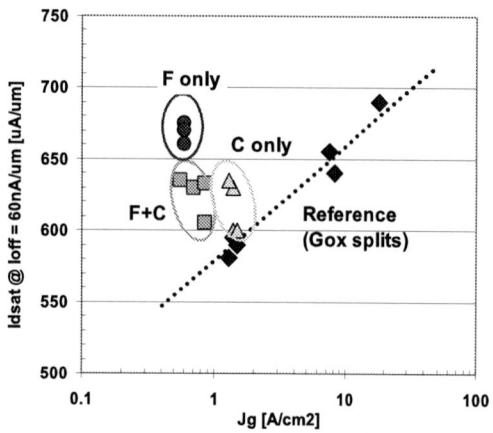

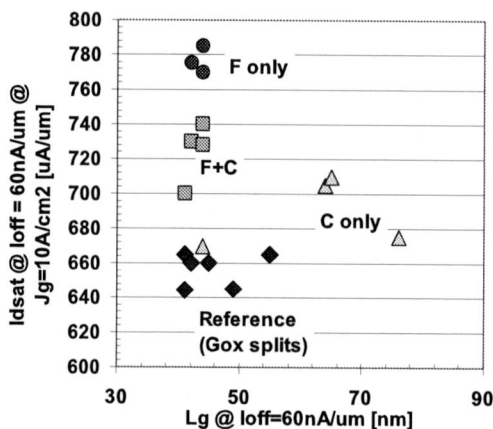

Figure 2. Drive currents measured at constant off-state current as function of gate leakage density for transistors with different processing conditions.

Figure 3. Drive currents extracted at constant J_G and off-state current plotted as function of L_G at constant off-state current.

It can be seen that the transistors with Fluorineine co-implantation are driving ~50uA/um more than those with F+C or C co-implantation and more than 90µA/µm than reference (no co-implantation) transistors. The gate leakage density of F co-implanted transistors is increased but only by half of the J_G of reference transistors.

In Fig. 3 drive currents at constant J_G and off-state current are plotted as function of L_G at constant off-state current. From this point of view, the Fluorineine co-implanted transistors are driving more than 20% more current at the same L_G than reference transistors. They also significantly outperform the C+F co-implanted transistors.

From comparison of SCE (Short Channel effect) performance presented in Fig. 4, Fluorine co-implanted transistors can improve the SCE performance by 30mV as compared to reference transistors.

8

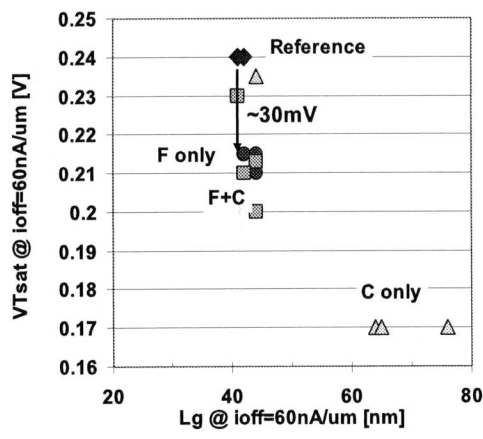
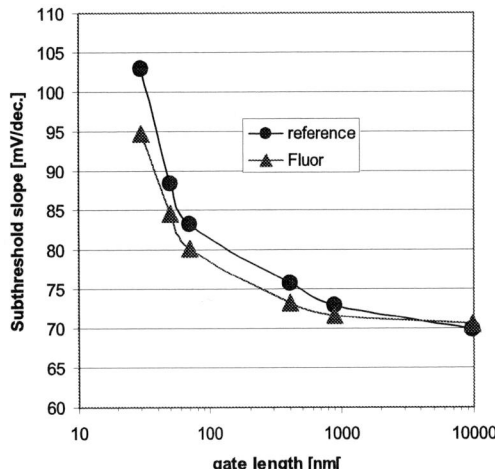

Figure 4. V_{TSAT}-L_G comparisons for different conditions (all with same gate dielectric).

Figure 5. Sub-threshold slope of plotted as function of the gate lengths for reference and Fluorine coimplanted transistors.

C_{INV} measurements on long channel transistors suggest that the improvement probably does not come from the reduced poly depletion. Neither the effect of improved channel mobility of series resistance was confirmed by electrical measurements. On the other hand improvement in sub-threshold slope and DIBL was observed. Sub-threshold slope as function of the gate length is plotted in Fig. 5 for reference transistors and transistors co-implanted with Fluorine.

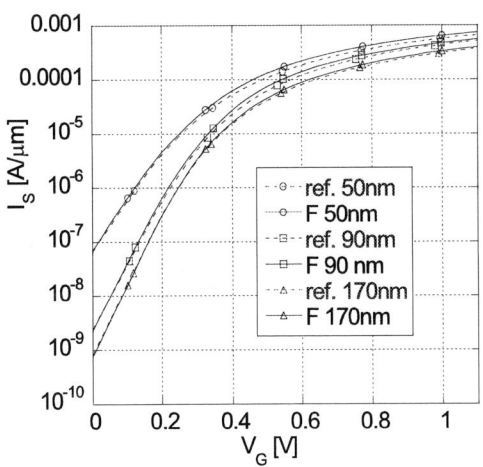
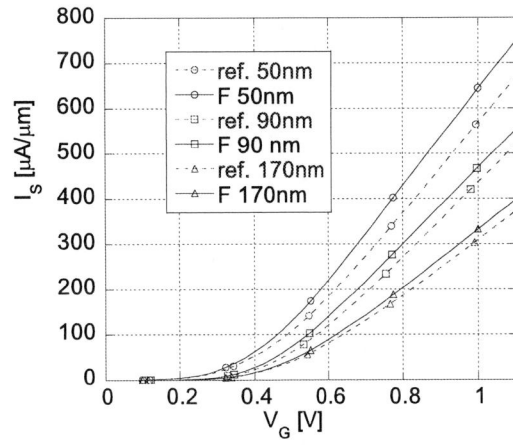

Figure 6. I_D-V_G curves of reference transistors and transistors with coimplanted Fluorine plotted in the logarithmic scale. Reference curvese were shifted so that Ioff coincide.

Figure 7. I_D-V_G curves of reference transistors and transistors with coimplanted Fluorine plotted in the log. Scale. Reference curvese were shifted so that Ioff coincide.

I_D-V_G curves of reference transistors and transistors with co-implanted Fluorine are plotted in the logarithmic and linear scale in Fig. 6 and 7. Reference I_D-V_G curves were shifted (to compensate for different V_T) so that Ioff coincide. From linear plots (Fig.7) it is clear that the

Fluorine co-implanted transistors outperform the reference ones. DIBL of reference transistors as function of gate lengths compared to transistors with co-implanted Fluorine is plotted in Fig. 8. As can be seen the Fluorine co-implanted transistors again outperform the reference ones. The improved SCE behavior, like better V_T roll-off and DIBL were already reported in [5] and it was attributed to the steeper Boron HALO profiles.

All the presented partial improvements lead to final improvement of I_{OFF}-I_{ON} curve of F co-implanted transistors as compared to other splits (reference, F+C and C co-implantation) as seen in Fig.9.

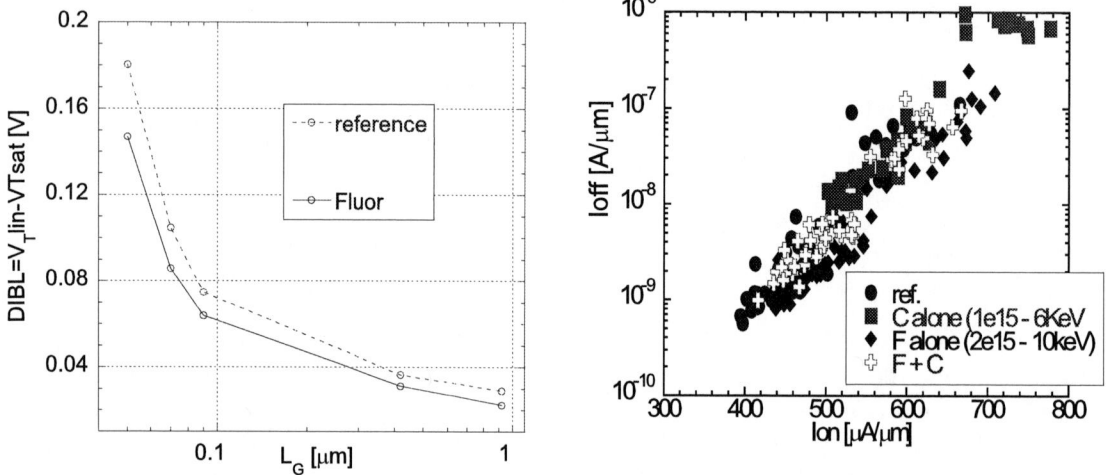

Figure 8.   DIBL of as function of gate lenght for reference transistors compared to that of transistors with coimplanted Fluorine.

Figure 9. I_{OFF}-I_{ON} curves for the transistors with different extension conditions.

IV. CONCLUSIONS

We have demonstrated that the benefit of the PAI and F co-implantation route applies at device level: not only are short channel effects better controlled, but also device drivability improves. For the first time, NFET transistors with gate lengths down to 40 nm are found to outperform As counterparts. To better understand the role of Fluorineine more fundamental studies are being carried on.

REFERENCES

[1] L-Å Ragnarsson et al., VLSI 2005

[2] S. Severi et al., SSDM 2005

[3] S. Severi et al., VLSI-TSA 2006, MRS symp. Proc. Vol.912

[4] S. Severi et al., MRS 2006, MRS symp. Proc. Vol.912

[5] N. Cagnat et al., MRS 2006, MRS symp. Proc. Vol.912

Formation of sharp-apex pyramids for active tips used in scanning probe microscopy

J. Šoltýs[a], D. Gregušová[a], R. Kúdela[a], A. Šatka[b], I. Kostič[c], and V. Cambel[a]

[a]Institute of Electrical Engineering, Slovak Academy of Sciences, Bratislava, Slovakia
[b]Faculty of Electrical Engineering and Information Technology, Slovak University of Technology, Bratislava, Slovakia
[c]Institute of Informatics, Slovak Academy of Sciences, Bratislava, Slovakia
e-mail: jan.soltys@savba.sk

This work is a part of a deeper study into technologies developed for the so-called active tips for use in scanning probe microscopy. We show that using an AlAs facet-forming sacrificial layer and a H_3PO_4, H_2O_2, H_2O based solution, symmetric pyramidal structures with tip diameter below 35 nm can be prepared. We have used such pyramidal objects for further MOCVD overgrowth. Finally, we have controlled the quality of the mesa sidewalls obtained using the SEM and AFM. Tthe pyramids are suitable for the next processing, and various semiconductor devices can be prepared on the pyramid.

1. Introduction

In the last decades, Scanning Probe Microscopy (SPM) has been used to investigate surface properties of materials. The family of SPM uses no lenses, but rather a probe that interacts with the sample surface. The type of interaction measured between the probe tip and the sample surface defines the type of SPM being used. Some examples are atomic force microscopy (AFM), electric force microscopy (EFM), scanning thermal microscopy (SThM), or magnetic force microscopy (MFM).

The above-mentioned microscopes and techniques have one feature in common: The tip and the cantilever function as the primary detector of force, charge, temperature or magnetism. The magnitude of a physical quantity measured is in general dependant on the deflection or deformation of the cantilever. The spatial resolution is given by the tip perimeter at its end (usually 5 – 20 nm). The quantity measured is at first transformed into the mechanical deflection of the cantilever. The motion of the cantilever is monitored by an optical way and the mechanical quantity is thus transformed to the measurement of electrical current of the photodetector. The read-out is further processed and the magnitude of the original physical quantity measured by the tip can be determined.

This evokes a question whether it would be possible to measure and process a physical quantity directly, i.e. without the conversion through a mechanical quantity. It has been already predicted that the development of scanning techniques will take a route towards the so-called laboratory on the tip. It is, of course, less complicated to define sensors on the cantilever itself, but this would inevitably result in a loss of spatial resolution. The tip is usually several micrometers long, and it considerably separates the cantilever from the sample. The only option is to define active elements closer to the tip or on the tip itself, i.e. on the side facets of the tip close to its pointed end. This approach has so far been used only rarely. It was applied in scanning Hall probe microscopy [1, 2], and in a scanning system to detect charge by means of a FET transistor [3].

As tips for SPM are usually prepared from Si, GaAs and InP, it is in principle possible to prepare active layers on facets of tips. The layers can be further processed into

1-4244-0396-0/06/$25.00 ©2006 IEEE

semiconductor sensors, amplifiers, etc. GaAs and InP are especially attractive for this purpose, as quantum heterostructure layers can be prepared on the materials using well-established epitaxial technologies.

The technology of III-V nonplanar structures is complicated due to two steps, the first, for the formation of smooth 3D objects with adjustable sidewall slope, the second, for the epitaxial overgrowth of the objects formed in the first step. Two ways have been used to prepare patterned III-V semiconductor substrates for the following epitaxial overgrowth – either by the direct patterning of 3D objects by wet-chemical etching [4, 5], or by the deposition and definition of masking materials for the selective area growth [6].

In this paper we report on the micromachining of pyramid structures using sacrificial AlAs layer and the subsequent MOCVD overgrowth of such 3D objects. The two technology processes represent the first two steps for the preparation the so-called active tips for scanning probe microscopes.

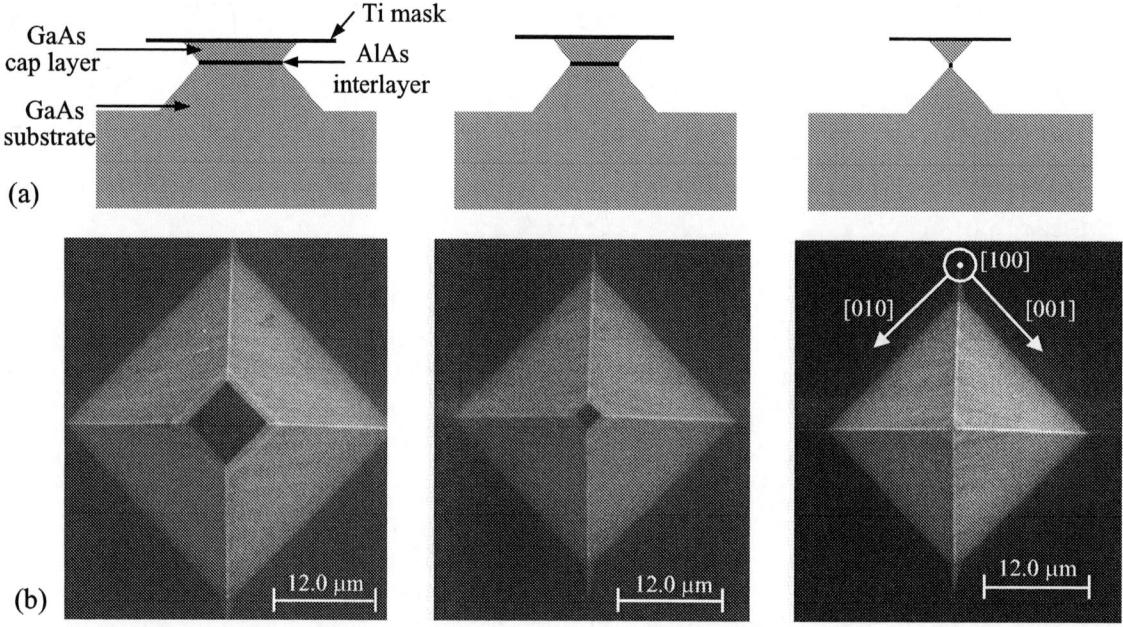

Fig. 1 (a) Schematic illustration of cross-sectional profile of the GaAs mesa etched using an embedded AlAs layer before the cap and AlAs are removed. The size of the mask pattern decreases from the left to the right.
(b) SEM images of the [001] -oriented mesas etched in 1 H_3PO_4 : 3 H_2O_2 : 8 H_2O for 7 min. The size of the mask patterns was 50, 46, and 44 μm from left to the right.

2. Experiment

The experimental structure consisted of n-type (100) GaAs substrate, a 0.1-μm-thick AlAs interlayer, and a 2-μm-thick GaAs cap layer prepared by MOCVD. A 100-nm-thick titanium layer was evaporated as an etching mask. Four sets of masking patterns with a pitch of 60 μm were used: squares, circles, lozenges, and rotated squares. The size of the mask patterns varied from 30 to 50 μm in steps of 2 μm. The squares had the edges in parallel with the [011] and [0$\bar{1}$1] directions; the rotated squares were oriented along [001] and [010]. The etching mask patterns were transferred into a Ti layer in a HF solution. The micromachining of the GaAs substrates was carried out in 1H_3PO_4 : 3H_2O_2: 8H_2O solution via the Ti mask patterns at room temperature (25°C) without agitation. The etched 3D objects were about 5 ÷ 7 μm high. On the top of the pyramids, the plateau or apex were formed, depending on the mask square side length. After etching in a H_3PO_4-based solution, the GaAs cap and the AlAs layer were removed using a solution based on HF in ultrasonic vessel. On the top of the 3D objects, the plateau or apex were formed, depending on the mask square side length. After the formation of the 3D objects with or without a plateau, the samples were overgrown by MOCVD to form GaAs about 350-nm-thick n-doped GaAs layer.

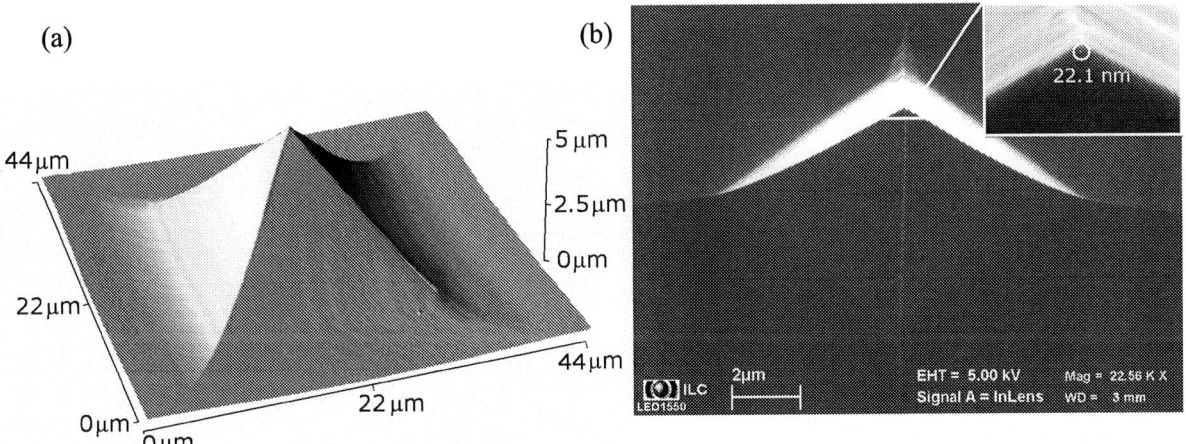

Fig. 2 (a) AFM and (b) SEM picture of the etched pyramid.

3. Results and discussion

The shape of the object etched depends on the mask topology as well as on the etching process. The AlAs layer controls the lateral etching rate and it influences the cross-sectional profile of the GaAs three-dimensional objects. On the basis of our experiments we found that the rotated squares were the most suitable mask patterns for the preparation of tips of pyramidal shape.

Fig. 1a presents a schematic cross-section of pyramidal structures depending on the size of the mask pattern. Figure 1b shows a top view of the pyramids which were etched during 7 min. As the size of the mask pattern decreases, the plateau of a pyramid diminishes, and it is eventually etched away, i.e. the pyramid is sharped into a sharp tip. The quality of the sidewalls, sidewall slope, and curvature of the tip were studied by AFM and SEM. In the Fig. 2 one can see that the mesa sidewalls are smooth, with a constant tilt for the selected directions, and all the crystallographic features and the diffusion profiles are suppressed. The etching technique applied provides the formation of symmetric GaAs pyramids with the sides, tilted at the same 51° slope. The etching rate is about 1.1 μm/sec. The inset of Fig.2b shows a detail at the apex of the pyramid. A typical diameter of curvature of the tip is 13 ÷ 35 nm.

In Fig. 3a SEM study of the overgrown pyramid is presented. One can see from Fig. 3a that the sidewalls are continuously covered with GaAs. We observe two interesting features at all pyramidal structures. The first feature is that the sidewall slope changes at the upper part of the pyramid from angle 51° to about 44° (arrow in Fig. 3a). We suppose that at upper part the layer was close to the (110) and (101) crystallographic planes which is at 45° to (100). We evaluated the surface roughness of the sidewalls using the AFM. The rms (root mean square) roughness in the lower part of pyramids was about 7 nm, whereas a roughness of 0.8 nm in the upper part was measured. This finding gives us new possibilities in the future. We will prepare pyramids with a sidewall slope of 45° which belong to (110) and (101) planes. The roughness of such sidewalls will be probably better after the overgrowth.

The second feature is observable at two of four edges when two sidewalls meet. Figure 3b shows that at the [011]- and [0 $\overline{1}$ $\overline{1}$]-oriented bottom corners of the pyramids small facets were formed. This is probably caused by an asymmetry of the surface, i.e. the surfaces can be Ga- or As-rich, and consequently this arrangement leads to the revelation of the faceting. The diameter of curvature of the tip is about 50 nm after overgrowth.

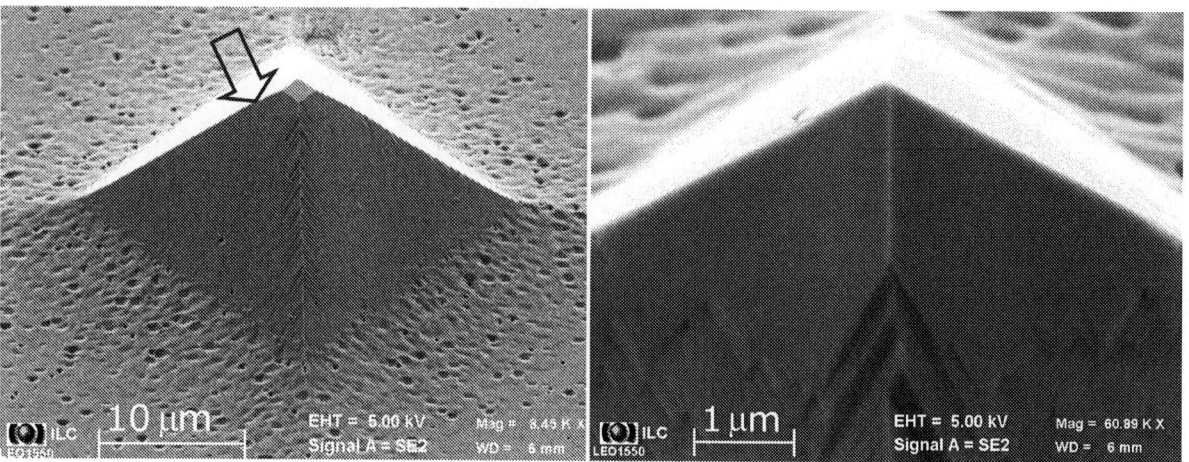

Fig. 3 (a) SEM picture of overgrown pyramid. The arrow indicate the place when sidewall slope decrease from angle 39° to circa 33°. (b) Detail at the upper part of the sharpened pyramid (not a detail of pyramid in Fig. 3a). Smooth surface probably represent a (203) crystallographic plane.

4.   Summary

This paper reports on the microfabrication of 3D structures that can be used as active tips for use in SPM. We found that the rotated squares were the most suitable mask patterns for the preparation of tips of pyramidal shape. The pyramids are overgrown with n-doped GaAs layer. The smooth objects prepared are suitable for next technological steps (lithography, metallization, ohmic contacts processing, passivation, etc) to prepare active tips that will incorporate active devices, such as transistors and sensors.

Acknowledgement
This work was sponsored by the Slovak APVV Agency, project n. APVV- 51-045705 and by the project n. VEGA /3108/06.

References

[1]   S. J. Bending, *Adv. Phys.* **48,** 449, 1999.
[2]   A. de Lozanne, *Supercon. Sci. Technol.* **12,** R43, 1999.
[3]   L. H. Chen, M.A. Topinka, B.J. LeRoy, R.M. Westervelt, K.D. Maranowski, and A.C. Gossard, *Appl. Phys. Lett.* **79**, 1202, 2001.
[4]   R. Shinohara, K. Yamaguchi, Y. Suzuki, W. Nabhan, *Jpn. J. Apl. Phys.* **37**, 7151, 1998.
[5]   V. Cambel, D. Gregušová, R. Kúdela, *J. Appl. Phys.* **94**, 4643, 2003.
[6]   G.J. Bauhuis, P. Mulder, H. van Kempen, *J. Crystal Growth* **240**, 104, 2002.

RF characterization and modeling of AlGaN/GaN based HFETs and MOSHFETs

A. Fox, M. Marso, G. Heidelberger and P. Kordoš*

Center of Nanoelectronic Systems for Information Technology and Institute of Bio- and Nanosystems, Research Center Jülich, D-52425 Jülich, Germany
* Institute of Electrical Engineering, Slovak Academy of Sciences, Bratislava, SK-84104 Bratislava, Slovakia, and Department of Microelectronics, Slovak Technical University, SK-81219 Bratislava, Slovakia
e -mail: A.Fox@fz-juelich.de

An increased RF-Performance of heterojunction field-effect transistor was found to be due to passivation and in addition a SiO₂ insulation underneath the gate matallization. This leads to an increase of cutoff frequency from 17 GHz up to 24 GHz for devices with 500 nm gate length. The RF output power increased from 4.1 to 6.7 W/mm at 7 GHz. RF simulation based on measured S-parameter showed a decrease of gate-source-capacitance and transconductance for devices with a dielectric layer underneath the gate metallization. The increase of the ratio g_m/C_{gs} of about 25% is in agreement with the measured cutoff frequency f_t.

1. Introduction

High power and high temperature electronics applications are one of the reasons for the enormous effort and progress which has been made in the development of III-nitride semiconductor material. A lot of investigations were done with regard to DC/RF dispersion since it influences the RF-performance of devices extremely. It results in a discrepancy between measured RF output power of the device and the calculated output power from DC characterization according to $P_{out}= (I_{max} * V)/8$, where I_{max} and V are maximum current and voltage swing at the output of the device. The RF-drain current has to be optimized to increase the power performance; however it is reduced by trapping effects at the exposed gate drain region. Silicon nitride passivation mainly reduces current slump by removing the surface charges [1]. Another method to reduce trapping effects and hence current slump is the addition of a p-GaN cap layer. In a further approach the introduction of a thin insulation layer underneath the gate to reduce the gate-source-capacitance and the gate leakage current leads to the MISHFET or MOSHFET. [2,3]. In this work we compare the two approaches passivation and MISHFET.

2. Device properties

The AlGaN/GaN material structures were prepared by LP-MOVPE growth on SiC substrates. Standard device processing steps were used for the passivated and unpassivated devices. Details are given in [1]. Additionally 5, 10 and 20 nm thick SiO₂ layer were deposited as a dielectric layer underneath the gate metallization for the MOSHFETs.

3. Small signal characterization

The influence of the passivation/insulation layer on device performance is demonstrated in the results of device modelling and equivalent circuit parameter extraction from the measured small signal S-parameters. The S-parameters were measured with the Vector Network analyzer system 8510XF (up to 110 GHz). The range for the drain voltage was 0 to 20V for

1-4244-0396-0/06/$25.00 ©2006 IEEE

all devices. The gate voltage for the HFET and the passivated devices varied from 1V to –5V and for the MOSHFET from 1V to –10V due to the higher negative threshold voltage.

The influence of the dielectric layer on the small signal performance is investigated by comparing h_{21} vs. frequency of the MOSHFET and the HFETs, as shown in Fig 1.The cutoff frequency f_t is defined as the frequency for zero gain value of h_{21}. The cutoff frequency f_t increased from 17 GHz for the HFET up to 24 GHz for the MOSHFET.

Due to different threshold voltages for HFET and MOSHFET devices the best cutoff frequencies were achieved with different gate source voltages at –3V and –6.5V, respectively, for a fixed drain source voltage of 20V. The cutoff frequency decreases for all types of devices with increasing gate length, as shown in Fig 2.

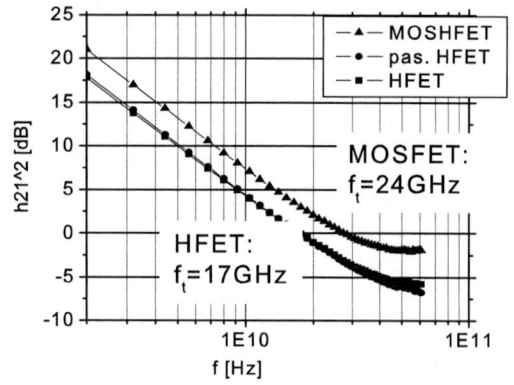

Fig 1: h_{21} vs. frequency for passivated-HFET, unpassivated-HFET and MOSHFET. Cutoff frequency f_t is determined @h_{12}=0dB. SiO$_2$ layer thickness: 10nm

Fig 2: Cutoff frequency f_T vs. gate length l_g for passivated-HFET, unpassivated-HFET and MOSHFET. SiO$_2$ layer thickness: 10nm

4. Modelling

Device modelling based on the measured S-parameter was done to evaluate the extrinsic and intrinsic parameter, e.g. g_m and C_{gs}, of the device model. Fig 3 shows the input capacitance C_{gs} of the three devices, while the extracted high frequency transconductance is depicted in Fig. 4. The cutoff frequency relevant ratio g_m to C_{gs} vs. the drain voltage is shown in Fig. 5. The calculated values of f_t from the ratio gm/C_{gs} are in agreement with the increase of cutoff frequency for the MOSHFET compared to unpassivated and passivated HFET.

5. Large signal characterization

Large signal measurements are performed with a load- and source- pull measurement system to match input and output impedances of the device to achieve maximum output power. P_{OUT} increases with the drain-source voltage V_{ds}. In our measurements we did not apply the highest possible drain source voltage to prevent the device from breaking through. Fig. 6 shows the output power for HFET and MOSHFET with different dielectric layer thicknesses vs. input power measured at 7 GHz. The maximum output power is found with a dielectric thickness of 5 nm.

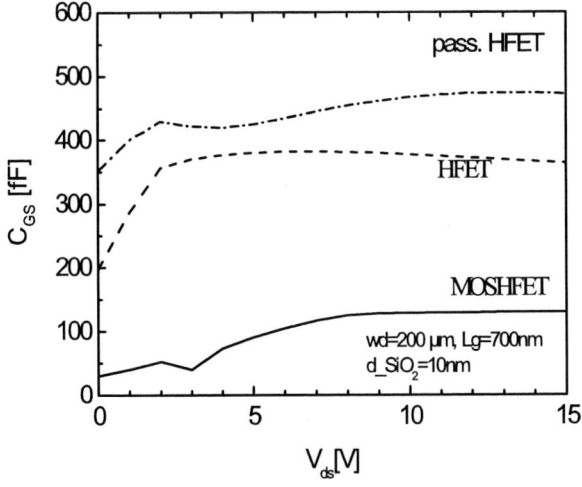

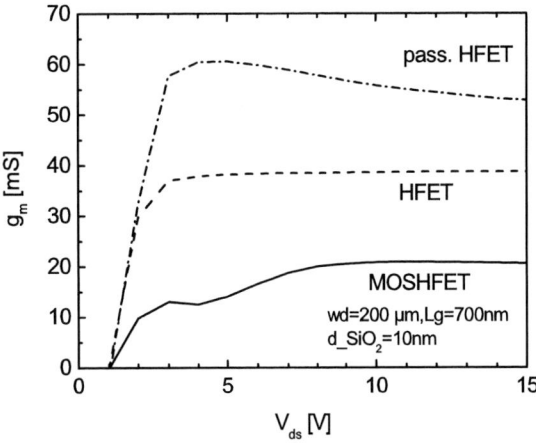

Fig 3: calculated gate source capacitance versus drain source voltage.

Fig 4: calculated transconductance versus drain source voltage.

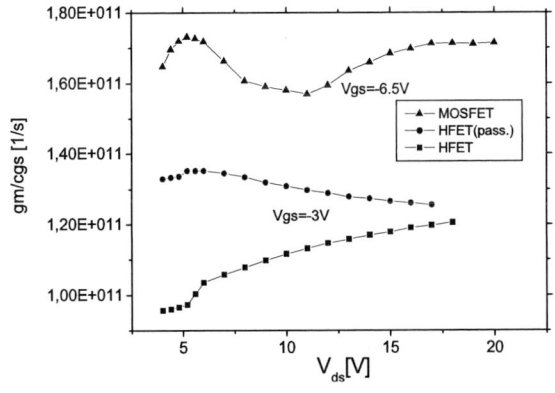

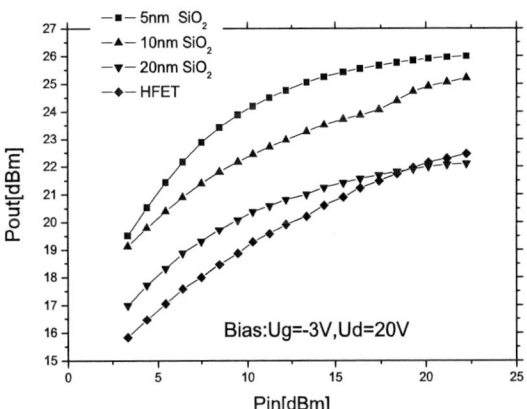

Fig 5: g_m / C_{gs} ratio vs. drain source voltage for HFET and MOSHFET. SiO$_2$ layer thickness: 10nm

Fig 6: P_{out} for HFET and MOSHFET with different SiO$_2$ dielectric layer.

6. Simulation

With the measured small signal S-parameter further simulations are performed by the simulation tool "Advanced Design System", ADS (Agilent Technologies). These simulations include all intrinsic and extrinsic parameters from the devices in a simulation-file created by the extraction program TOPAS (IMST, Kamp-Lintfort, Germany). A linear simulator for DC simulations shows the basic behaviour of the model which is comparable with the measured DC- data with respect to the extrinsic parameter. The S-parameter are simulated by the linear S-parameter simulator to compare the quality of simulated data with measured data. For large signal investigations the non-linear simulation tool "harmonic balance simulator" is used to perform the load- and source- pull simulations which give the impedance matching at the in-and output of the devices. With the matched impedances at input and output the maximum output power, power added efficiency PAE and gain are simulated at different bias points and frequencies of interest. Figs 7. and 8. show first results comparing simulation and measurements of output power, gain and PAE. The PAE shows slight deviations since it depends on input matching. Further investigations could be done for gain stability, noise behaviour and higher order harmonic components of the output signal which were brought up by the device. The generated simulation data file can be used in ADS as circuit design tool e.g. for designing schematics for mixer-, filter- or amplifier design

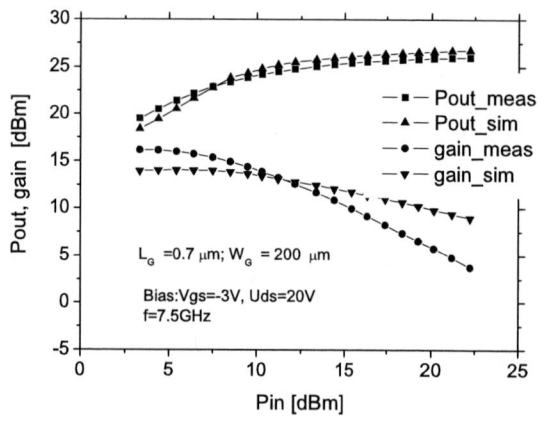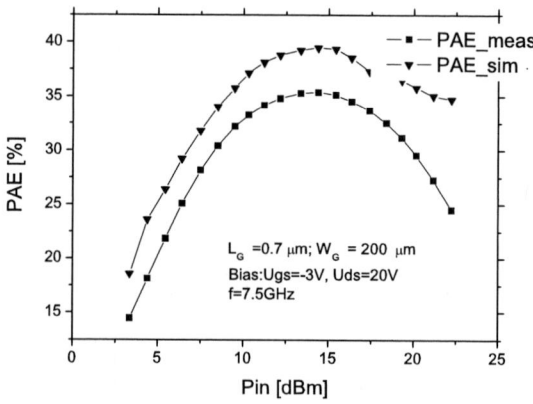

Fig. 7:.Comparison of simulated and measured
  P_{out} and gain for MOSHFET.
    SiO$_2$ layer thickness: 10nm

Fig. 8: Comparison of simulated and
  measured PAE for MOSHFET.
  SiO$_2$ layer thickness: 10nm

7. Conclusion

From literature it is known that performance of AlGaN/GaN HFET increases by passivation of the devices. In this work we report an increase of RF- performance by introduction of the MOSHFET- technology. This is demonstrated by the increase of cutoff frequency with the small signal behaviour and the increase of output power with large signal performance. The simulations are in accordance to the large signal measured results and are also confirmed by the S-parameter measurements. Further investigation will be done with different dielectric materials underneath the gate.

References

[1] J. Bernát, P. Javorka, A. Fox, M. Marso and P. Kordoš, "Influence of Layer Structure on Performance of AlGaN/High Electron Mobility Transistors before and after passivation. Journal of Electronic Materials, 33 (2004) 436

[2] M. Marso, G. Heidelberger, M. Indlekofer, J. Bernat, A. Fox, P. Kordoš and H. Lüth, "Origin of Improved RF Performance of AlGaN/GaN MOSHFETs Compared to HFETs, IEEE Transactions on Electron Devices 53 (2006) 1517

[3] P. Kordoš, G. Heidelberger, J. Bernat, A. Fox, M. Marso and H. Lüth, "High-Power SiO$_2$/AlGaN/GaN metal-oxide-semiconductor heterostructure field effect transistors, Applied Physics Letters, 87, 143501 (2005).

Electronic transport in carbon nanotubes: From individual nanotubes to thin and thick networks

V. Skákalová, A. B. Kaiser[*], Y.-S. Woo and S. Roth

Max Planck Institute for Solid State Research, Heisenbergstr. 1, 70569 Stuttgart, Germany
[*]MacDiarmid Institute for Advanced Materials and Nanotechnology, SCPS, Victoria University of Wellington, P O Box 600, Wellington, New Zealand

PACS number_s_ : 73.63._b

We measure and compare the electronic transport properties of individual single-wall carbon nanotubes (SWNTs) and SWNT networks of varying thickness [1]. Study of electronic transport through an individual metallic SWNT suggests Luttinger liquid mechanism. Arrhenius plot shows (Fig. 1) no sign of Schottky barrier at the contacts between the SWNT and the metal leads.

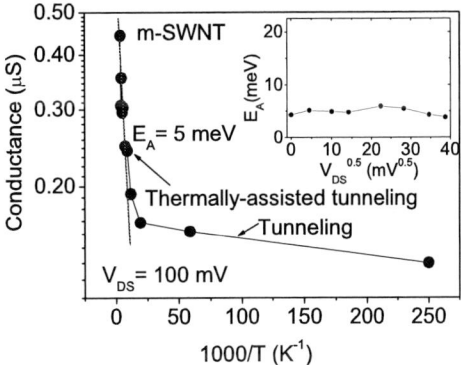

Fig. 1. Conductance G at 100mV (using a logarithmic scale) as a function of inverse temperature, showing two mechanisms: tunneling at low temperatures and activated behavior with energy E_A= 5 meV at high temperatures.

This is very different from an individual semiconducting SWNT, where Schottky barrier obviously dominates the electronic transport (Fig. 2).

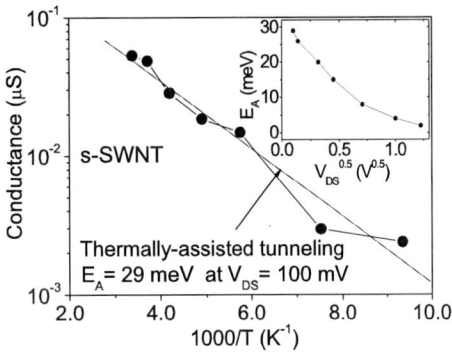

Fig. 2. Conductance G at 100mV (using a logarithmic scale) as a function of inverse temperature, showing a good agreement with activated behavior with activation energy E_A= 29 meV.

The thinnest SWNT networks, like the individual semiconducting SWNTs, show nonlinear current-voltage (*I-V*) characteristics at low temperatures (Fig. 3). Schottky barriers are formed at the interface of the metal contacts as well as at the junctions of semiconducting and metallic tubes, similar to the case of the individual s-SWNT presented above. We also demonstrate a field effect transistor (FET) device with a current that can be tuned by a gate voltage (Fig. 4). At room temperature, the "on-off" ratio of the device made of the highly transparent SWNT network (Net 1) is about 10000.

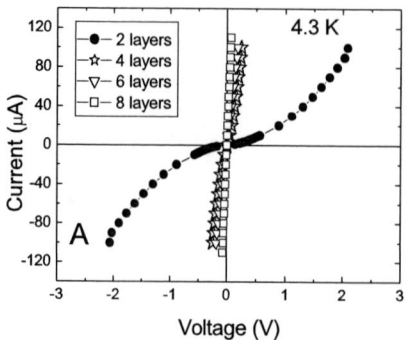

Fig. 3. I_D–V_{DS} characteristics of thin SWNT networks Net 1 (thinnest) to Net 4 at 4.3 K

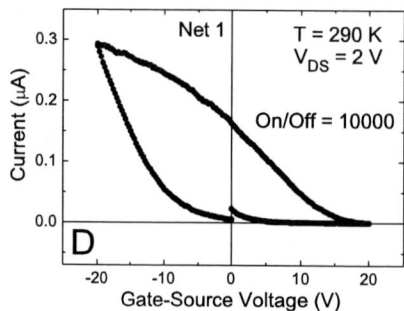

Fig. 4. The effect of a gate voltage V_{GS} on the current I_D of Net 4 at temperatures of 4.3 K (A) and at 290 K (B) and of Net 1 at temperatures of 4.3 K (C) and at 290 K (D).

The overall temperature dependence of conductance in the transparent networks changes systematically as the thickness of the network increases and is consistent with hopping conduction (Fig. 5). On the other hand, the thickest SWNT networks (free-standing film) show more metallic behaviour: their *I-V* characteristics are linear with no gate-voltage effect, and a large fraction of their conductivity retained at very low temperatures, consistent with tunnelling through thin barriers separating metallic regions. In the present fit we varied the value of T_m to achieve the best fit to the free-standing film data in Fig. 5, obtaining a value of 1830 K (corresponding to 160 meV or 1230 cm^{-1}). This value agrees with the band of energies 1000 – 2000 cm^{-1} expected for the in-plane zone boundary phonons in a SWNT [2,3] This agreement leads us to suggest that it is backscattering by these phonons that causes the turnover to metallic sign of the conductivity T dependence for the free-standing film sample. We note that in a complementary experiment investigating the dependence of conductance on bias voltage rather than temperature, Yao et al. [3] found that optical and zone boundary phonons played the major role in limiting the conductance as bias voltage reached large values for SWNTs with low-resistance electrode contacts.

We make a comparison with individual MWNTs, which in some cases show even greater retention of conductance at very low temperatures, but (unlike the thickest SWNT networks) no changeover to metallic temperature dependence at higher temperatures.

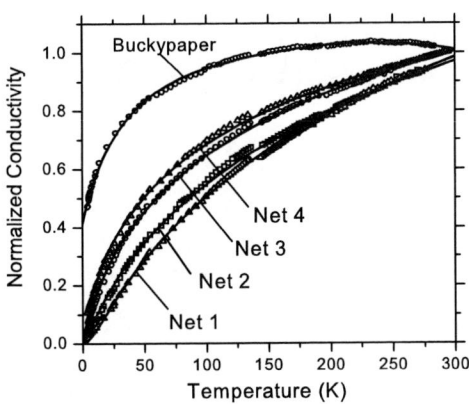

Fig. 5. Temperature dependences of the electrical conductance of SWNT buckypaper and four thin networks normalized to their values at room temperature $\sigma(T)/\sigma(292\ K)$

[1] V. Skákalová, A. B. Kaiser[*], Y.-S. Woo and S. Roth, *Accepted to Phys. Rev. B* 74 (2006).
[2] M. S. Dresselhaus, G. Dresselhaus, and P. C. Eklund, *Science of Fullerenes and Carbon Nanotubes* (Academic, San Diego, 1996).
[3] Z. Yao, C. L. Kane, and C. Dekker, Phys. Rev. Lett. **84**, 2941 (2000).

Electron-electron Interaction Induced Parabolic Negative Magnetoresistance in Two-dimensional Electron Gas in InGaAs/InP

B. Pődör[1,2], and G. Reményi[3]

[1] Budapest Tech, Kandó Kálmán Faculty of Electrical Engineering,
Institute of Microelectronics and Technology, Budapest, Hungary

[2] Hungarian Academy of Sciences, Research Institute for Technical Physics
and Materials Science, Budapest, Hungary

[3] CNRS Centre de Recherches sur les Tres Basses Températures et Laboratoire
des Champs Magnétiques Intenses, Grenoble, France
E-mail: podor.balint@kvk.bmf.hu

We have studied the parabolic negative magnetoresistance in modulation-doped $In_{0.53}Ga_{0.47}As/InP$ heterostructures below 4.2 K temperature. It is shown that the observed negative magnetoresistance in the two-dimensional electron gas can be explained in terms of electron-electron interaction in two-dimension.

1. Introduction and theoretical bacground

The magnetoresistance is one if the most frequently studied characteristics of the two-dimensional electron gas (2DEG). In the Drude-Boltzmann theory the resistivity of a degenerate 2DEG does not depend on the magnetic field, $\rho_{xx}(B) = \rho_0 = m^*/(e^2 n_s \tau_t) = (h/e^2)/(k_F l_e)$, where n_s is the 2DEG density, τ_t is the transport scattering time, k_F is the Fermi wave number and l_e is the electron mean free path. There are several sources of non-trivial magnetoresistance (MR), which reflect the rich physics of the 2DEG [1]. Besides quasi-classical effects two types of quantum corrections, i.e. weak localization and electron-electron interaction induced negative MR.

Two types of quantum corrections, such as weak localization [2] and electron-electron interaction [3] affect the transport in disordered two-dimensional electron gas. These perturbative corrections are of the order of $(k_F l_e)^{-1}$. In the lowest order the weak-localization and electron-electron interaction corrections are additive, and in zero magnetic field they give a correction to the Drude conductivity, which is logarithmic in the temperature [4, 5]

$$\sigma_{xx}(T) = \sigma_0 + \Delta\sigma_{xx}(T) = \sigma_0 + (e^2/2\pi^2 h)[\alpha p + (1-3F^*/4)\ln(kT\tau/h)] \qquad (1)$$

here σ_0 is the Drude conductivity, the second and the third term in the right-hand side is from the weak-localization effect and from the electron-electron interaction, respectively. For an orbitally non-degenerate electron gas $\alpha = 1$, p is the exponent in the temperature dependence of the phase-breaking relaxation time $\tau_\phi \propto T^{-p}$, F^* is the renormalized screening parameter. Finally τ is the elastic scattering time, and the pre-factor is $e^2/2\pi^2 h = 1.233 \times 10^{-5}$ ohm^{-1}.

In nonzero magnetic field, the weak-localization correction is quenched for fields above a maximum value given by $B_{tr} = h/2el_e^2$ [5, 6]. The electron-electron correction is essentially independent of the magnetic field. However the Zeeman effect can give rise to an additional correction for fields $B_Z > kT/g^*\mu_B$ (here μ_B is the Bohr magneton and g^* is the effective electron g-factor). So for a wide enough magnetic field range $B_{tr} < B < B_Z$ only the electron-electron interaction is significant and we have

$$\sigma_{xx}(T) = \sigma_0 + \Delta\sigma_{xx}{}^{eei}(T) = \sigma_0 + (e^2/2\pi^2 h)g(F^*)\ln(kT\tau/h) \qquad (2)$$

where $g(F^*) = (1-3F^*/4)$, is the order of unity.

Theory also predicts that there is no correction to the Hall conductivity due to the electron-electron interaction, $\Delta\sigma_{xy}{}^{eei} = 0$ [4, 7]. Then inversion of the conductivity tensor yields for the longitudinal resistivity in function of the magnetic field [7, 8]

$$\rho_{xx}(B) = \sigma_0{}^{-1} + [1 - (\omega_c\tau)^2]\Delta\sigma_{xx}{}^{eei}/\sigma_0{}^{-2} \qquad (3)$$

where ω_c is the cyclotron frequency. Eq. (3) predicts a negative magnetoresistance proportional to B^2 even though the correction to the conductivity is independent of B. An important consequence of Eq. (3) is that when $\omega_c\tau = \mu B = 1$ then $\rho_{xx} = \sigma_0{}^{-1} = \rho_0 = \rho_{Drude}$, i.e. the $\rho_{xx}(B)$ curves measured at different temperatures should cross each other at this magnetic field [9].

Similarly, for the low-field Hall coefficient $R_H = \rho_{xy}/B$ we have

$$\Delta R_H/R_H = \Delta\rho_{xy}/\rho_{xy} = 2\Delta\rho_{xx}/\rho_{xx} = 2\Delta\sigma_{xx}/\sigma_{xx} \qquad (4)$$

We report here our results on the parabolic negative magnetoresistance due to electron-electron interaction in 2DEG in $In_{0.53}Ga_{0.47}As/InP$ heterostructures. Some aspects of this work have already been published in [10].

2. Experimental details and sample characterization

The samples were liquid phase epitaxial (LPE) modulation-doped $In_{0.53}Ga_{0.47}As/InP$ heterostructures with 2DEG density and mobility of $(0.3-4)\times10^{11}$ cm^{-2} and $(2-6)\times10^4$ cm^2/Vs respectively [10-12]. The low-temperature mobility in the 2DEG in these samples, corresponding to $k_F l_e = 3$ to 40, is mainly limited by alloy scattering. The samples were characterized by the ratio of the transport to quantum scattering time, τ_t/τ_q, deduced from the Shubnikov-de Haas oscillations [10, 13]. The values of τ_t/τ_q decreased with decreasing 2DEG concentration, and were typically ~1.5, 2 to 3, and 6 to 8 respectively for concentrations of $(4-5)\times10^{10}$, $(1-2)\times10^{11}$, and $(3-4)\times10^{11}$ cm^{-2} respectively, reflecting the dominant role of small-angle scattering, i.e. of alloy disorder scattering and of scattering on interface irregularities at low 2DEG concentration. Persistent photoconductivity was used to control the 2DEG density. Magnetotransport coefficients (ρ_{xx} and ρ_{xy}) were measured in the temperature range from 40 mK to 4.2 K on photolithographically defined double cross Hall bars. Both conventional dc and low-frequency lock-in techniques were used.

3. Experimental results and discussion

Fig. 1 shows typical data for the temperature dependence of the longitudinal and Hall resistivity, plotted as relative changes $\Delta\rho_{xx}/\rho_{xx}$ and $\Delta\rho_{xy}/\rho_{xy}$, respectively, referred to their corresponding values at 4.2 K temperature. The logarithmic dependence of $\Delta\rho_{xx}/\rho_{xx}$ on the temperature, characteristic for both types of quantum corrections is clearly demonstrated. The experimental values of the parameter g(F*) in Eq. (2) were found to be 0.6 to 3.0 in various samples. These are in accord with the values found in other material systems, notably in AlGaAs/GaAs [14]. Weak-localization would yield $\Delta\rho_{xy}/\rho_{xy} = 0$, i.e. no or vanishingly small correction to the Hall resistivity, in contrast to the experimental observations. For the samples investigated the ratio $(\Delta\rho_{xy}/\rho_{xy})/(\Delta\rho_{xx}/\rho_{xx})$, i.e. the ratio of the relative corrections to the Hall resistivity to the relative corrections to the longitudinal resistivity was in the range 1 to 2.5, in an acceptable agreement with the theoretical value of 2 expected on the basis of electron-electron interaction theory.

Representative data for the dependence of the longitudinal resistivity on the magnetic field at various temperatures are presented in Fig. 2. At low magnetic fields, before the onset of Shubnikov-de Haas (ShdH) oscillations, a parabolic negative magnetoresistance was clearly observed, which can be ascribed to electron-electron interaction effects [13]. In intermediate fields ShdH oscillations were observed, followed by the quantum Hall effect regime.

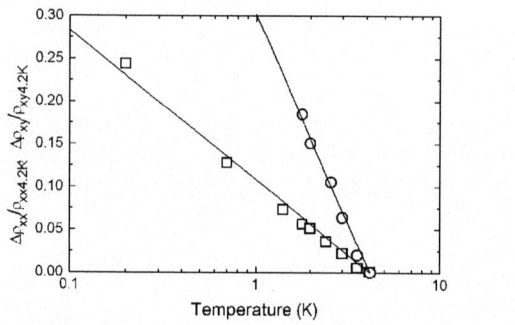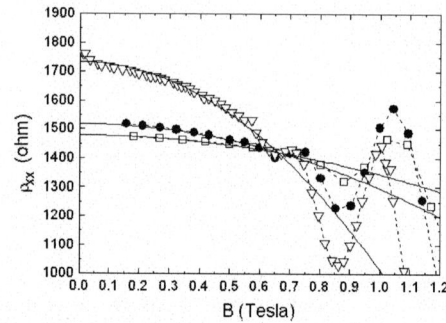

Fig. 1. Relative changes of the longitudinal resistivity $\Delta\rho_{xx}/\rho_{xx}$ and of the Hall resistivity $\Delta\rho_{xy}/\rho_{xy}$ versus the temperature.

Fig. 2. Longitudinal resistivity ρ_{xx} vs. the magnetic field at temperatures 4.2 K (squares), 1.8 K (full circles), and 0.3 K (down triangles). Full lines – fitted component quadratic in B.

The experimental $\rho_{xx}(B)$ curves were fitted by the appropriate theoretical relation. The results for the fitted parabolic contribution are shown in Fig. 2. It can be seen that the smoothly varying parabolic contributions to the $\rho_{xx}(B)$ resistivity for different temperatures cross each other in one point in the $\rho_{xx}(B)$-B plane as stipulated by the electron-electron interaction theory. The relaxation time determined from the crossing point of the magnetoresistance curves ($\omega_c\tau = 1$) corresponded within about 20 percent to the quantum relaxation time τ_q, deduced from the decay of the Shubnikov-de Haas oscillations.

The values of the electron-electron interaction induced correction to the Drude conductivity, $\Delta\sigma_{xx}^{eei}$, decuced from the magnetoresistance curves using Eq. (3) exhibited the characteristic logarithmic temperature dependence too.

All the above results are consistent with each other and with the model for the quantum corrections due to electron-electron interaction.

4. Conclusions

To sum up, we have observed a large negative parabolic magnetoresistance in 2DEG in modulation-doped $In_{0.53}Ga_{0.47}As/InP$ heterostructures. The magnitude of this negative magnetoresistance as well as its dependence on the temperature can be satisfactorily interpreted invoking electron-electron interaction effects in the 2DEG.

Acknowledgments

The measurements were performed in the laboratories of the CNRS in Grenoble, in part under projects SE1295 and SE2198. Supported in part from the Hungarian National Research Fund (OTKA) grants No. 031763 and 048696.

References

[1] I. V. Gornyi and A. D. Mirlin, *Physica E* **22**, 260, 2004.
[2] E. Abrahams, P. W. Anderson, D. C. Licciardello, and T. V. Ramakrishnan, *Phys. Rev. Lett.* **42**, 673, 1979.
[3] B. L. Althsuler, A. G. Aronov, and P. A. Lee, *Phys. Rev. Lett.* **44**, 1288, 1980.
[4] B. L. Altshuler and A. G. Aronov, in *Electron-Electron Interaction in Disordered Systems*, Ed. A. L. Efros, M. Pollak, Elsevier, Amsterdam, 1985, p. 1.
[5] P. A. Lee and T. V. Ramakrishnan, *Rev. Mod. Phys.* **57**, 287, 1985.
[6] K. K. Choi, *Phys. Rev. B* **28**, 5774, 1983.
[7] A. Houghton, J. R. Senna, and S. C. Ying, *Phys. Rev. B* **25**, 2196, 1982.
[8] M. A. Paalanen, D. C. Tsui, and J. C. M. Hwang, *Phys. Rev. Lett.* **51**, 2227, 1983.
[9] Yu. G. Arapov, G. I. Kharus, O. A. Kuznetsov, V. N. Neverov, and N. G. Shelushkina, *Fiz. Tekh. Poluprovodn.* **33**, 1073, 1999.
[10] B. Pődör, I. G. Savel`ev, Gy. Kovács, and G. Reményi, in *Proceedings of the International Workshop on Semiconductor Nanostructures, SEMINANO2005*, Budapest, Hungary, 2005, Ed. B. Pődör, Zs. J. Horváth, P. Basa, Vol II, p. 341.
[11] L. V. Golubev, A. M. Kreshchuk, S. V. Novikov, T. A. Polyanskaya, I. G. Savel`ev, and I. I. Saidashev, *Fiz. Tekhn. Poluprovodn.* **22**, 1948, 1989.
[12] I. G. Savel`ev, A. M. Kreshchuk, S. V. Novikov, A. Y. Shik, G. Remenyi, Gy. Kovács, B. Pődör, and G. Gombos, *J. Phys. Cond. Matter* **8**, 9025, 1996.
[13] B. Pődör, S. V. Novikov, I. G. Savel`ev, and G. Gombos, *Acta Phys. Hung.* **74**, 147, 1994.
[14] L. Li, Y. Y. Proskuryakov, A. K. Savchenko, E. H. Linfeld, and D. A. Ritchie, *Phys. Rev. Lett.* **90**, 076802, 2003.

Terahertz-Radiation Photomixers on Nitrogen-Implanted GaAs

M. Mikulics,[a,b] M. Marso,[a] S. Stanček,[c] E. A. Michael,[d] and P. Kordoš[e,f]

[a] Institute of Bio and Nanosystems, Research Centre Jülich, D-52425 Jülich, Germany
[b] Institut für Hochfrequenztechnik, Technische Universität Braunschweig, D-38106 Braunschweig, Germany
[c] Slovak University of Technology, Department of Nuclear Physics and Technology, SK-81219 Bratislava, Slovak Republic
[d] 1. Physics Institute, University of Cologne, D-50937 Cologne, Germany
[e] Department of Microelectronics, Slovak Technical University, SK-81219 Bratislava
[f] Institute of Electrical Engineering, Slovak Academy of Sciences, SK-84104 Bratislava, Slovak Republic

We have fabricated and characterized photomixers based on high energy nitrogen-ion-implanted GaAs. For material optimization and annealing dynamics in MSM photodetector structures, we used 400 keV implantation energy with an ion dose of 1×10^{16} cm^{-2}. For photomixer structures we used 3 MeV energy to implant N^+ ions into GaAs substrates, with an ion concentration dose of 3×10^{12} cm^{-2}. The N^+-implanted GaAs photomixers exhibit improved output power in comparison to their counterparts, photomixers fabricated on low-temperature-grown GaAs. The highest output power was 2.6 μW at 850 GHz and about 1 μW at 1 THz. No saturation of the output power with increased bias voltage and optical input power was observed. These characteristics make N^+-implanted GaAs the material of choice for efficient high power sources of terahertz radiation.

1. Introduction

The low temperature grown GaAs (LT-GaAs) fabricated by molecular-beam epitaxy has been recognized for more than a decade for its subpicosecond photocarrier trapping time and acceptably high carrier mobility, and so it is the mostly used material for tunable THz sources [1,2]. However, efficiency limits of the LT-GaAs material are already reached and therefore further performance improvement of photomixer devices can be obtained only by alternative material systems with subpicosecond carrier lifetime. One such approach is the implantation of various ions, such as O, Si, Ga, As, into GaAs [3,4]. It has been used to reduce the photocarrier trapping time and thus to achieve high-speed and broad-band performance of GaAs-based photodetector and photomixer devices [5,6]. The purpose of this paper, is to describe fabrication process, material optimization and annealing dynamics in 400 keV nitrogen implanted GaAs studied on MSM photodetector structures, as well as the properties of traveling-wave photomixers fabricated on high energy 3 MeV N^+-implanted GaAs substrates and to demonstrate the performance improvement of these photomixers compared to those fabricated on LT-GaAs [1].

1-4244-0396-0/06/$25.00 ©2006 IEEE

2. Device fabrication

At first, we have prepared the photomixer and photodetector material by implantation of nitrogen ions with energies of 400 keV (ion dose $1 \times 10^{16}/cm^2$) and 3 MeV (ion dose $3 \times 10^{12}/cm^2$), respectively, into semi-insulating GaAs (001) substrates with a resistivity $> 10^7 \ \Omega cm$ at 300K. Then, the whole surface except the designated area for the MSM structures was coated with 400 nm SiO_2. On top of this insulator layer 10 mm long and 15 μm wide Ti/Au coplanar strip lines (CPS) with thickness of 50/600 nm and 10 μm gap were fabricated using conventional photolithography and lift-off technique (Fig. 1). For photomixer devices we used electron beam lithography for fabrication of interdigitated MSM structure integrated with bow-tie antenna in CPS (Fig. 2). For the sake of comparison the same photomixer geometry was fabricated on LT-GaAs.

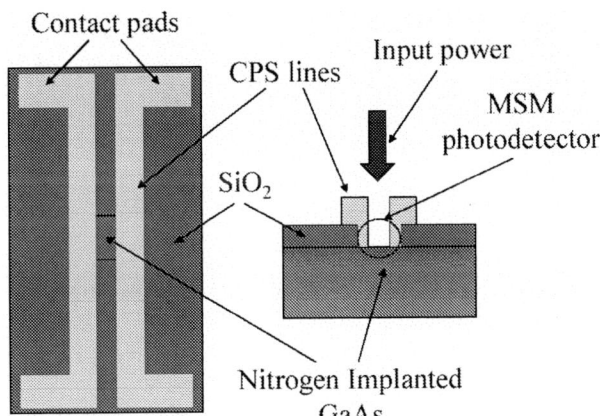

Figure 1: Schematics of photodetector structure integrated with coplanar strip lines (CPS) and contact pads.

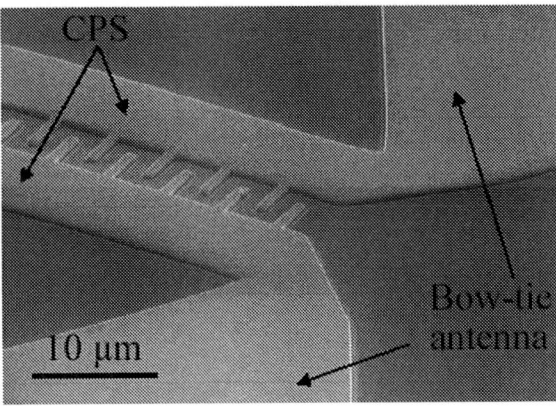

Figure 2: Traveling-wave photomixer structure integrated with coplanar strip lines and bow-tie antenna.

3. Annealing dynamics

Annealing of the N^+–implanted GaAs samples at various temperatures from 200°C to 600°C for 10 min in nitrogen ambient was performed in order to evaluate the annealing dynamics of nitrogen ions implanted into GaAs. The lifetime of the photogenerated carriers in the annealed nitrogen implanted GaAs was studied using femtosecond time resolved reflectivity measurements (Fig. 3). Carrier lifetimes of less than 400 fs can be obtained for the

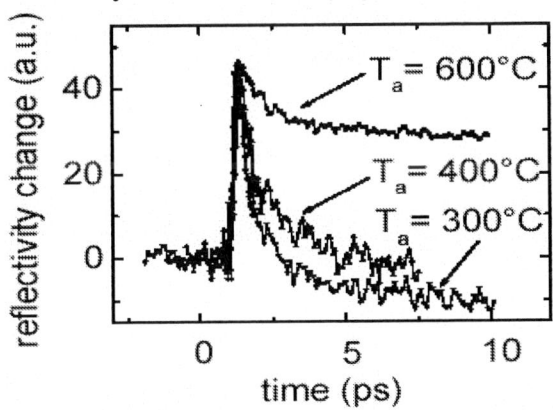

Figure 3: Reflectivity measurement on annealed GaAs:N samples.

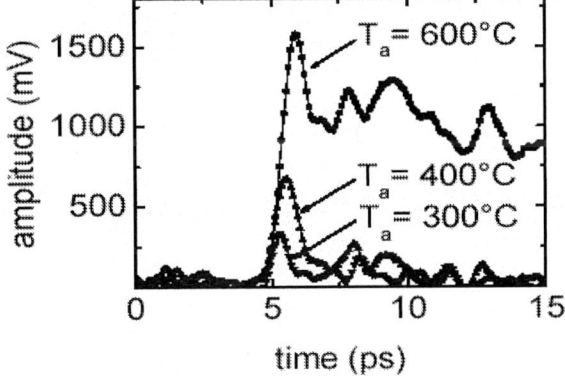

Figure 4: Transient photoresponse of annealed GaAs:N implanted samples.

samples annealed up to 400°C. Similar results were measured by an electro-optic sampling system on photodetector structures (Fig. 4). An increase of carrier lifetime with increased annealing temperature indicates that two processes are involved in the annealing dynamics of high-energy (400 keV) and high-dose (1×10^{16} cm^{-2}) implanted GaAs:N. The first one is characterized by low influence on the carrier lifetime, i.e. the carrier lifetime increases slowly with the annealing temperature. In the second process the thermally activated traps have longer relaxation time but are created in sufficient density only at higher annealing temperatures. The traps created at low annealing temperatures are responsible for the sub-ps carrier lifetimes. On the other hand, the traps generated by the second process reduce the dark current and the responsivity in the high temperature annealed samples (Fig. 5, 6), but their longer relaxation time is also responsible for the increase of the carrier lifetime.

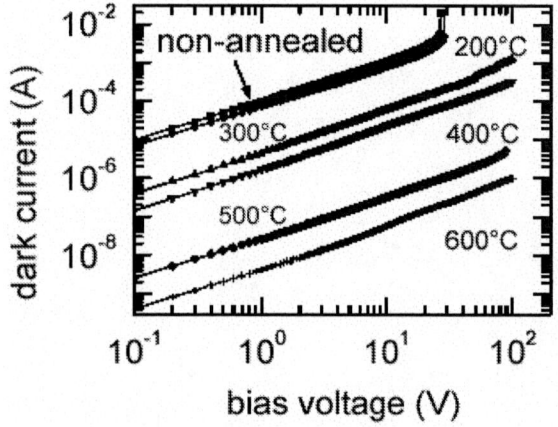

Figure 5: Current-voltage characteristics in the dark for MSM photodetectors on non-annealed and annealed GaAs:N.

Figure 6: Bias voltage dependence of responsivity for MSM photodetectors on annealed GaAs:N ($\lambda = 850$ nm, $P_{in} = 170\ \mu$W).

4. *Photomixing measurements*

The THz power spectrum of a traveling wave photomixer based on 3 MeV nitrogen implanted GaAs was measured with a calibrated magnetically tuned 4.2 K InSb-bolometer, which has a cutoff around 1.4 THz. Two heterodyned laser beams with different frequencies (~780 nm wavelength) were used for photomixer pumping [7]. Figure 7 compares the output power vs. frequency characteristics of a photomixer based on nitrogen implanted GaAs with

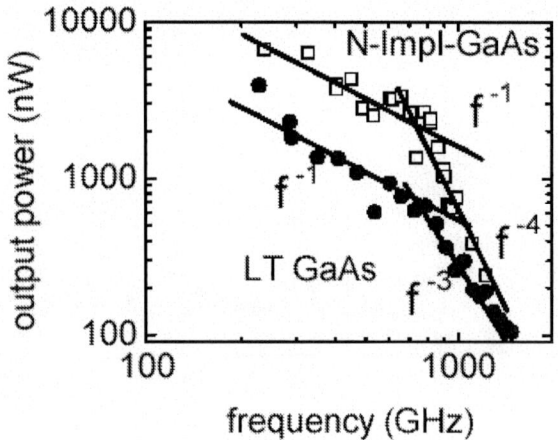

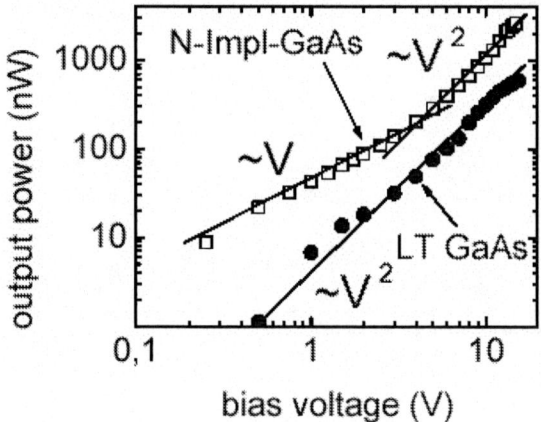

Figure 7: THz output power spectrum for a traveling-wave photomixer based on 3 MeV nitrogen-implanted GaAs and LT-GaAs.

Figure 8: Output power at 850 GHz vs. bias voltage for traveling-wave photomixers on 3 MeV nitrogen-implanted GaAs and LT-GaAs.

one on LT-GaAs material. The total input power was 400 mW and the bias voltage was 15 V. Figure 8 shows the output power as a function of the bias voltage for a photomixing frequency of 850 GHz. Better performance of the nitrogen implanted GaAs photomixer in the whole range of applied bias voltage is evident. For lower biases (up to ~4 V) a more than 300% increase of the output power is obtained, when compared to the LT-GaAs photomixer, with linear dependence on the bias voltage. At higher biases the output power increases quadratically with the bias voltage for both materials, according to the theory [1]. The output power of the nitrogen implanted GaAs based photomixer at the highest applied bias level (15 V) and 850 GHz is 2.64 µW, which is more than three times higher than that for LT-GaAs. Further, no indication of the output power saturation, that would be typical for LT-GaAs photomixers, is observed.

5. Conclusion

To conclude, we have fabricated and tested traveling-wave THz photomixers on nitrogen implanted GaAs as well as ultrafast MSM photodetector structures for material and device properties evaluation. Pump-probe experiments and DC characterization of ultrafast photodetectors based on nitrogen implanted GaAs with different annealing temperatures show the annealing dynamics resulting in changes of the dark current and photocurrent characteristics as well as carrier lifetime. It is clear that with the choice of the optimal annealing temperature, it is possible to achieve material suitable for the fabrication of ultrafast photodetectors and high efficiency photomixers. As compared to the photomixers fabricated on LT-GaAs, the GaAs:N-based devices show 200%-higher output power in the whole frequency range and up to 300% enhancement at 850 GHz for lower bias levels. The output power of 2.64 µW was obtained at 850 GHz and an input power of 400 mW and 15 V bias voltage. The nitrogen implanted GaAs interdigitated traveling-wave photomixers show no evidence of output power saturation that is typical for the LT-GaAs based photomixers. All these advantages are accompanied with the potentially higher attraction for industrial applications major for lower cost reasons and higher reproducibility of material parameters for implantation technology in comparison to the low-temperature MBE process. A further improvement of the device performance is expected by the next optimization of the dose and energy of the nitrogen ions implanted into the GaAs material.

References

[1] E. R. Brown, F. W. Smith, and K. A. McIntosh, J. Appl. Phys. **73**, 1480 (1993).
[2] E. Peytavit, S. Arscott, D. Lippens, G. Mouret, S. Matton, P. Masselin, R. Bocquet, J. F. Lampin, L. Desplanque, and F. Mollot, Appl. Phys. Lett. **81**, 1174 (2002).
[3] B. Salem, D. Morris, V. Aimez, J. Beerens, J. Beauvais, and D. Houde, J. Phys.: Condens. Matter **17**, 7327 (2005).
[4] A. Krotkus and J.-L. Coutaz, Semicond. Sci. Technol. **20**, S142 (2005).
[5] Tze-An Liu, M. Tani, M. Nakajima, M. Hangyo, Ci-Ling Pan, Applied Physics Letters, **83**, 1322 (2003).
[6] B. Salem, D. Morris, V.Aimez, J. Beauvais, D. Houde, Semicond. Sci. Technol. **21**, 283 (2006).
[7] E. A. Michael, B. Vowinkel, R. Schieder, M. Mikulics, M. Marso, and P. Kordoš, Appl. Phys. Lett. **86**, 111119 (2005).

Simulation of Advanced Tunneling Devices

A. Heigl and G. Wachutka

Institute for Physics of Electrotechnology, Munich University of Technology,
Arcisstrasse 21, 80290 Munich, Germany.
e-mail: heigl@tep.ei.tum.de and wachutka@tep.ei.tum.de

The progressive shrinking of the physical gate length of MOS devices involves problems such as punch-through, DIBL, and roll-off. In order to cope with short channel effects in nanoscale MOS devices, alternative device concepts, among others the tunnel FET, have been suggested. The proper description of the operation of these devices requires the consideration of two mechanisms of electron tunneling through the bandgap: Defect-assisted and phonon-assisted tunneling. Both prove to be crucial for predictive device simulation, making it possible to obtain realistic device characteristics of tunneling transistors.

1. Introduction

The everlasting gradual decrease of the physical gate length of MOS devices leads to well-known problems such as punch-through, roll-off, and DIBL. To some extent these could be coped with by down-scaling the vertical dimensions of the devices. However, a number of problems are remaining due to leakage currents, which are caused by parasitic tunneling. A radical possibility to reduce these effects consists in going to alternative device concepts like the tunneling transistor [1]. Here, tunneling is not an indesired parasitic effect, but part of the working principle. We studied such structures using the device simulator DESSIS. As for many nanometer MOS devices, it is essential to properly include tunneling through the bandgap in the physical device model. In case of indirect semiconductor materials, the most relevant processes are phonon-assisted and defect-assisted tunneling, because an exchange of energy as well as of momentum is necessary.

2. Modelling

The implementation of tunneling currents in state-of-the-art device simulators can conveniently be done by implementing generation rates, which translate a tunneling event into the equivalent generation of an electron-hole pair in the middle of the bandgap.

2.1 Defect-assisted Tunneling

An easy-to-use, but yet realistic physical model of tunneling via traps in the bandgap can be obtained on the basis of the effective-mass Schrödinger equation under the assumption of a linear potential [2]. This leads to an equivalent recombination rate of Shockley-Read-Hall-type as follows

$$R_{dat} = \frac{np - n_i^2}{\frac{\tau_p}{\Gamma_p + 1} \cdot (n + n_1) + \frac{\tau_n}{\Gamma_n + 1} \cdot (p + p_1)} . \tag{1}$$

with the recombination-lifetimes $\tau_{n,p}$ and the field-dependent functions $\Gamma_{n,p}$. The quantities

n_1 and p_1 depend on the difference between the trap energy levels and the intrinsic energy, and they scale with the intrinsic carrier density n_i. In the case of low electric fields, one obtains $\Gamma_{n,p} \ll 1$ and arrives, in the limit, at the classical SRH recombination-rate.

2.2 Phonon-assisted Tunneling

Setting up a model for phonon-assisted tunneling requires the calculation of the transmission probabilities; a popular approach is the WKB-approximation [3]. With the particle energy E and the electric field $\vec{F}$ as control variables, we arrive at the following expression for the tunneling rate:

$$ R_{pat} = -B \cdot \left| \vec{F} \right|^{\alpha} \cdot D\left(\vec{F}, E\right) \cdot \exp\left[-\frac{F_0}{\left| \vec{F} \right|} \right]. \tag{2}$$

where $\alpha = 7/2$ holds for indirect semiconductor materials [4] and $D = f_v - f_c$ denotes the Fermi-Dirac distribution functions related to the difference of intrinsic energy and the corresponding Fermi-energy levels. For use in numerical simulations, this expression is rewritten in an alternative form:

$$ D\left(\vec{F}, E\right) = f_v - f_c = \frac{n_i^2 - np}{\left(n + n_i\right) \cdot \left(p + n_i\right)}. \tag{3}$$

3.   Problem Description

We investigated a tunneling transistor (Fig. 1) consisting of a 10^{20} cm^{-3} arsenic-doped drain region, a source region that is 10^{20} cm^{-3} boron-doped and a 10^{16} cm^{-3} boron-doped channel of 90 nm length. A considered structure considered consists of a MOSFET with a Zener diode (ZFET) in series. The source region is split in a $4 \cdot 10^{19}$ cm^{-3} arsenic- and 10^{20} cm^{-3} boron-doped layer (Fig. 2). The working principle of the TFET is gate-controlled tunneling through a heavily doped pn-junction. Applying a positive gate voltage causes band bending so that an electron channel is formed. In case of high gate voltages, the tunneling barrier becomes independent of V_{DS}, resulting in an excellent saturation behaviour.

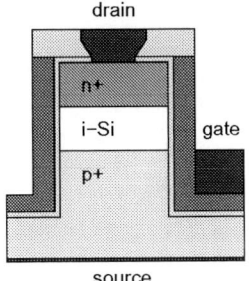

Fig. 1. Schematic view of simulated TFET structure.

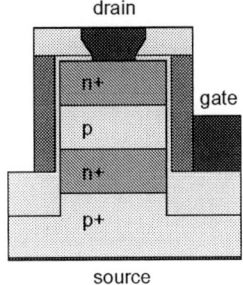

Fig. 2. Schematic view of simulated ZFET structure.

4.   Numerical Analysis

The simulations presented in the following were performed using the device simulator DESSIS. In Fig. 3 the simulated conduction and valence band energies of the ZFET are shown at different source-drain voltages. At the pn-junction on the source side, a very small barrier is formed and, thus a reasonably high tunneling current in the channel is observed. The simulated transfer characteristics of this device is shown in Fig. 4. We identify three different regions: For small V_{GS}, the device behaviour is dominated by the MOS channel. Then in a transition region, the barrier height is controlled by the gate voltage, and at high gate voltages, we obtain a very good saturation of the current. The latter effect results from the maximum tunneling current induced by the applied source-drain voltage. It should be noted that this device is governed by the Zener diode and, therefore, a further improvement is expected for a higher doping of the source region.

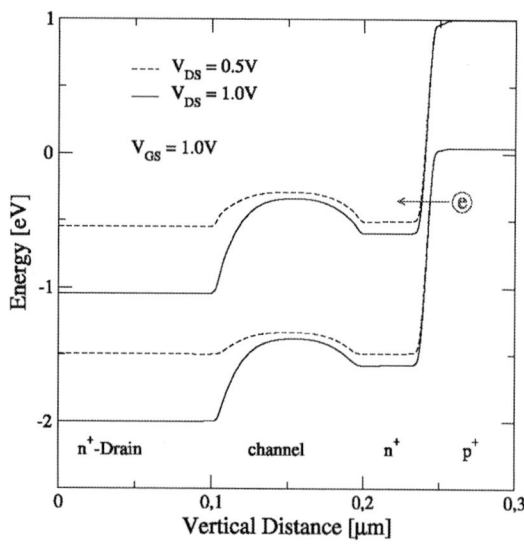

Fig. 3. Band diagram of the ZFET at $V_{GS} = 1.0V$.

Fig. 4. Simulated transfer characteristics of the ZFET.

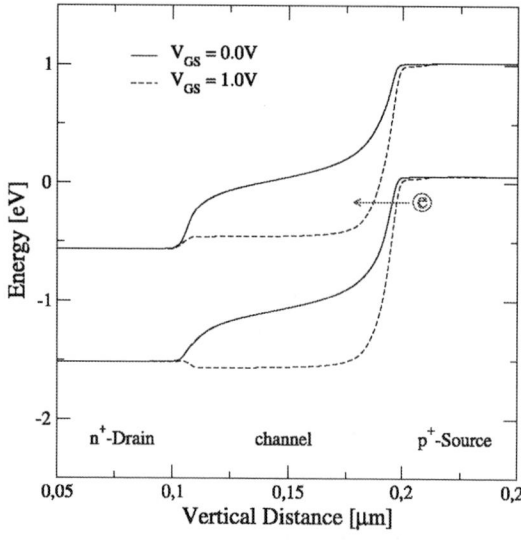

Fig. 5. Band diagram of the TFET at $V_{DS} = 0.5V$.

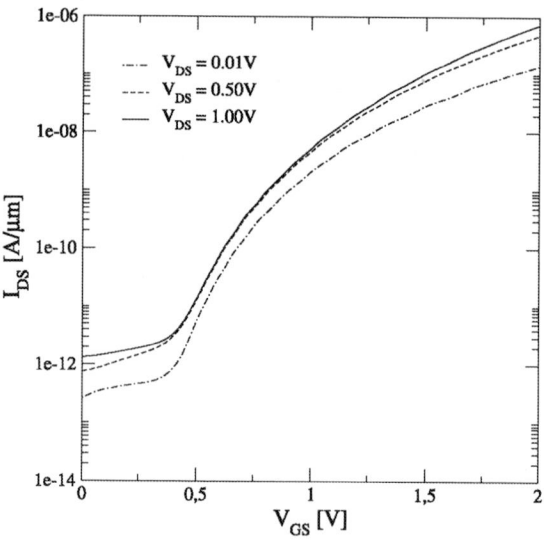

Fig. 6. Simulated transfer characteristics of the TFET.

Fig. 5 depicts the conduction and valence band energies as simulated for the TFET at different gate voltages. The most relevant finding is that the increase of V_{GS} is able to reduce the barrier width such that it causes a sufficient tunneling current in the channel. The transfer characteristics is shown in Fig. 6. Evidently the TFET has a significant on-current generated by phonon-assisted tunneling. Tunneling via traps is only in the case of negative V_{DS} of importance. A very important feature is the extremely low off-current of the tunneling FET and the resulting high on-off ratio. In the case of high gate voltage, the tunneling barrier becomes independent of V_{DS} and, therefore, we obtain an excellent saturation behaviour of the drain current in the forward output characteristics (Fig. 7). By using the models presented above it is possible to simulate the complete device characteristics of the TFET. Even the excess current typical of a highly doped tunneldiode can be observed (Fig. 8).

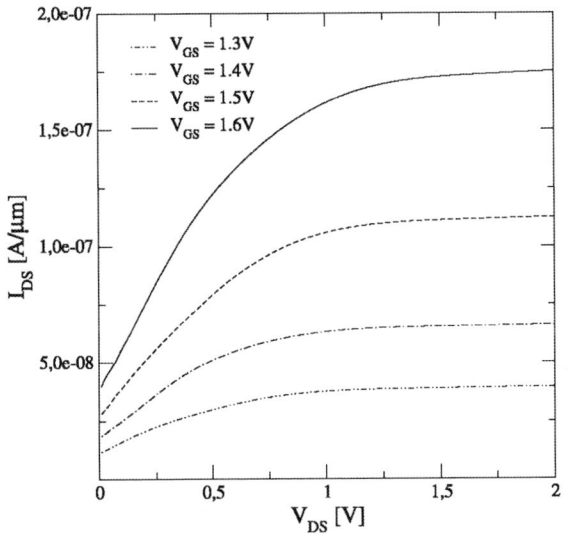

Fig. 7. Simulated forward output characteristics of the tunneling field-effect transistor.

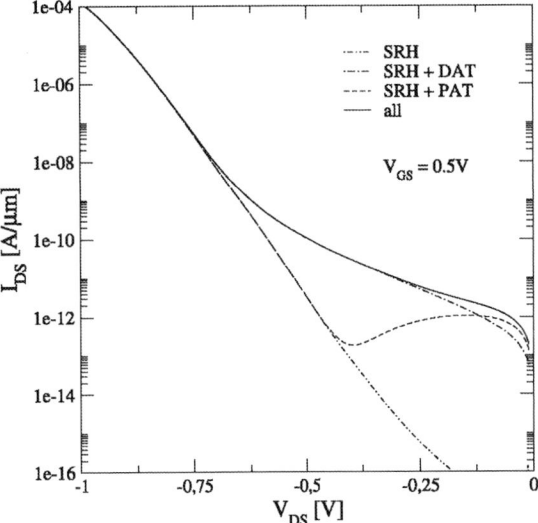

Fig. 8. Current components in the TFET reverse output characteristics at $V_{GS} = 0.5V$.

5. Summary

We proposed a model to describe tunneling effects within the framework of semiclassical device simulation. Two mechanisms for tunneling through the bandgap are identified as being crucial: Defect-assisted and phonon-assisted tunneling. The model can be used for the predictive simulation of various tunneling devices. For the first time, it showed to be possible to obtain the full device characteristics of a TFET via simulation.

Acknowledgement

This project is funded by the Bavarian Science Foundation within the research consortium for nanoelectronics FORNEL.

References

[1] W. Hansch, C. Fink, J. Schulze and I. Eisele, "A vertical MOS-gated Esaki tunneling transistor in silicon", *Thin Solid Films* **369**, 2000, pp.387-389.
[2] G. Hurkx, D. Klaassen and M. Knuvers, "A new recombination model for device simulation including tunneling", *IEEE Trans. Elec. Dev.* **39**(2), 1992, pp. 331-338.
[3] E.O. Kane, "Theory of tunneling", *J. Appl. Phys.* **32**(1), 1961, pp. 83-91.
[4] A. Schenk, "Advanced Physical Models for Silicon Device Simulation", *Springer-Verlag*, Wien, 1998.

Simulation of the p-type RTD

J. Voves

Department of Microelectronics, CTU FEE Prague,
Technická 2, 16627 Praha 6, Czech Republic,
voves@fel.cvut.cz

The simulation of I-V characteristics of p-type AlAs-GaAs resonant tunneling diodes (RTD) is presented. P-type RTDs are important for the possible spintronic applications. The modification of nonequilibrium Green function (NEGF) based 1D quantum transport simulator Wingreen is used for the hole transport simulation. Our results are analyzed from point of view of RTD geometrical and physical parameters.

1. Introduction

Current-voltage (I-V) characteristics of the n-type resonant tunneling diodes (RTD) are understood very well together with all the electron related phenomena [1]. Advent of spintronic applications of p-type RTDs stimulated the analysis of the hole transport phenomena in the resonant tunneling structures. Klimeck's group [2] studied the hole transport by means of the sp3s* tight-binding model. They performed numerical calculations by the program NEMO based on the non-equilibrium Green's functions (NEGF). Their approach is very complex and time consuming.

2. The model and results

Our approach is based on the NEGF as well. We used the program Wingreen [3]. By means of this program we simulated n-type RTD successfully [4]. The model implemented in Wingreen originally covers only electron tunneling. We adopted the carrier parameters to enable simulation with the single band model apart for the heavy and light holes. We simulated the structure described in [5]: AlAs-GaAs double barrier RTD grown by molecular beam epitaxy with two identical 5.1 nm AlAs barriers, 4.2 nm GaAs well and 5.1 nm undoped GaAs spacer layers (RTD1). Other two structures differs with the well width 8.4 nm (RTD2) and barrier width 7.1 nm (RTD3) respectively. Valence band offset is assumed 0.53 eV and hole effective mass $m_{LH} = 0.09\ m_e$ and $m_{HH} = 0.45\ m_e$. In Fig. 1 we compare the hole transmisivity for 10 band model (similar as in [2]) with the transmisivity for the heavy hole (HH) and light hole (LH) single band model. When we summarize the HH and LH transmisivity, no significant difference between 10 band and single band model is observed. The transmisivities and the local density of states (Fig. 2) are calculated for the zero transversal momentum and zero applied bias. Resonant states are clearly visible for heavy and light holes in the quantum well. I-V characteristics based on the HH tunneling for RTD1-3 are shown in the Fig. 3. Higher barrier width shifts the resonant voltage to higher value. Similar effect produces the higher well width. The current is lower for the broader resonant structures RTD2 and RTD3. The comparison of the I-V characteristics based on the HH resp. LH tunneling for RTD1 in the Fig. 4 shows, that very insignificant resonance appears for the light hole tunneling in this case. The HH3 state seems to be the most important level responsible for the appearance of the negative differential resistance region on the simulated IV curve.

1-4244-0396-0/06/$25.00 ©2006 IEEE

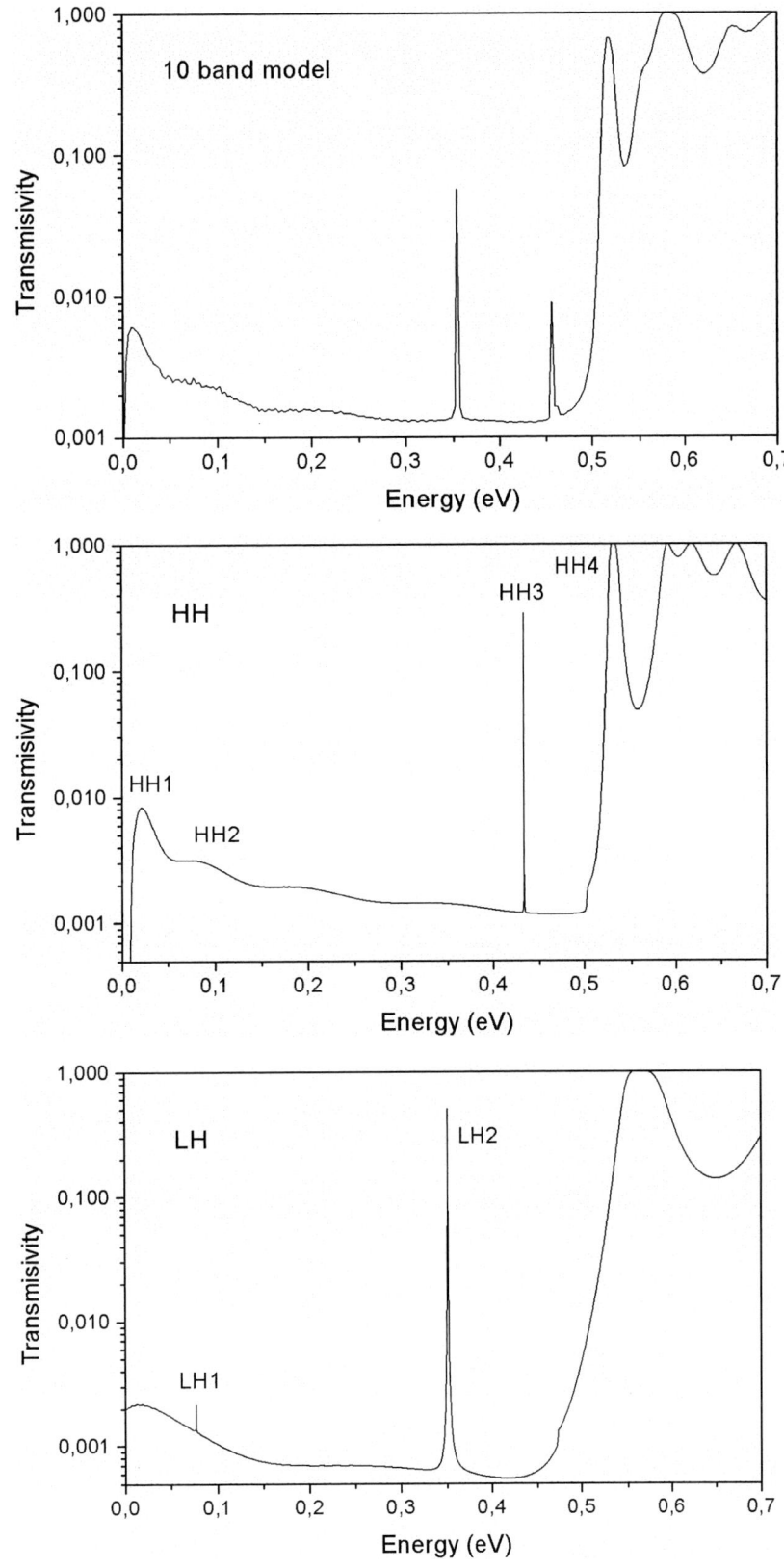

Fig. 1 Hole transmisivity for 10 band model and HH, LH transmisivity of RTD1 for the single band model.

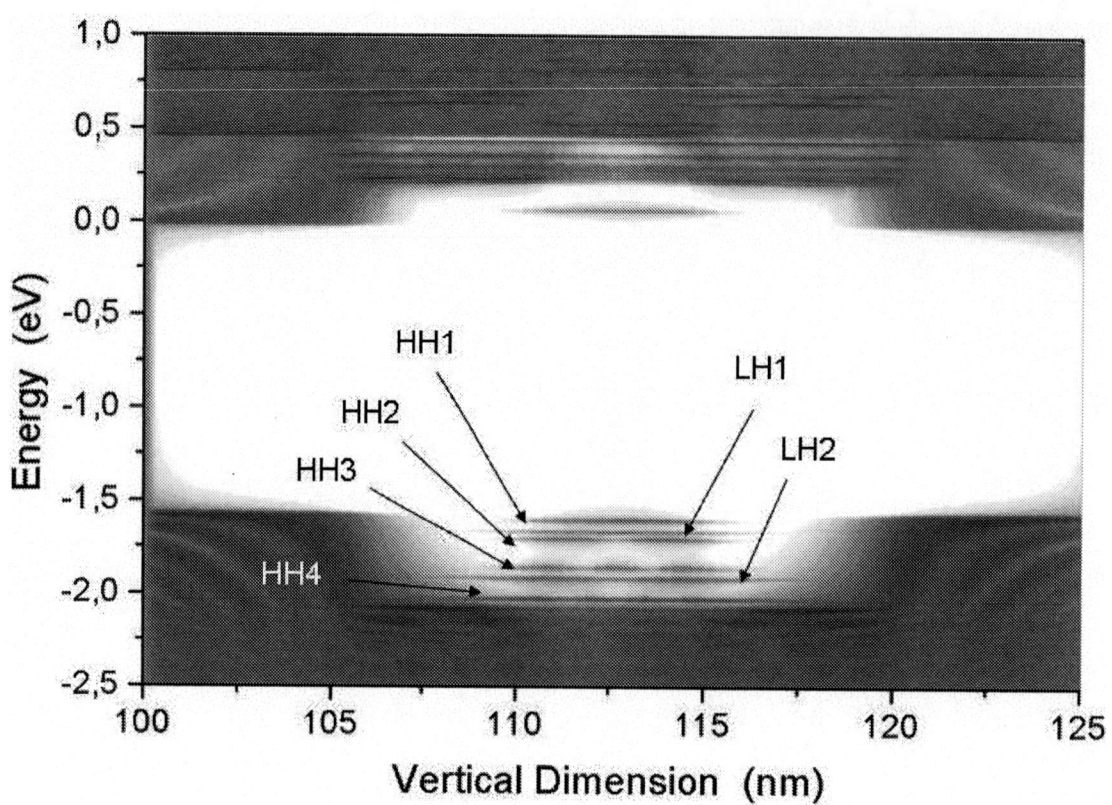

Fig. 2 Local density of states of RTD1 for the 10 band model.

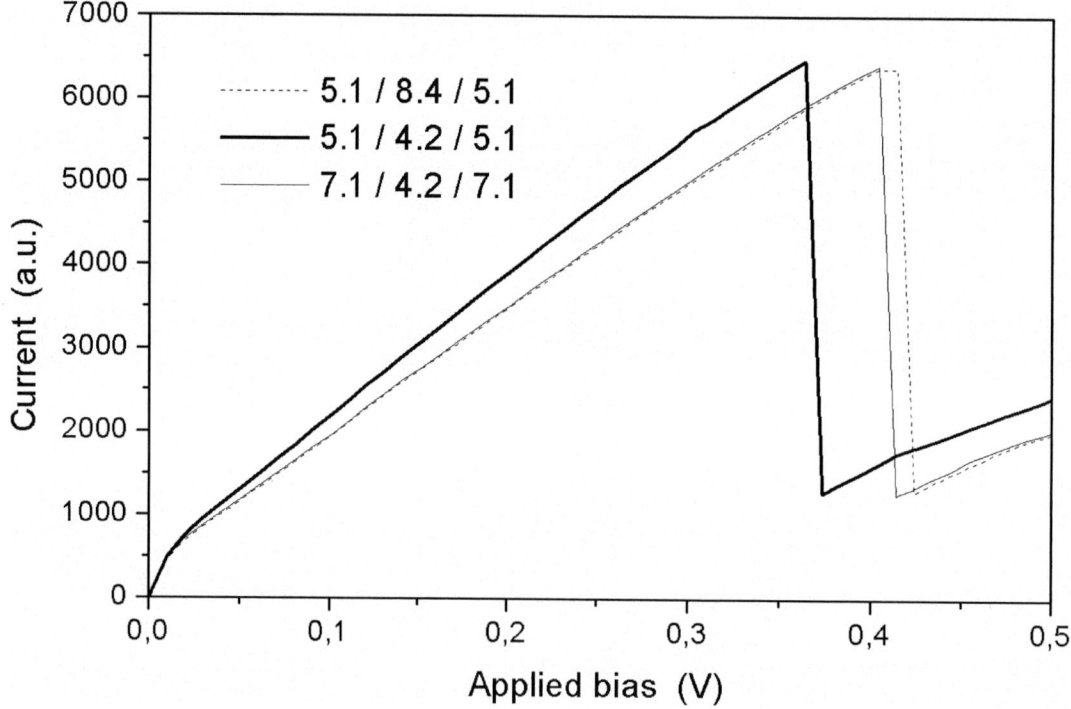

Fig. 3 I-V characteristics of RTD 1-3 calculated by the single band model (heavy holes), T=77K.

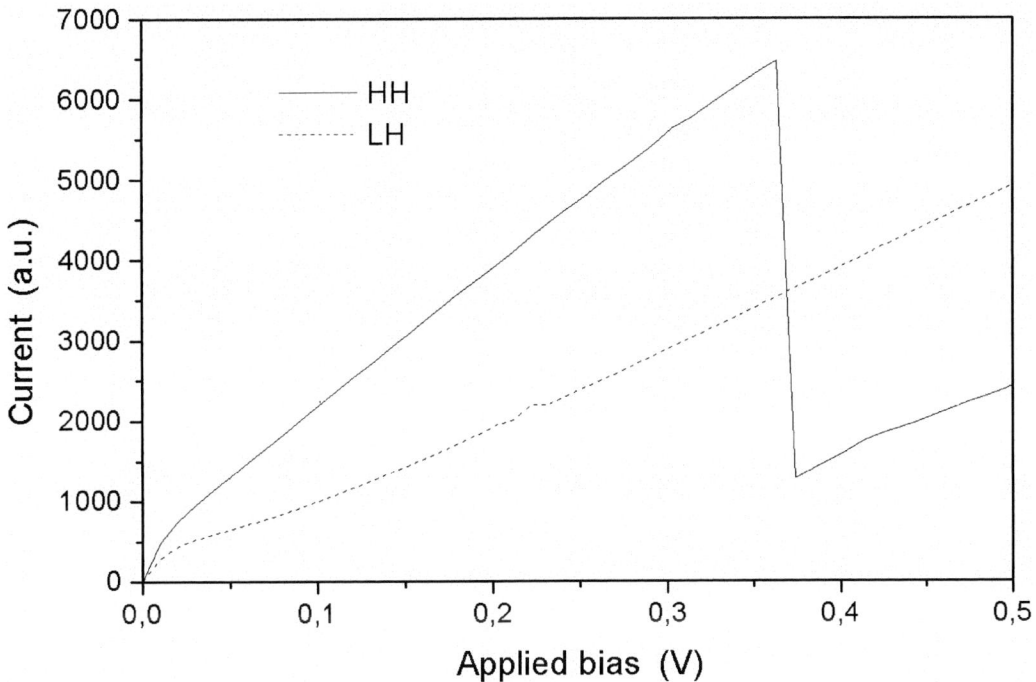

Fig. 4 I-V characteristics of RTD 1 calculated by the single band model (heavy holes and light holes) by 77K.

3. Conclusions

The single band model does not take into account mixing of the LH and HH states and the off-zone-center hole transport. It could be important by the precise analysis of the structure. In the coincidence with [2] the single band model shows the dominance of HH injection in the resonant tunneling structure. Our simulations can serve as the relatively fast estimate of the p-type RTD behavior. Further analysis and the comparison with experimental results will be performed.

Acknowledgement

The work was supported by the grant No. 102/06/0381 GACR and by the grant No. MSM 6840770014 The Ministry of Education CR.

References

[1] C.L. Gardner, G.Klimeck, Ch. Ringhofer, *J. of Comp. Electronics* **3**, 95, 2004
[2] G. Klimeck, R. Ch. Bowen, T. B. Boykin, *Superlatt. and Microstructures* **29**, 187-216, 2001
[3] K. M. Indlekofer, J. Malindretos, http://www.fz-juelich.de/isg/mbe/software.html
[4] J. Voves, T. Třebický, R. Jackiv, *Journal of Computational Electronics,* in print
[5] R. K. Hayden, T. Takamasu, N. Miura, M. Henini, L. Eaves, G. Hill, *Semicond. Sci. Technol.* **11**, 1424-1428, 1996

Investigation of Si delta-doped InGaAs/GaAs QW MSM photodetectors

M. Florovič[1], J. Kováč[1], R. Srnánek[1], J. Jakabovič[1], J. Chovan[2],
B. Ściana[3], D. Radziewicz[3], M. Tlaczała[3]

[1]Department of Microelectronics, Faculty of Electrical Engineering and Information
Technology, Slovak University of Technology, Ilkovičova 3, 812 19 Bratislava, Slovakia
[2]International Laser Centre, Ilkovičova 3, 812 19 Bratislava, Slovakia
[3]Faculty of Microsystem Electronics and Photonics, Wroclaw University of Technology,
Janiszewskiego 11/17, 50-372 Wroclaw, Poland

The aim of this work is to point on the electrical and optical properties of interdigital MSM photodetectors containing Si delta-doped $In_{0.22}Ga_{0.78}As/GaAs$ QW in dependence of different QW depth under UD GaAs cap layer. For this purposes the delta-doped structure was bevelled. The I-V characteristics in dark and under illumination as well as time response dependence on cap layer thickness were measured and evaluated.

1. Introduction

The advanced growth technologies enable the growth of nano-dimensional structures (quantum wells, quantum wires and quantum dots). In delta-doped QW electronic system carriers could be not only confined by the V-shaped potential caused by delta-doping [1], but also by the barriers formed by QW heterostructure [2], which can be described by a combination of triangular and rectangular QW. The optical methods such as photoluminescence [2], Raman spectroscopy [4] and photocurrent spectroscopy [5] are preferably used for investigation of optical properties of nano-dimensional structures.

This work focuses on the electrical and optical properties (I-V characteristics, spectral characteristics, time response) of interdigital MSM photodetectors containing Si delta-doped $In_{0.22}Ga_{0.78}As/GaAs$ (QW) in dependence of different QW depth under UD GaAs cap layer. For this purposes the delta-doped structure was bevelled as described in [6].

2. Experimental

Test structures were grown by MOVPE on n-doped GaAs substrate using an atmospheric pressure AIX200 R&D Aixtron reactor at Wroclaw University of Technology. The TMGa, TMIn, AsH$_3$, DEZn and SiH$_4$ (20 ppm in H$_2$, growth temperature 670°C) were used as growth and dopant precursors, respectively. Interdigital MSM photodetectors with different thickness of UD GaAs cap layer were realized on the beveled structure as shown in Fig. 1. For cap layer thinning the sample was dipped into the etchant solution $H_3PO_4 : H_2O_2 : H_2O = 3 : 1 : 10$. Bevel angle of $\sim 1.55.10^{-5}$ rad (M $\sim$ 65 000) was determined from the profile measurements. Along the beveled structure interdigital Au Schottky contacts (*finger:gap=2μm:2 μm*) were prepared using standard vacuum evaporation. Dark and light I-V characteristics were measured using halogen lamp and source meter Keithley 2400. The photocurrent spectroscopy measurements on MSM photodetectors were realized using standard lock-in technique and halogen source. For the time response measurements the laser diode of wavelength 780 nm and 850 nm (output power $\sim$ 400 μW) was employed as a light source and digital oscilloscope Le Croy 9362 for the photoresponse measurements.

1-4244-0396-0/06/$25.00 ©2006 IEEE

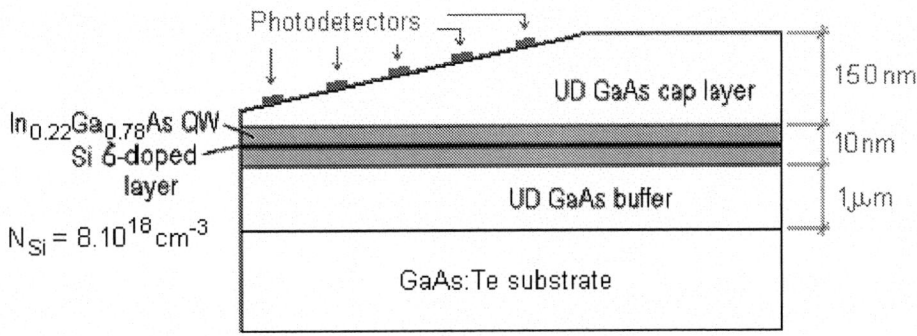

Fig.1 Bevelled Si delta-doped $In_{0.22}Ga_{0.78}As/GaAs$ QW structure

3. Results and discussion

Dark and light I-V characteristics of MSM photodetectors with different cap layer thickness (~ 35 and 60 nm) are shown in Fig. 2a,b. The high concentration of electrons in Si delta-doped $In_{0.22}Ga_{0.78}As/GaAs$ QW act as a conductive channel near the surface and contribute to the dark current increase with increasing applied voltage in dependence on cap layer thickness. With increasing of UD GaAs cap layer thickness the turning point voltage on the dark current I-V characteristics shifts to higher value with increased applied voltage (Fig.2b). The measured dark current I-V characteristics were compared at the same current (100 nA, 100μA) in forward and reverse applied voltage in dependence on the thickness of UD GaAs cap layer as shown in Fig.3a. For halogen lamp illumination the light current level at I-V characteristics is almost constant for applied voltages under the photocurrent turning point (d-l voltage, Fig.3a), where photocurrent merge the dark current value.

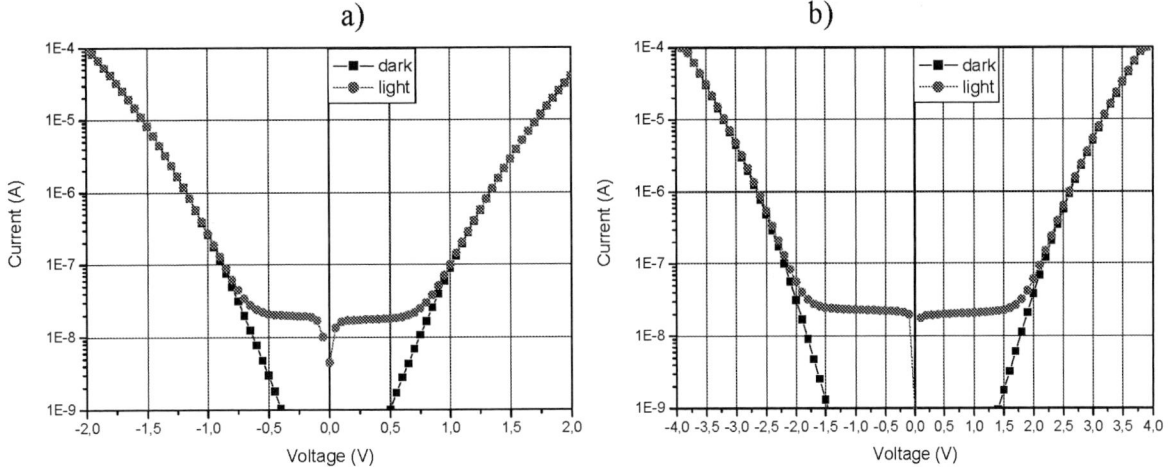

Fig.2 I-V characteristics of delta-doped QW MSM photodetector
with different cap layer thickness
a) 36 nm   b) 60 nm

In the photocurrent spectra of MSM photodetectors (Fig. 3b) an absorption peak corresponding to the optical transition e2-hh2 with energy ~ 1.37 eV (λ ~ 905 nm) is resolved and shifted with the bias voltage due to the Stark effect. As the thickness of UD GaAs cap layer decreases the intensity of the absorption peak (optical transition e2-hh2) increases in proportion to the GaAs absorption edge in photocurrent spectra of MSM photodetectors.

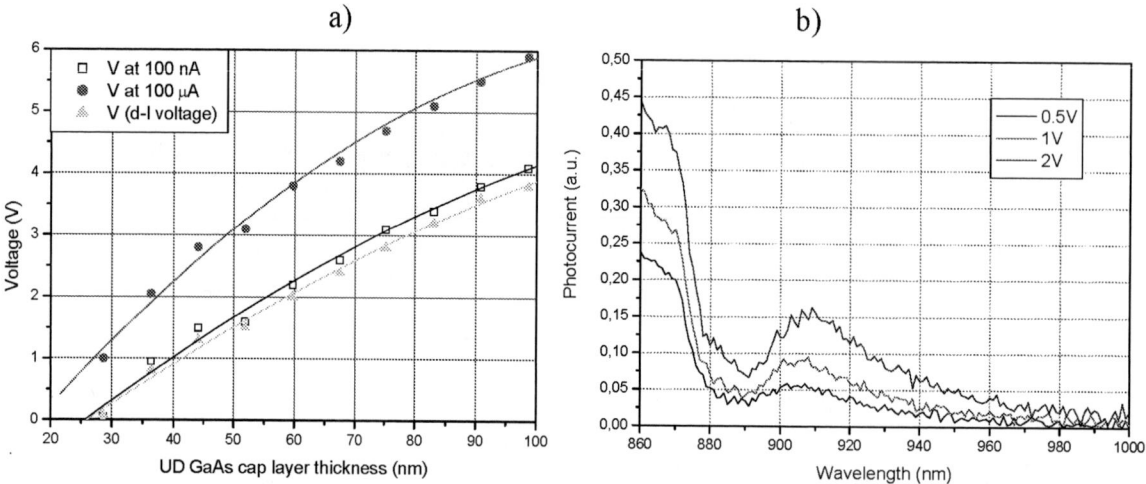

Fig. 3 a) Voltage turning points in dependence on cap layer thickness
b) Spectral characteristics of MSM photodetector (cap layer thickness 60 nm)

Time measurements were realized by using pulse laser light source of wavelength 850 nm (energy) and optical power ~ 400 µW. The energy of the laser light (~1,46 eV) is higher than e2-hh2 transition energy of delta-doped QW (~1,37 eV) and GaAs bandgap energy (~1,42 eV) and the incident photons are absorbed in both $In_{0.22}Ga_{0.78}As$/GaAs QW and GaAs layers. Under the Schottky contacts the space charge region is formed in dependence of applied voltage. The higher electrical field causes more rapid separation of generated free carriers in structure. The time response of investigated delta-doped QW MSM photodetector with different cap layer thickness (45 and 90 nm) is shown in Fig. 4a,b. As the electrical field increases, rise and fall time decreases with rising applied voltage. The rise and fall time of particular MSM photodetectors in dependence of applied voltage is shown in Fig. 5a,b. With increasing thickness of UD GaAs cap layer the device attributes limit to the undoped GaAs MSM photodetector properties while the rise and fall time generally decreases and depends less on the applied voltage.

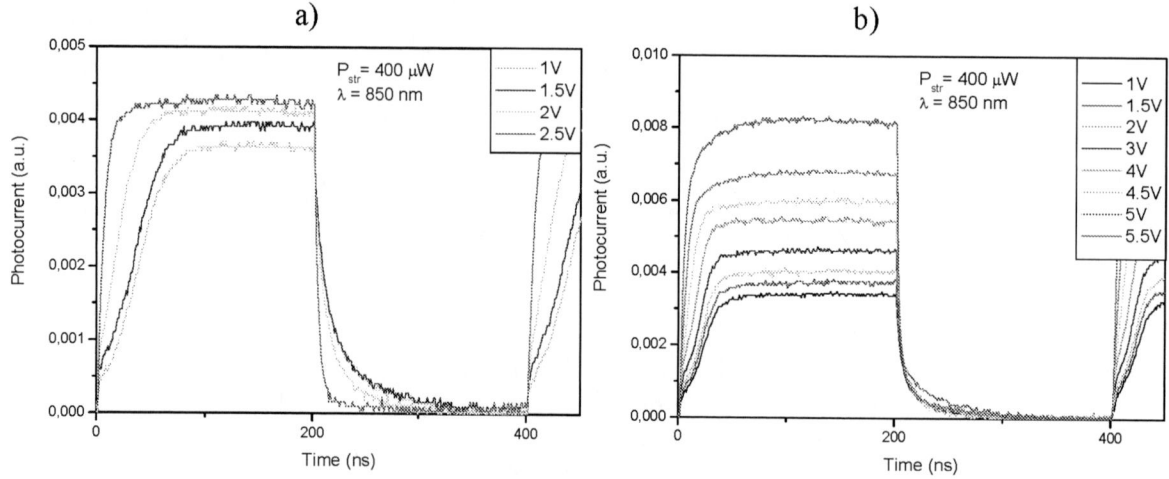

Fig. 4 Time response of delta-doped QW MSM photodetector
with different cap layer thickness
a) 44 nm   b) 90 nm

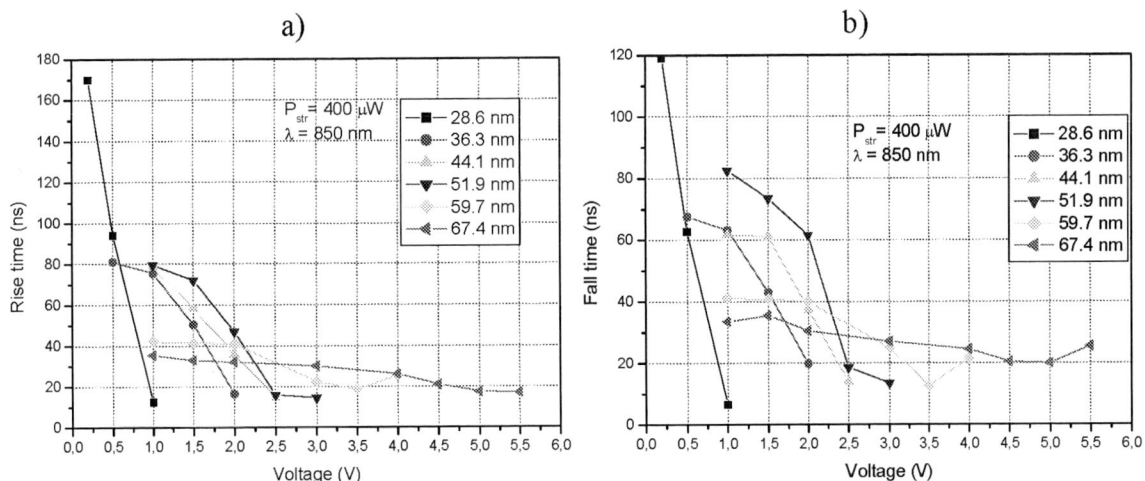

Fig. 5 a) Rise time and b) fall time voltage dependence of MSM photodetector

4. Conclusion

It was experimentally confirmed that the UD GaAs cap layer thickness has considerable influence on Si delta-doped $In_{0.22}Ga_{0.78}As/GaAs$ QW interdigital MSM photodetector electrical, optical and time response properties. With decreasing of UD GaAs cap layer the turning point voltage shifts to lower value for measured dark current I-V characteristics. In the same time the QW optical transition e2-hh2 ($\lambda \sim 905$ nm) is better resolved in measured photocurrent spectra and time measurements show strong rise and fall time dependence with rising applied voltage. As the thickness of UD GaAs cap layer increases the device attributes limit to the undoped GaAs MSM photodetector properties while the rise and fall time generally decreases and depends less on the applied voltage. For practical purposes it's important to optimize the UD GaAs cap layer thickness for required applications.

Acknowledgement

This work was supported by Slovak Grant Agency projects 1/3108/06, 1/3076/06, 1/3111/06, bilateral cooperation project Pol/Slov, the Wrocław University of Technology statutory grant.

References

[1] Schubert, E.F.: Delta-doping of Semiconductors, Cambridge University Press, 1996.
[2] Frensley, W.R.: Heterostructures and Quantum Devices, Microstructure Science, Academic Press, San Diego, 1994, Ch. 1.
[3] Florovič, M., Kováč, J., Srnánek, R., Irmer, G., Sćiana, B., Radziewicz, D., Tlaczala, M., Geurts, J.: Optical properties of delta-doped layers, EDS 2003, Brno, pp. 174-177.
[4] Srnánek, R.,Gurník, O., Harmatha, L., Gregora, I.: Diagnostics of Si-multi-delta-doped GaAs layers by Raman spectroscopy on bevelled structures, Applied Surface Science, Vol. 183 (1-2), 2001, pp. 86-92.
[5] Florovič, M., Kováč, J., Jakabovič, J., Sciana, B., Radziewicz, D., Tlaczala, M., Kučera, M.: Optical properties of $In_xGa_{1-x}As/GaAs$ MQW and delta-doped QW structures, Wocsdice 2004, Smolenice Castle, May 17-19, 2004, pp. 21-22.
[6] Srnánek, R., Novotný, I., Hotový, I., Gomati, M. E.: Chemical bevelling of GaAs-based structures, Materials Science and Engineering B47, Elsevier, 1997, pp. 127-130.

Micro-Raman scattering: A versatile tool for characterization of semiconductor structures

G. Irmer

Institute of Theoretical Physics, TU Bergakademie Freiberg
Leipziger Strasse 23, D-09596 Freiberg, Germany
e-mail: irmer@physik.tu-freiberg.de

As a non-destructive and relatively rapid method, Raman spectroscopy is well established for the characterization of semiconductor substrates, heterostructures and devices. Some applications of micro-Raman spectroscopy to measurements of stress, crystallinity and charge carriers in semiconductor structures are discussed with emphasis on the scope and the limitations of the information that can be obtained with Raman spectroscopy.

1. Introduction

Raman spectroscopy is well established as a non-destructive optical method to characterize semiconductor substrates, heterostructures and devices. Raman scattering yields information about vibrational and electronic properties, impurities, carrier concentration, composition, crystal structure and crystal orientation.

The talk is focused to the influence of stress fields, crystallinity and charge carriers onto the micro-Raman spectra of phonons in semiconductor structures.

One of the most interesting applications of Raman scattering in microelectronics processing is its sensitivity for local strain. By monitoring the phonon frequencies at different positions on the samples, a strain-map with local resolution of about 1 μm and a stress sensibility of about 5 MPa can be obtained at different levels: in wafers, chips and IC devices. Applications of stress measurements for different semiconductor structures and MEMS structures (micro mirror arrays, sensors, membranes) on the basis of Si, SiC and III-V compound semiconductors will be discussed. Limitations of the method will be shown. The Raman method is also applicable to investigate strain fields near defects like microcracks or dislocations and in the vicinity of indentations on substrates.

Secondly, phonon shifts can also be induced by changes of the crystallinity. They are often accompanied by a halfwidth broadening of the phonons. Due to the small scattering volume of a nanocrystalline particle the wavevector transferred in the scattering process is uncertain, and with decreasing particle size phonons with larger wavevectors and smaller frequencies contribute to the light scattering process. This confinement effect can be used to determine the size of nanocrystals. The confinement also enables the size determination with observation of acoustic phonons and observation of Fröhlich modes. Examples of applications are given.

Thirdly, in polar compound semiconductors the LO-phonons can be shifted and broadened by free charge carriers due to the LO-phonon-plasmon coupling. The Raman spectroscopic characterisation of intrinsic and photoinduced charge carrier distributions in nanostructurized compound semiconductors (e.g. columnar nanostructures in GaN, honeycomblike structures in InP) and thin layers will be discussed. The technique of etching bevels of layered structures with layer thicknesses of some nm gives access to single layers for Raman measurements.

1-4244-0396-0/06/$25.00 ©2006 IEEE

2. Strain measurements

Strain in crystalline semiconductors affects the phonon frequencies, and in this way the position of the Raman peak. In general, compressive stress will result in an increase of the phonon frequency and tensile stress results in a decrease. The relation between stress and frequency shift is simply linear for uniaxial or biaxial stress. However, in the case of more complicated stress fields, often expected in microelectronic devices, the situation is more complex because all non-zero strain tensor components influence the phonon positions. The Raman data have then to be fitted by theoretical stress models.

Applications of Raman scattering for strain measurements in the microelectronics industry date back to 1980, but only for uniform films: The strain in silicon films on sapphire substrates was measured [1]. Around 1989 the first stress measurements were reported on local oxidation of silicon(LOCOS) samples [2]. The first measurements were performed at single points. With the development of new micro-Raman spectrometer with high optical throughput, cooled CCD detectors and computer controlled X-Y-Z stages automatic scans with high local resolution became possible. It was proved that the Raman technique could be used to measure local stresses in silicon devices and became popular in microelectronics research and industry. The influence of different isolation techniques, of silicides, nitrides and metals on stress in Si have been investigated. Overviews are given in [3] and references therein.

Applications to compound semiconductors are not so numerous, but the Raman technique was successfully used to determine the residual strain between substrates and epilayers or in order to measure strain near defects or indentations in substrates with high local resolution.

The local resolution depends on the wavelength and the microscope objective used, resolutions in the order of the wavelength of the exciting laser can be obtained, with immersion objectives values about 0.3 μm are possible. The information depth in opaque semiconductors is $\Box 1/\alpha$, where α is the absorption coefficient. Information about variation of the stress with the depth can be obtained by changing the probing depth using different wavelengths of the exciting laser. This works for instance in the case of Si and for compound semiconductors with small penetration depth for visible or UV light. Recently, the characterization of strained Si for sub-100 nm MOSFET's with UV laser excitation has been reported [4]. For transparent semiconductors like GaN or SiC the technique of confocal imaging can be used in connection with autofocus equipment in order to obtain depth profiles of stress fields.

Below some typical results of local stress measurements are shown. They demonstrate possibilities for application to different semiconductors.

2.1 Silicon based structures

In Fig. 1 a typical Raman spectrum of crystalline silicon is shown. The Raman scattering frequency of the optical phonon is 520.6 cm^{-1}. The sharp line which can also be seen in the spectrum comes from a Hg lamp which was used for calibration. The light of the calibration lamp was coupled through a fiber optic cable into the optical path by means of a beamsplitter positioned before the entrance slit of the spectrometer. In this way, the calibration signal height can be kept always constant during the stress measurement. The frequencies of the phonon band and of the calibration line are obtained after background correction with fit to a Lorentzian and a Gaussian band, respectively.

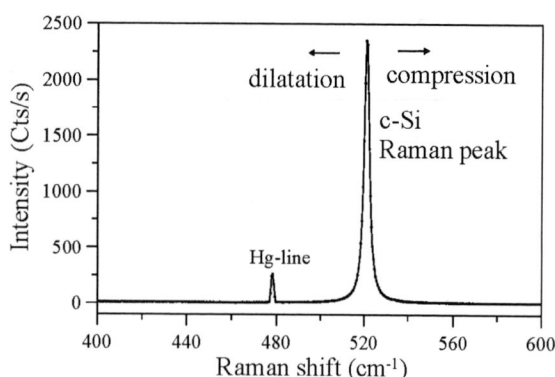

Fig.1
Typical Raman spectrum of crystalline silicon. Compression of the lattice shifts the phonon frequency upwards and dilatation downwards.

Fig.2
Stress induced by amorphisation of Si: The area shown was sputtered with Ga ions (30 keV), penetration depth about 8 nm. The graph shows the shift of the c-Si phonon (circles). The stress was calculated with the edge force model (solid line).

Mechanical stress results in a shift of the phonon frequency. By monitoring the phonon frequency at different positions on the sample, a strain map can be produced. In general, compressive stress results in an increase of the frequency and tensile stress in downshift of the frequency. For backscattering from a [001] surface of silicon with uniaxial in-plane stress σ the relation is $1\,cm^{-1} \square -434\,MPa$ (Raman shift measured in cm^{-1} and stress in MPa, $\Delta\omega > 0$ is obtained for compressive stress $\sigma < 0$). However, in more complex cases, for instance near edges, trenches or in components of microelectromechanic systems (MEMS), more complicated relations between the frequency shift and the stress tensor components have to be used and some prior knowledge about the stress and strain distribution in the sample is necessary. An example is shown in Fig.2 . The surface of a silicon wafer was sputtered with Ga ions (30 keV) in a localized area with penetration depth of about 8 nm. Scans of the c-Si Raman signal indicate compressive stress below the damaged area and tensile stress near the edges. The stress field was modelled with an edge force model (solid line in Fig.2). Due to absorption the c-Si signal from below the amorphous layer is diminished, in the Raman spectra the phonon band at about 480 cm-1 of amorphous Si appears besides the c-Si phonon band. The thickness of the amorphous layer can be estimated using the intensities of the two bands.

2.2 Compound semiconductor structures

In compound semiconductors several optical phonons of different symmetry can be observed. Therefore, in dependence on the scattering geometry, in general different stress tensor components are involved in the stress measurements. Fig. 3 shows results of Raman measurements characterising the crystal quality near the surface of micro holes in AlGaN/GaN transistors on SiC, drilled by an UV laser using nanosecond pulses [5]. The

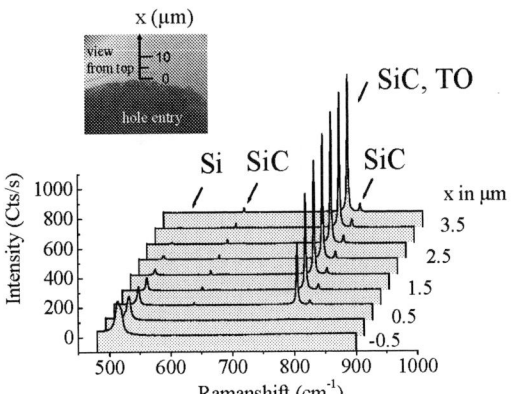

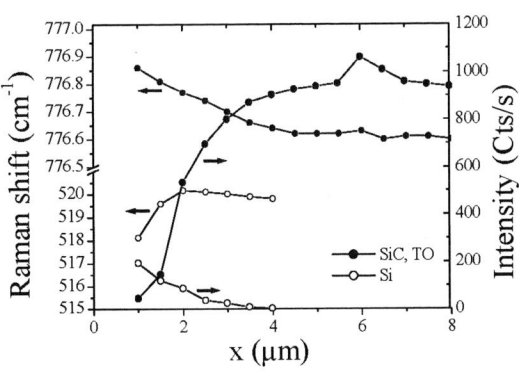

Fig.3
UV laser micro drilling of via holes in AlGaN/GaN transistors on SiC: The inset shows an optical micrograph of the cross section of the hole entrance (SiC side). Raman spectra are taken at different distances from the rim of the hole entrance. The indicated SiC TO mode has a Raman shift of 776.6 cm^{-1} and the Si mode of 520.6 cm^{-1} in bulk material.

Fig.4
Frequency shift and peak intensity of the SiC TO-phonon and the Si phonon in dependence on the distance to the hole.

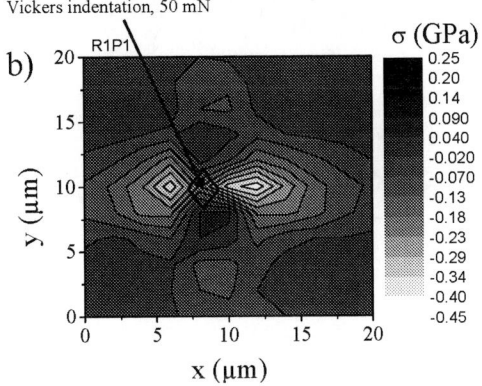

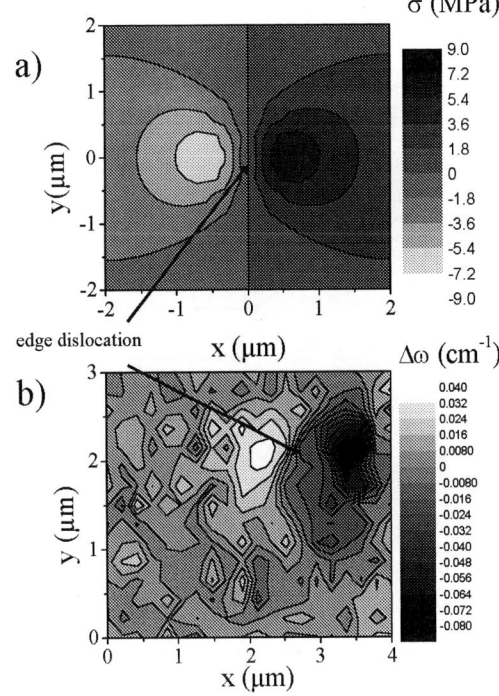

Fig.5
Raman analysis of the stress fields around a nanoindentation on a (100) GaAs wafer.
(a) AFM image. The depth of the indentation is about 500 nm.
(b) Stress around the indentation calculated from the Raman shift of the LO-phonon using the biaxial stress model.

Fig.6
Stress field around a single edge dislocation line intersecting the (100) surface of a GaAs wafer .
(a) Calculated stress filed
(b) Raman shifts of the LO phonon

45

micro-Raman spectra indicate a change of internal strain due to laser drilling on the hole entrance (SiC) side and evidence the formation of nanocrystalline Si. On the GaN side no variation indicating significant change of crystal perfection or strain in the hole vicinity could be found.

For investigations of residual stress in semiconductor wafers photoelastic investigations are state of the art. However, the lateral resolution of these methods is not sufficient to investigate stresses with lateral resolution down to about 1 μm. Figs. 4 and 5 show examples of Raman measurements on GaAs wafers with high resolution in the vicinity of a nanoindentation and in the vicinity of a single edge dislocation. In the last case the method is near its detection limit of about 5 MPa [6].

3.   Crystallinity

Frequency shifts of Raman peaks can be caused also by other effects. Raman scattering is sensitive to probe the crystallinity. The Raman bands of optical phonons of nanocrystals are frequency shifted. The size dependent shift is accompanied by a peak broadening. This effect can be used for size determinations. By a method described e.g. in [7] the average particle

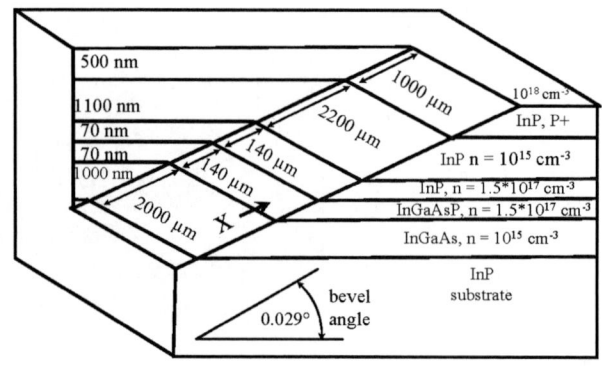

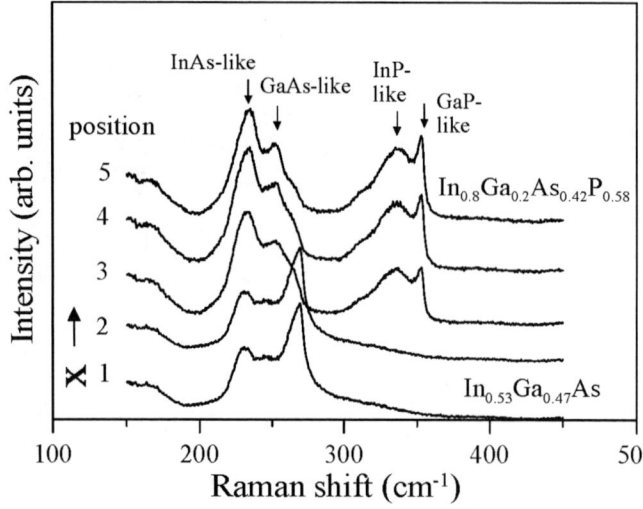

Fig.7
Spatially resolved Raman measurements of bevelled InP/InGaAsP/InGaAs heterostructures.
Measurements along the interface of InGaAs/InGaAsP are shown.

size of the Si nanoparticles found near the hole surface in the example shown in Fig. 3 can be estimated from the frequency and the peak width of the Si mode to be about 4 nm near the surface. Amorphous Si could also be detected.

Indication of amorphous GaAs was found near the nanoindentation centre in the example shown in Fig.5.

4. Charge carriers

Free charge carriers in compound semiconductors couple with LO phonons via their accompanying electric fields. The interaction results in coupled LO phonon plasmon modes (PPM). The analysis of the Raman spectra of the PPM can be used to determine concentration and mobility of free carriers with high local resolution.

In order to get access to single layers of semiconductor nanostructures a bevelling technique was developed and successful applied to layered semiconductors at the TU Bratislava (for a review, see [8]). By scanning the laser beam along the bevelled surface doping profiles and interface characterisation with very high resolution can be obtained.

As an example Fig. 7 shows Raman measurements across the interface between InGaAs and InGaAsP of a structure used for diodes. Observed frequency shifts could be interpreted as caused by charge carriers rather than by stress.

Acknowledgement

Examples used in this talk came from cooperation with different colleagues. The author especially thanks R. Srnanek, M. Herms, T. Wernicke, B. Köhler, M. Jurisch , A. Shaporin, J. Leschner and G. Richardson for successful cooperation.

References

[1] T. Englert, G. Abstreiter and J. Pontcharra, *Solid State Electronics* **23**, 31, 1980.

[2] K. Brunner, G. Abstreiter, B. O. Kolbesen and H. W. Meul, *Appl. Surf. Science.* **39**, 116, 1989.

[3] I. de Wolf, *Semicond. Sci. Technol.* **11**, 139, 1996; *J. Raman Spectrosc.* **30**, 877, 1999.

[4] K. Rim, L. Shi, K. Chan, J. Ott, J. Chu, D. Boyd, K. Jenkins, D. Lacey, P. M. Mooney, M. Cobb, N. Klymko, F. Jamin, S. Koester, B. H. Lee, M. Gribelyuk and T. Kanarsky, *Semiconductor Application Notes* , www.jyhoriba.co.uk

[5] T. Wernicke, O. Krüger, M. Herms, J. Würfl, H. Kirmse, W. Neumann, G. Irmer, T. Behm and G. Tränkle, submitted to *Appl. Surf. Science*

[6] G. Irmer and M. Jurisch, submitted to *J. Appl. Phys.*

[7] G. Irmer, J. Monecke and P. Verma, in: *Encyclopedia of Nanoscience and Nanotechnology (Ed. H. S. Nalwa)*, American Scientific Publishers, California, USA, 2004, 561

[8] R. Srnanek , G. Irmer , J. Geurts , J. Kovac , D. Donoval and L. Peternai , in: *New Research on Thin Solid Films (Ed. Maria G. Benjamin)*, Nova Science Publishers, Hauppauge NY, USA (ISBN 1-60021-454-1, **Status:** Announced)

Measurement of the Germanium fraction in strained and relaxed SiGe by Spectroscopic Ellipsometry

J. Moers[1], D.M. Buca[1], M. Goryll[1], R. Loo[2], M. Caymax[2] and S. Mantl[1]

[1]Institute of Bio- and Nanosystems and Center of Nanoelectronic systems
for Information Technology
Research Centre Jülich, D-52425 Jülich, Germany

[2]IMEC, Kapeldreef 75, B-3001, Leuven, Belgium

The further development of MOSFET devices is mandatory for the development of information technology. Besides the optimization of lateral layouts and introduction of alternative device architectures, new materials were introduced, too. To increase the on current I_{on} the carrier mobility is increased by straining the silicon channel layer. To get strained silicon layers on insulator (SSOI) the so called Jülich-Process was developed, which uses an ion implantation assisted relaxation of thin $Si_{1-x}Ge_x$ layer to get a high-quality surface to grow pseudomorphically strained silicon. To characterize the $Si_{1-x}Ge_x$, layers non-destructively by spectroscopic ellipsometry (SE), in this work a parametric expression of the dielectric function was established, which depends only on the Ge fraction x. With that function, SE can be used as powerful tool to characterize thickness and composition homogeneity of the epitaxial $Si_{1-x}Ge_x$, layers.

1. Introduction

Since three decades MOSFET devices are developed further. Their dimensions were scaled from several μm down to sub 50 nm, which is production standard today. The physical problems arising from that scaling, the short channel effects, were addressed by improvements of the lateral layout. For the next generations, alternative structures as ultra thin body devices and multi-gate MOSFETs were discussed to maintain the development. But there is still need for improvement of device performance. Besides the integration of alternative high-k dielectrics to decrease the gate leakage current and different materials for metallisation to minimize parasitic series resistances and adjust threshold voltage V_{th}, new materials for the channel are discussed, too. It is favourable to increase the carrier mobility in the channel to increase I_{on}. By stressing the normal bulk silicon, the six fold degenerated conduction band minimum is split. The now energetically favourable energy states have a higher mobility. The same is true for the holes at valence band maximum. Therefore, integrating strained silicon into the channel of MOSFETs will increase device performance.

Growing silicon pseudomorphically strained on relaxed SiGe buffers offers the possibility to adjust the strain by the Ge fraction in the buffer. The problem is to realize a high quality surface of the relaxed SiGe. Defects will proceed into the epitaxial silicon layer and hence will deteriorate the properties of the material. One widely used method to get high quality relaxed SiGe surfaces is to grow thick layers with varying Ge fraction (graded buffer). These buffers are very thick, which is unfavourable in terms of growth time (which means high costs) and further processes to remove that layer.

In the so called "Jülich process" [1] a thin (~200 nm) SiGe layer is grown pseudomorphically strained on Si (100) substrates. After the layer growth He is implanted below the Si/SiGe interface. In the subsequent annealing step the high density of dislocation sources in the substrate originated by the ion implantation enables efficient strain relaxation and a low dislocation density in the relaxed layer. Hence the surface has low defect density.

To characterize the epitaxial layer, it is mandatory to determine the Ge fraction x, the thickness d and its homogeneity across the wafer. Spectroscopic ellipsometry (SE) is a fast and non-destructive method for this purpose. For SE, however, it is essential to know the dispersion of the absorption $k(\lambda)$ and the refractive index $n(\lambda)$ and their dependence on Ge fraction x and strain. If this is known, the measured values for $n(\lambda)$ and $k(\lambda)$ can be used to determine the strain and Ge fraction x.

In this work we developed a data set for the dispersion of SiGe, which depends only on the Ge fraction and implemented it into the Ellipsometer Software. Therefore the Ge fraction is now accessible to direct fit and the SE is a powerful tool to characterize the epitaxial layers non-destructively.

2. Experimental

The ellipsometry is an optical method to measure thin film properties with high accuracy and declines to P. Drude [2]. Linearly polarized light is lead under an angle of 45° to the plane of incidence on the surface and is reflected. During the reflexion the light penetrates the sample partially and interacts with the materials. Depending on the dielectric function of the materials, the state of polarization is altered and the reflected light is polarized elliptically. The phase difference Δ between the light components parallel and perpendicular to the plane of incidence and the ratio $\tan(\psi)$ of the maximum amplitudes are measured. The Δ and ψ depend on the angle of incidence φ, the film thicknesses d_i and the dielectric function $\varepsilon(\lambda)$ of the materials, and hence on the wavelength λ.

If the dielectric function and the thicknesses are known, $\Delta(\lambda)$ and $\psi(\lambda)$ can be calculated theoretically using the Fresnel reflection coefficients. If $\varepsilon(\lambda)$ is given in a parametric form, the theoretical spectrum can be fitted to the measured spectrum by adjusting those parameters and the film thicknesses.

$Si_{1-x}Ge_x$ layers on (100) Silicon wafers with x from 19% to 29% and different thickness (80 nm to 200 nm) were grown by CVD to investigate the dependence of $\varepsilon(\lambda)$ on the

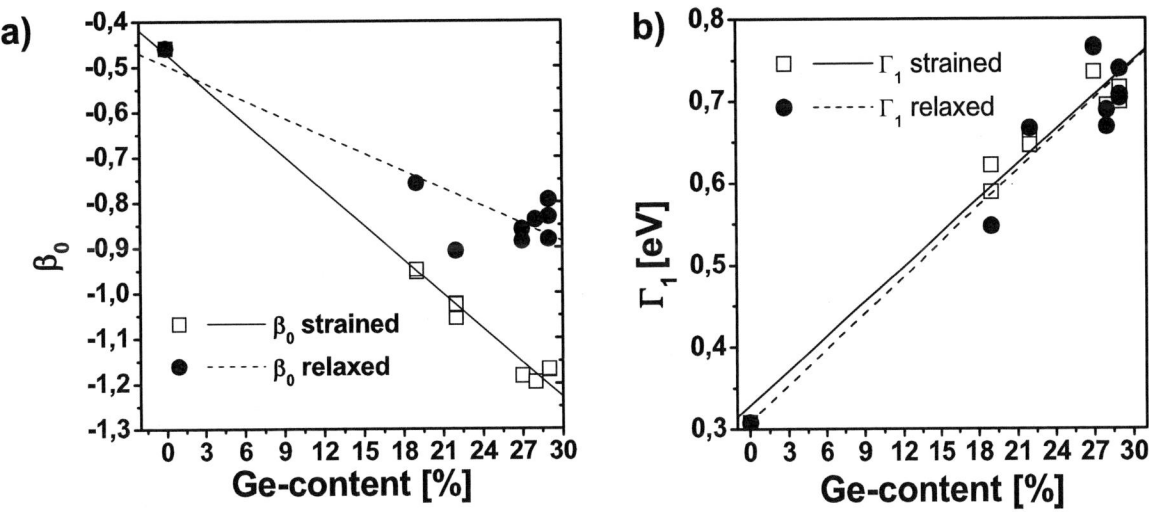

Fig. 1: Dependence of (a) β_0 and (b) Γ_1 on Ge-content and strain.

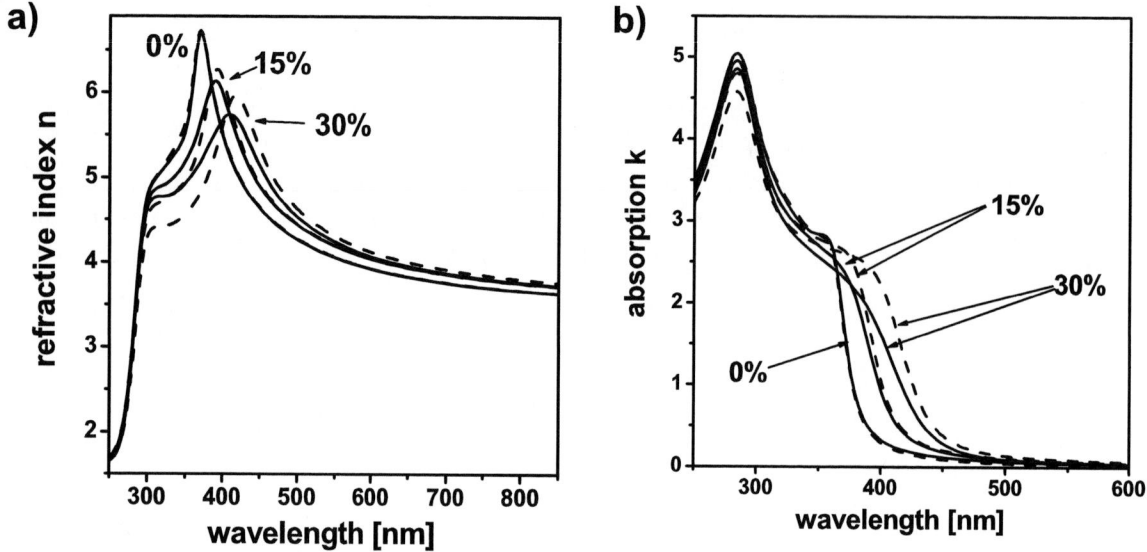

Fig. 2: (a) Refractive index n(λ,x) and (b) absorption k(λ,x) for x = 0%, 15% and 30% and strained (solid lines) and relaxed (dashed lines) layers.

Ge fraction x and the strain. From each wafer some parts were relaxed using the "Jülich process", so strained and relaxed layers from the same epitaxy were available. The measurements were performed with a SE800 from SENTECH GmbH in the wavelength range from 250 nm to 850 nm at three different angles of incidence φ (50°, 60° and 70°).

A harmonic oscillator formulation [3] is used for the parameterized description of the dielectric function of $Si_{1-x}Ge_x$. In this description the dielectric function is described by four oscillators. Each oscillator has 5 parameters (Energy level E_{gi}, amplitude C_i, broadening γ_i, phase β_i and order μ_i), which are combined with five global parameters (ε_∞, k_0, x_0, μ and thickness d). This adds up to 25 parameters to model the $Si_{1-x}Ge_x$-layers. Furthermore a native oxide was assumed at the surface of the samples. This was modelled as a thermal oxide with known dispersion and its thickness as fitting parameter.

To decrease the number of 26 free parameters and identify reliable dependencies of these parameters on strain and Ge fraction x, each spectrum was fitted with more and more free parameters. Here at first the parameters with the biggest impact on $\varepsilon(\lambda)$ were chosen for the fit. The evolution of each parameter was watched carefully, whether there was a clear trend or whether the values were random.

3. Results and discussion

It was found, that the global parameters are constant for all samples, as well as the parameters of two oscillators. The other two oscillators shifted to lower energies with increasing Ge fraction, reflecting the decreasing direct band gap at the centre of the Brillouin zone. The broadening parameters Γ_0 and Γ_1 of those oscillators increase with increasing x, indicating a higher spin orbit split in those excitations. For even higher x, those oscillators have to be modeled as two oscilators, each. For all parameters of the dielectric function a clear dependence on x could be established, but it is difficult to distinguish between strain and different germanium content. Hence two sets of parameters for SiGe were compiled, one for relaxed and one for strained layers. For example Fig. 1 shows the dependence of β_0 and Γ_1 for relaxed and strained layers. While Γ_1 varies with x, but is independent of the strain , β_0 varies with x and strain. Fig. 2a shows calculated refractive indices and Fig. 2b the absorption

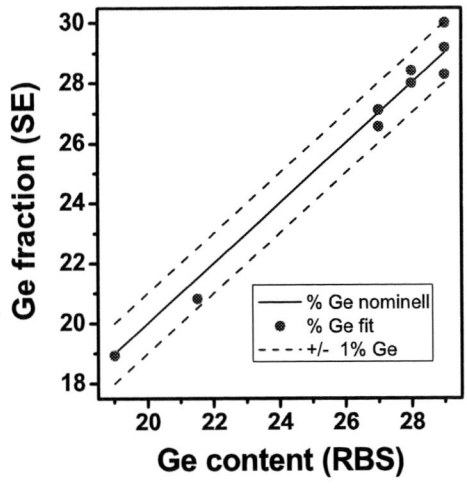
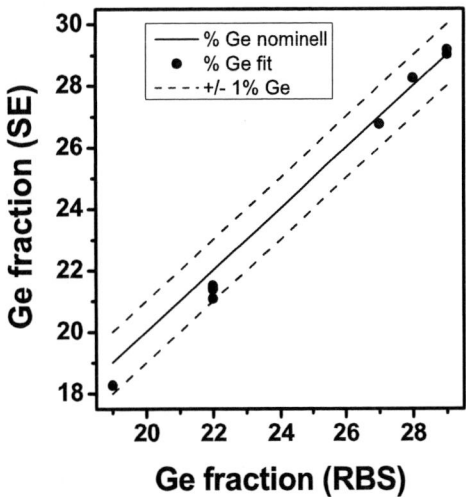

Fig. 3a: Relaxed $Si_{1-x}Ge_x$:SE measured x versus RBS values

Fig. 3b: Strained $Si_{1-x}Ge_x$:SE measured x versus RBS values

for strained (solid lines) and unstrained (dashed lines) $Si_{1-x}Ge_x$. The peak position of refractive index n and the absorption edge is shifted to higher wavelength with increasing x. For high wavelength (>500 nm) $k(\lambda,x)$ does not depend on x or λ, while there are only small differences in $n(\lambda,x)$. So strained and relaxed $Si_{1-x}Ge_x$ can be modeled with only two parameters, the thickness d and Germanium fraction x, and hence those parameters are measurable with SE.

Fig. 3a shows the fitted Germanium fraction x versus the RBS-measured values for relaxed layers. It can be seen that it is reflected very well within an accuracy of ±1% Ge. In case of strained layers (Fig. 3b) the deviation is in the range of 1%, too, but there is the tendency of underestimating the Germanium fraction for low values. But still an accuracy of ±1% Ge is achieved in the range from 19% to 29% Ge.

4. Conclusion

We developed two representation of the dielectric function of SiGe, one for strained and one for unstrained layers, which depends only on the Ge fraction x within the range from 19% to 29% Ge fraction. The fitting results done with those representations are very stable and enable measurement of x with an accuracy of 1%-Ge-fraction. However, the relative measurement results are still more accurate, so SE is a powerful tool to measure the thickness and compositional homogeneity of epitaxial SiGe layers non-destructively.

References

[1] D. Buca, B. Holländer, H. Trinkaus, S. Mantl, R. Carius, R. Loo, M. Caymax, and H. Schäfer, *Appl. Phys. Lett.* **85** (2004) 2499 and articles therein.

[2] P. Drude, *Ann. Phy. Chem.* **36** (1889) 532 ff

[3] J. Leng, J. Opsal, H. Chu, M. Senko, D.E. Aspens, *Thin Solid Films* **313-314** (1998) 132-136

X-ray diffraction characterization of Low Temperature grown GaAs/InP epilayers

C. Ferrari[a], F. Dubecky[b], R. Kudela[b], J. John[c], R. Srnanek[d],

[a]IMEM-CNR Institute, Parco area delle Scienze 37/A, 43010 Parma, Italy,
[b]IEE-SAS Institute, Dúbravská cesta 9, 841 04 Bratislava, Slovakia
[c]IMEC, Kapeldreef 75, B-3001 Leuven, Belgie

[d]Faculty of Electrical Engineering and Information Technology , University of Technology, Ilkoviova 3, SK-812 19, Bratislava, Slovakia

e-mail: ferrari@imem.cnr.it

Abstract

In the present work low temperature GaAs epitaxial layers grown by molecular beam epitaxy on InP semi-insulating substrates and annealed at different temperatures have been extensively characterised by high resolution X-ray diffraction and X-ray topography to determine the crystal quality and the residual strain of the highly mismatched GaAs epilayer.

The results of the X-ray characterization indicate that in all the samples a negligible residual strain is detected except in a thin region near the GaAs/InP interface and that the GaAs lattice parameter and the full width at half maximum (FWHM) of the GaAs 004 peak decrease when increasing the annealing temperature from 360 to 640 °C, indicating an improvement of the epilayer quality.

1. Introduction

The possibility of integrating electronic and optoelectronic properties on the same device has stimulated the research in highly mismatched heterostructures such as GaAs/InP ([1], Y. Takano et al.). Very defective epilayers with X-ray diffraction peak widths of the order of several minutes of arc and with a large density of misfit and threading dislocations are generally obtained. Furthermore due to the different thermal expansion coefficients of GaAs and InP (α_{GaAs}=4.6x10^{-6} K^{-1}, α_{InP}=5.0x10^{-6} K^{-1}) residual strains after cooling from growth temperature may be introduced.

In the past much attention has been devoted to the study of residual strain in highly mismatched heterostructures (see for instance C. Songyan, L. Yudong., S. Hongbo, P. Yuheng, L. Shiyong [2], L. H. Avanci et al. [3], M. Ichimura, Y. Moriguchi, A. Usami, M. Tabuchi, A. Sasaki [4]). In the paper of Ichimura et al. [4] the authors measured the residual strain as a function of the layer thickness by means the micro-Raman technique. Due to the presence of As precipitates the LT grown GaAs has interesting electronic properties, as for instance a very high resistivity and a very short carrier lifetime and thus can be used as buffer layers in several devices. Due to the As excess in the compound an increase of the lattice parameter was found in LT GaAs (see for instance M. Leszczynski and J. F. Walker [5]).

1-4244-0396-0/06/$25.00 ©2006 IEEE

In the present study we have characterized the residual strain, the lattice parameter and the crystallographic quality of LT grown GaAs/InP epilayers as a function of annealing temperature.

2. Experimental

400 nm thick GaAs layers have been grown at low temperature (Tg = 250 °C, beam equivalent pressure = 15) by molecular beam epitaxy on InP semi-insulating substrates. Four samples have been then annealed in a MOCVD reactor in arsine overpressure at temperatures of 360, 470, 560 and 640 °C respectively.

The samples have been characterized by high resolution X-ray diffraction in the vicinity of CuKα symmetric 004 (θ_B=31.67° for InP) and asymmetric 224 (θ_B=40.02° for InP) reflections for both the InP substrate and GaAs epilayer. Reciprocal lattice maps have been measured in the 224 asymmetric geometry. The comparison of results between symmetrical and asymmetrical geometry allows the unambiguous determination of residual strain and of the absolute lattice parameter of the GaAs epilayers. The double crystal X-ray topography in the asymmetric 335 Bragg was used to verify the presence of extended defects in the heterostructure.

To study the possible variation of strain versus the strain thickness the GaAs epilayer were partially removed by means of selective etching and variable etching time, in order to form a bevel with decreasing thickness. Several diffraction profiles were then taken in different points of the sample.

3. Discussion

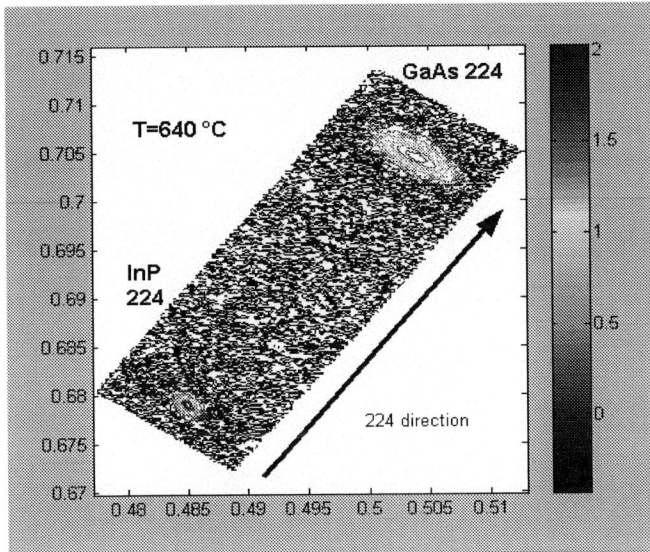

Fig. 1: reciprocal lattice map of the 640 °C annealed sample taken in the vicinity of InP and GaAs 224 reflections. The alignment of the peaks demostrates the fully relaxed state of the heterostructure.

In fig. 1 the reciprocal lattice map (RLM) near the 224 diffraction peak of the sample annealed at 640 °C is reported. In all the maps the main peak at lower values of the RL coordinates corresponds to the InP 224 substrate whereas the broader peak at larger RL coordinated represents the 224 peak of the GaAs epilayer.

The asymmetric 224 geometry at grazing incidence geometry was chosen in order to distinguish between lattice deformation (strain) parallel and perpendicular to the (001) interfaces, as deduced from the 004 rocking curves. All the maps show a 224 GaAs reciprocal lattice (RL) vector aligned with the 224 RL vector of the InP substrate. This indicates that the corresponding planes are almost parallel and the GaAs lattice is undeformed. It is then concluded that the GaAs apilayers is almost completely relaxed.

53

According to this assumption the lattice parameter can be derived by the position of the Bragg peaks in the 004 symmetrical rocking curves by using the simple formula:

$$\frac{\Delta d}{d} = -\Delta\theta \cdot \cot g\theta_B \qquad (1)$$

in which $\Delta\theta$ is the peak separation, $\Delta d/d$ the lattice mismatch and θ_B is the substrate Bragg angle.

The lattice parameter dependence as a function of the annealing temperature is shown in fig. 2. Within the experimental error the lattice parameter decreases by increasing the annealing temperature, in agreement with previous results on LT GaAs. The annealing contributes to solve the As precipitates that are responsible of the lattice expansion in the LT GaAs.

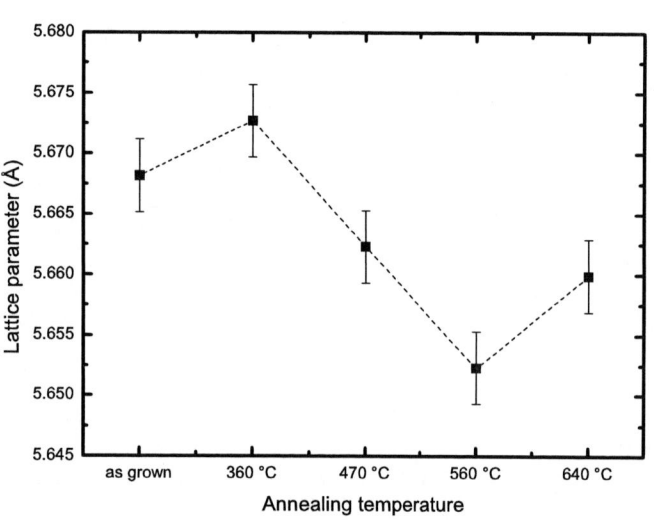

Fig. 2: lattice parameter of the LT grown epitaxial GaAs as a function of the annealing temperature

In fig. 3 the Full Width at Half Maximum of the GaAs 004 peak is reported for all the annealed samples and the as grown sample. A FWHM decrease from 1800 to 800 arc seconds is observed by increasing the annealing temperature. The FWHM values are larger than those observed in LT homoepitaxial GaAs (see for instance M. Leszczynski and J. F. Walker [5]), demonstrating that the large lattice mismatch between GaAs and the InP substrate is the main factor affecting the crystallographic quality but also that As precipitates or interstitial plays a significant role. Double crystal topographs made by using the 335 substrate reflection did not show defects related to the mismatched GaAs, but only threading dislocations from the InP substrate. This is not surprising since the average spacing between misfit dislocations needed to accommodate the lattice mismatch is by far much larger than the resolution of the technique, which is of a few microns in the present case. In fig. 4 several 004 diffraction profiles obtained from zones of different thickness of the bevelled sample are shown. The diffraction peaks do not shift or change shape, demonstrating the uniformity of the GaAs epilayer through the epilayer thickness. This finding in partially in disagreement with the results of M. Ichimura, Y. Moriguchi, A. Usami, M. Tabuchi, A. Sasaki [4] which found a variation of residual strain in the first 100 nanometres of the GaAs epilayer using the micro-Raman technique in bevelled GaAs/InP samples.

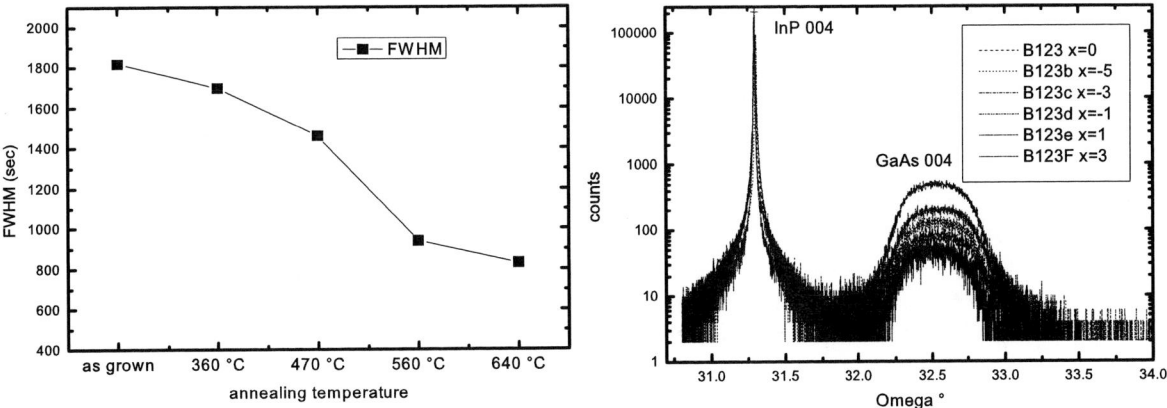

Fig. 3: FWHM of the 004 diffraction peaks as a function of annealing temperature

Fig. 4: 004 diffraction profiles of the sample annealed at T= 470 °C, taken at different thicknesses of the GaAs epilayers

The X-ray diffraction profiles of fig. 4 did not evidence any asymmetry even for the thinnest layer stating that the lattice mismatch must be completely released in a very thin layer close to the interface.

4. Conclusions

LT grown GaAs epilayers on InP substrates exhibit FWHM of X-ray diffraction peaks as large as 1800 arc seconds, mainly due to the large lattice misfit between GaAs and InP ($m=-3.67x10^{-2}$) and the presence of a high density of misfit dislocations at the interface that are not visible by means of low resolution techniques, as X-ray topography. As already seen in homoepitaxial LT grown GaAs the lattice parameter and the full width at half maximum decrease after the annealing, due to the decrease of number and size of As precipitates. Furthermore a negligible residual strain is evidenced in the as grown as well as in the annealed samples.

Acknowledgements

The present work has been supported by the CNR-SAV project: *"Applications of semiconductor single crystals for X-ray optics, monolithic X-ray detectors and high efficiency solar cells"*.

References

[1] Y. Takano, T. Sasaki, Y. Nagaki, K. Kuwahara, S. Fuke, T. Imai, *J. of Cry. Growth*, **169,** 621, 1996.

[2] C. Songyan, L. Yudong., S. Hongbo, P. Yuheng, L. Shiyong, *Mat. Sci. Eng. B* **35**, 133 1995.

[3] L. H. Avanci, M. A. Hayashi, L. P. Cardoso, S. L. Morelhao, F. Riesz, R. Rakennus, T. Hakkarainen, *J. of Cry. Growth* **188**, 220, 1998

[4] M. Ichimura, Y. Moriguchi, A. Usami, M. Tabuchi, A. Sasaki, J. of Cry. Growth **149**, 167, 1995.

[5] M. Leszczynski and J. F. Walker, Appl. Phys. Lett. **62**, 1484, 1993

Investigation of Nickel Silicide Contact Layers for Power Diodes

A. Šatka[1,2], R. Srnánek[1], A. Vincze[2], D. Donoval[1], G. Irmer[3], J. Ková č[1,2]

[1] Slovak University of Technology, FEI, Department of Microelectronics,
SK 812 19 Bratislava, Slovakia
[2] International Laser Centre, Ilkovičova 3, SK 812 19 Bratislava, Slovakia
e-mail: alexander.satka@stuba.sk
[3] TU Bergakademie Freiberg, Institute of Theoretical Physics, Leipziger-Strasse 23,
D-09596 Freiberg, Germany

We report our investigations of the nickel silicide based contact layers prepared for silicon power diodes by electroless nickel plating followed by furnace annealing and subsequent electroless deposition of contact layers. Selected properties of the final structure were studied by the SEM / EDS, micro-Raman spectroscopy and TOF SIMS. The distribution of the species in contacts, quality of interfaces and the role of technological conditions to the formation of the nickel silicide layers were examined in details.

1. Introduction

Silicon semiconductor devices require of low-resistance and reliable ohmic contacts. Nickel silicide (NiSi) has long been used for preparation of low-resistance ohmic contacts for silicon power devices for industrial and automotive applications. Recently NiSi has been proposed as a suitable silicide for the salicidation process in ULSI ICs [1] and Si nanostructures [2] due to its low resistivity comparable to $TiSi_2$ and $CoSi_2$, superior morphological stability and low film/silicon stress. NiSi formation includes Si surface pre-treatment, Ni deposition, and NiSi formation during the heat-treatment. Conventional furnace or rapid thermal annealing is used for the transformation of Ni film to Ni_2Si, NiSi and $NiSi_2$ in respective temperature windows at $250\div300\ °C$, $300\div650\ °C$ and above $700\ °C$. Evaporation or sputtering methods of Ni deposition often used in IC technology and low-count sample preparation are not cost-effective for commercial mass production of large-area power devices as they require oil-free high vacuum and expensive target materials. Ni electroless plating is therefore re-examined as a relatively cheap and reliable alternative for preparation of Si ohmic contacts [4] and nanostructures [5]. Despite of long-term experiences with electroless NiSi formation, some typical problems still remain when NiSi contacts are formed on both sides of the diode structures at the same time. The morphological and phase inhomogeneities in metallization layers, voids in NiSi layers, sensitivity of the Ni layer coating and adhesion to the surface treatment and activation, peeling of the metal layers from the Si substrate or silicide have to be studied to improve yield and reliability of the devices.

2. Experimental

Samples under study were Si (111) power diode wafers with p-n junction formed by a standard single step, both sides diffusion from solid diffusion sources. After the diffusion, surfaces of the wafers were chemically cleaned to remove oxides and rests from the diffusion process to prepare clean Si surfaces. Standard Ni plating solution was used for electroless plating of Si [6]. It comprised of a mixture of $NiSO_4.6H_2O$ nickel salt, which is reduced to Ni at the Si surface by a $NaH_2PO_2.H_2O$ chemical reducing agent buffered by CH_3COONa. Since the Si surface does not have satisfactory catalytic activity to the hypophosphite ions, the Si surface was catalytically activated prior to the plating by deposition of Pd from a solution of $PdCl_2$ in HCl. The Ni silicide was formed by furnace annealing performed at 630°C for 30 min in a forming gas. The Ni contacts were finalised by plating of Ni+P and thin Au layers.

The SEM equipped with EDS X-ray microanalysis, micro-Raman scattering spectroscopy (RSS) and TOF SIMS methods were used for structure and material characterization. The investigations were made on pre-treated substrates, sintered NiSi layers and final diode contacts. Eventually, cross-sections of the samples were prepared by sample cleaving followed by mechanical polishing and lapping.

3. Results and discussion

Surface of the formed silicide depicted in secondary electrons after the etch-off of the remaining Ni (Fig.1a) reveals homogeneous but not smooth coverage of the Si surface, typical for approx. 500 nm thick electroless silicide. Phase transformation of Ni to NiSi has been

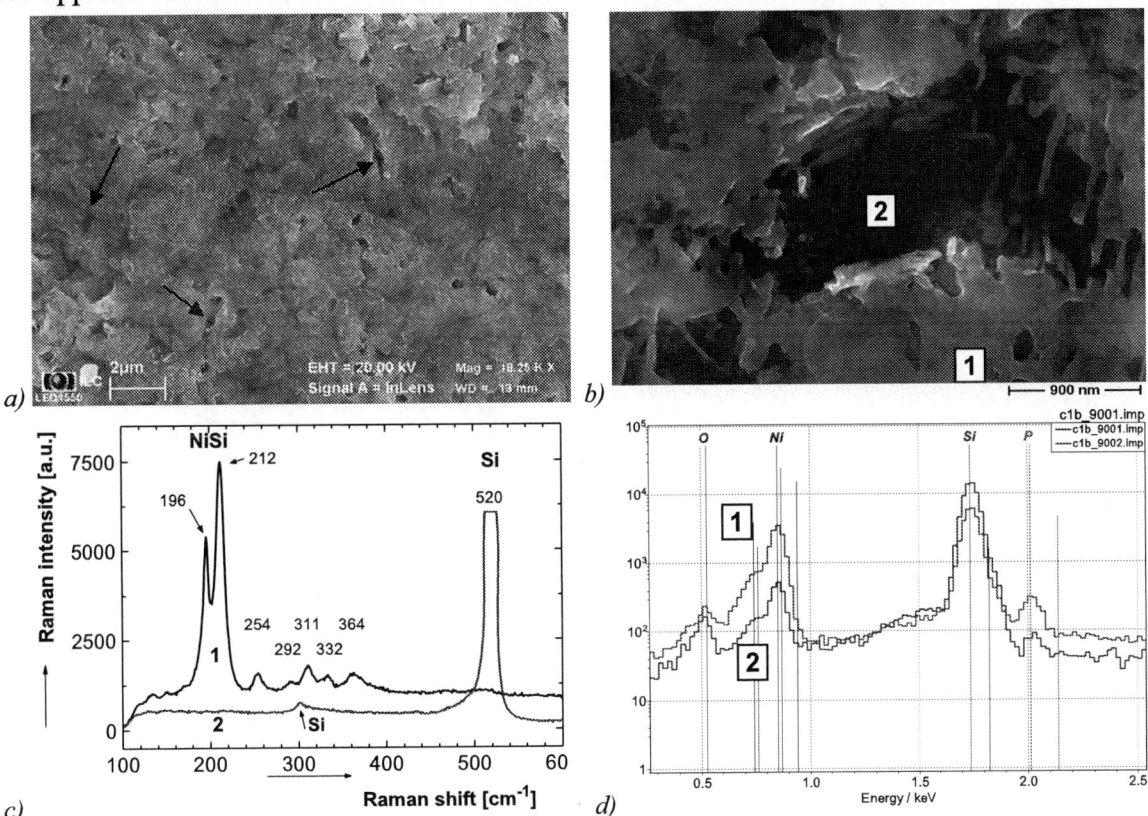

Fig.1: (a) Surface of NiSi film with micropores indicated by arrows, (b) voids in NiSi layers, (c) RSS spectra from the NiSi layer (1) and from the void (2), (d) X-ray EDS from NiSi (1) and void (2) at 5keV beam energy.

confirmed from the RSS spectra (e.g. line #1 in Fig. 1c), from which main peaks at 196 and 212 cm^{-1} and minor peaks at 254, 292, 332 and 364 cm^{-1} attributed to the NiSi phase [7] are clearly resolved. Slight red shift of the peaks can be attributed to a misfit strain in the NiSi. No other silicide phases e.g. $NiSi_2$ could be resolved from the spectra confirming completion of the NiSi formation. Detail SEM inspection of the NiSi surface reveals small micropores threading from the Si surface and voids in the NiSi films (Fig.1a, b). From the RSS taken from the voids in the silicide layer (e.g. line #2 in Fig. 1c), formation of NiSi is suppressed in the voids. From the EDS spectrum #1 in Fig.1d, Ni, Si, P were identified as main constituents in the NiSi layer in which P comes from the electroless plating process. The intensity of Ni L_α relative to Si K_α slightly varied in dependence on the P content in the layer. On the other hand, remarkable low, up to zero Ni L_α intensity is measured at the voids in the layers proving no silicide formation at the voids. The EDS spectra both from the compact silicide and voids reveal oxygen peak at the O K_α line with relatively stronger intensity at the voids. This reveals the presence of the oxide layer at the Si surface in the voids as a barrier for the silicide formation. The TOF-SIMS element depth profiles in Fig. 2 measured from the compact silicide layer confirm formation of the 500÷600 nm thick NiSi layer containing NiO_x and PO_x oxides at the surface formed during the sintering process. Therefore the oxygen O K_α peak in EDS spectra of the compact NiSi layer (Fig. 1) originates from the surface oxides on the silicide layer in contrary to EDS from the voids, where it originates from the oxides at the Ni/Si interface. Concentration of the P inserted from the hypophosphite reducing agent peaks at the silicide surface and practically follows the unreacted Ni during the silicide formation. Relatively slow decrease in Ni, NiSi and P signals corresponds to the inhomogeneous milling process of rough NiSi layer/interface during the

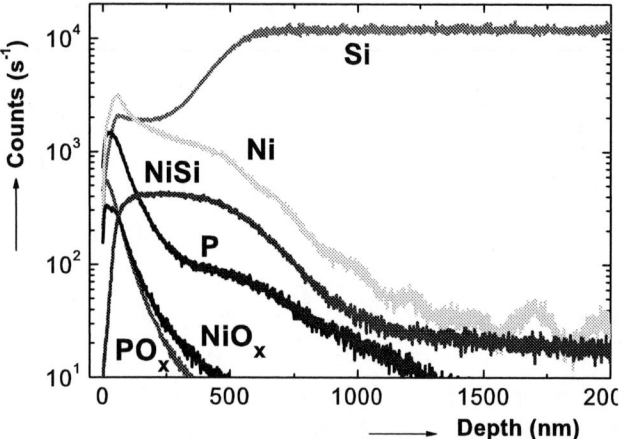

Fig. 2 SIMS elements depth profiles from the silicide layer.

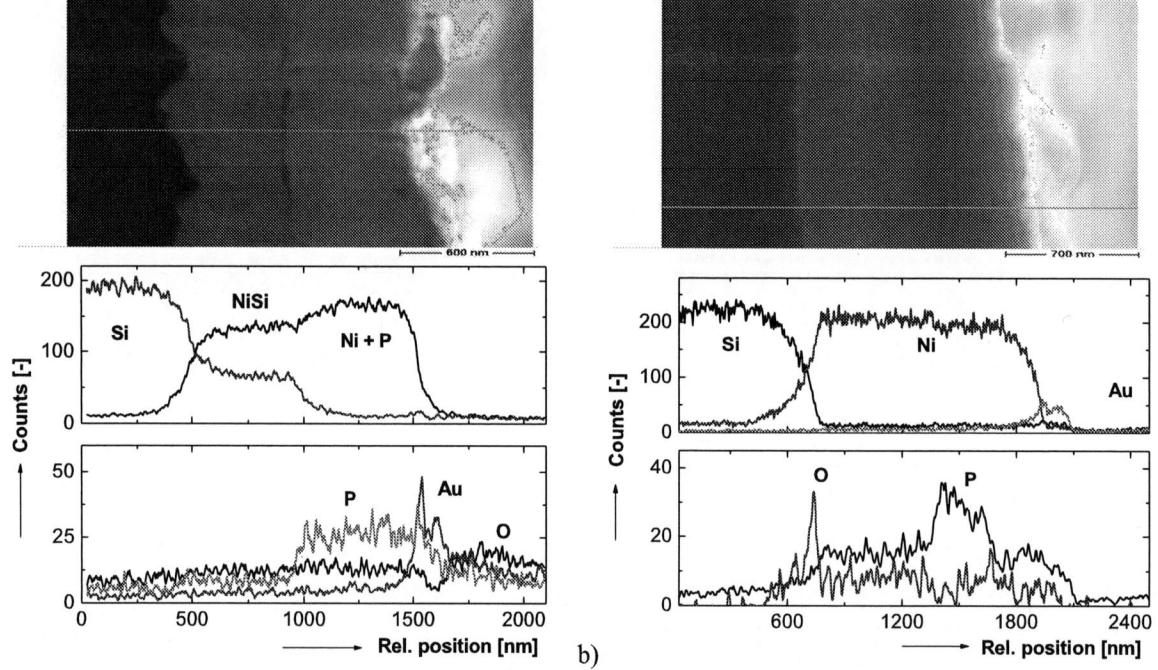

a) b)

Fig.3: SEM images of cleavage edges of two different diode contacts and corresponding Si K_α, Ni L_α and Au M_α EDS line profiles measured across straight line in the SEM images: a) NiSi layer completed with Ni+P contact layer and Au cover layer, b) failure silicide formation in presence of a thin oxide layer at the Ni/Si interface.

SIMS measurements.

The role of O on the NiSi/Ni contact formation could be better understood from the comparison of two diode contacts with formed and unformed silicide layer. Fig. 3a shows the cleaved edges of the diode contacts with silicide layer followed by Ni and Au layers. The

58

NiSi thickness varies between 400 and 600 nm. From the corresponding EDS element line profiles taken at 5 keV primary electron beam energy, Ni contact layer contains remarkable amount of introduced P from the hypophosphite reducing agent. P content in silicide is remarkably lower as a result of the P dissolution conditions at the temperature of NiSi formation. Relative intensity of the Ni L_α to Si K_α line profiles in the silicide is considerably different from nearly half to half values expected for the NiSi atomic composition confirmed by the RSS. Taking into account a strong reabsorption of the Si K_α radiation by Ni atoms adjacent to the Si atoms followed by the additional emission of the X-rays at the Ni L_α, total Ni L_α yield is nearly 30% higher of its value expected for NiSi. The reabsorption of the Si K_α radiation affects also the lateral resolution of the element line profiles, which could be better for Si than for Ni at the NiSi interface. Intensity of the Ni L_α in Ni+P layer is influenced by a relative high amount of P in the layer. The Au M_α signal peak at the surface of the contact corresponds to plating of Ni+P layer by thin Au layer. From the line profiles, both Ni and NiSi layers contain small amount of O. Intensity of the O K_α in both silicide and Ni+P layers is at the detection limit of the EDS system, so that the O profile could not be exactly evaluated. Cleaved edge of the defective diode contact prepared by the same electroless deposition and heat treatment processes is shown in Fig. 3b. Both Ni L_α and Si K_α EDS line profiles exhibit steep transition at the Ni/Si interface revealing no silicide layer formation. The O K_α signal peak at the Ni/Si interface confirm that thin oxide layer prevents the silicide formation. It is interesting to note no negligible oxygen signal in Ni layers, and slight decrease of the O signal with increasing P K_α signal.

4. Conclusions

Properties of the NiSi-based contact layers were studied by the SEM / EDS, RSS and SIMS. The distribution of the species in the contacts, quality of the interfaces and technological conditions of the formation of the NiSi layers were examined in details. The experiments confirm critical role of the silicon oxides in formation of the NiSi layers with inhomogeneities and holes. The sintering process was identified as a source of the oxides at the silicide surface. The electroless Ni plating process and preceding technological operations were identified as a source of oxides in the voids at the Ni/Si interface acting as a barrier for formation of NiSi and subsequent Ni plating.

Acknowledgement

The authors would like to thank M. Predmerský and R. Záluský for technical assistance. This work was partially supported by grants AV 4/0022/05, APVV-20-055405, VEGA 1/3111/06 and 1/3076/06.

References

[1]   C. Fitz et al.: *Microelectronic Engineering* **82**, 460-466, 2005.
[2]   Y. Wu et al.: *Nature* **430**, 61, 2004.
[3]   A.U. Ebong et al.: *Solar Energy Materials and Solar Cells* **31**, 499, 1994.
[4]   D. S. Kim et al.: *J. Korean Phys. Soc.* **6**, 1208, 2005.
[5]   D. Niwa et al.: *Electrochimica Acta* **48**, 1295, 2003.
[6]   S. Furukawa et al.: *Sensors and Actuators* **A56**, 261, 1996.
[7]   F. F. Zhao et al.: *Microelectronic Engineering* **B21**, 862, 2003.

Reliability issues in advanced High k/metal gate stacks for 45 nm CMOS applications

G. Groeseneken[1,2], M. Aoulaiche[1,2], S De Gendt[1,3], R. Degraeve[1],
M. Houssa[1,2], T. Kauerauf[1,2], L. Pantisano[1]

[1]IMEC, Kapeldreef 75, B-3001 Leuven, Belgium, Phone +32-16-281269
Email: Guido.groeseneken@imec.be
[2] also at KU Leuven, ESAT-Department, Leuven, Belgium
[3] also at KU Leuven, Chemistry-Department, Leuven, Belgium

Some recent insights in reliability issues of high k/metal gate stacks for the 45 nm CMOS node and beyond are discussed. The problem of transient charging effects leading to threshold voltage instability is illustrated. It is shown that nitridation of Hf-silicate layers leads to severe degradation of the Negative-Bias-Instability (NBTI) lifetime. Some insights in the mechanisms of Time-dependent-Dielectric breakdown (TDDB) are discussed and illustrated.

1.   Introduction

Due to the exponential increase in leakage current when scaling down the gate oxide thickness of MOSFET's, there is an urgent need to replace SiO_2-based dielectrics by alternative dielectrics with high dielectric constants. Also due to poly-depletion effects, the use of metal gates will become necessary. Although the requested decrease in gate leakage current can be relatively easily realized, as is shown on Fig. 1, other parameters such as drive current and mobility are more difficult to achieve [1]. Moreover, it is very important to guarantee sufficient reliability for these new types of gate stacks.

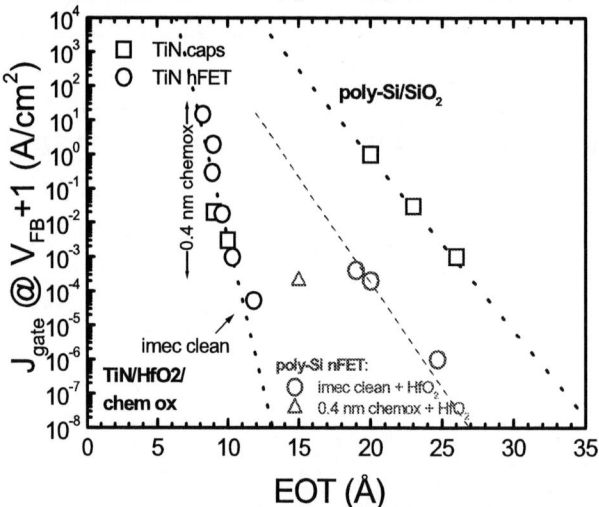

Fig. 1 – Gate leakage current vs. EOT for high k gate stacks with poly-Si and metal gates. A 0.8 nm EOT stack can be achieved by using a 0.4 nm chemical oxide interfacial layer, and a HfO_2/TiN gate stack [1]

As a result it is very important that already from the start of the technology development, reliability issues such as TDDB and NBTI are investigated to understand better the degradation and breakdown mechanisms of the high k gate stacks.

In this paper we will review some of the recent work on high k/metal gate stack reliability for 45 nm CMOS technologies and beyond. In the first part we will discuss the effects of defects inside the high k stacks on Vt-instability. The third part of the paper will deal with Bias-Temperature-Instability effects (BTI) in Hf-silicate gate stacks. The last part will discuss the recent insights in Time-Dependent-Dielectric-Breakdown (TDDB) in sub 1 nm gate stacks.

2.   Threshold voltage instability

The electrical stability of CMOS devices with conventional gate dielectrics is commonly studied using static (DC) measurement techniques. By applying the same methods to MOS devices with alternative gate dielectrics, it has been shown that alternative gate stacks suffer from severe charge trapping and that the trapped charge is not stable, leading to fast transient charging components and threshold voltage hysteresis. More particularly HfO_2 layers suffer from a defect band close to the conduction band, especially when poly-Si gates are used. These defects give rise to Vt-instability and C-V hysteresis [2,3,4]. New time-resolved measurement techniques down to the µs time range have to be applied to capture the fast transient component of the charge trapping observed in $SiO2/HfO_2$ dual layer gate stacks. An example of such measurement is shown in Fig. 2.

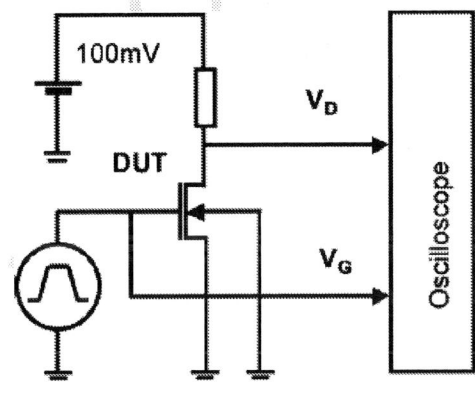

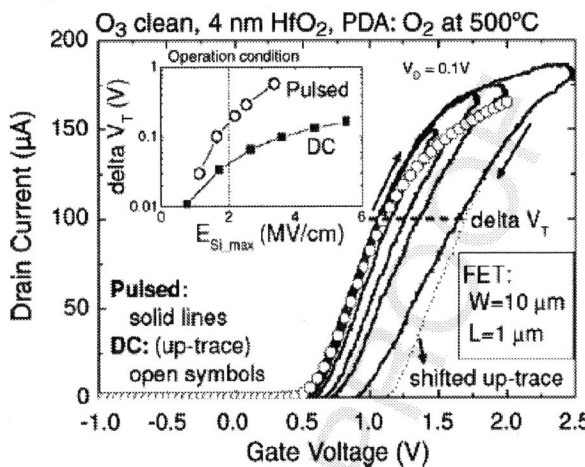

Fig. 2-a   Schematic of setup used for pulsed ID–VG measurements.

Fig. 2-b I_D–V_G traces measured with the setup of Fig. 1.a. The open symbols show an uptrace measured by conventional quasi dc technique. The inset shows the VT-shift as a function of the maximum Si-field for both measurement techniques.

These defects can also be assessed by charge pumping measurements, which are able to sense defects inside the bulk of the high k layers, by changing either the amplitude or the frequency of the charge pumping signal. Using metal gates and Hf-silicate layers, and by scaling down the high-k physical thickness the Vt-instability can be reduced to acceptable levels.

3.   Bias-Temperature Instabilities

Negative bias temperature instability (NBTI) has recently become a very important reliability issues for heavily nitrided SiON dielectrics. Also for high k/metal gate stacks this problem needs attention. For high k/metal gate stacks both PBTI and NBTI are important, although NBTI is the lifetime limiting degradation mechanism. PBTI can be attributed to trapping in pre-existing defects, whereas NBTI is caused by generation of new defects under influence of the presence of holes at the Si/high k interface. The NBTI problem was mainly investigated for SiOx(N)/HfSiO(N)/TaN based pMOSFETs [5, 6]. NBTI is found to be reduced when the N-content in the gate stack is reduced and when Deuterium is used instead of Hydrogen during the FGA anneal. It is shown that nitrogen-incorporation in the gate stack (either by NH_3 anneals or decoupled plasma nitridation - DPN) result in much enhanced NBTI, as is shown on Fig. 3

Device degradation is mainly due to fast (interface) state generation in the non-nitrided stacks, while a substantial contribution of the defects produced in the nitrided stacks are slow (bulk) states. When extrapolating the lifetime, stacks annealed in N_2 or O_2 "pass" the reliability criteria fixed at $(V_G-V_{th})=1V$ for 10 years lifetime, whereas the nitrided stacks do not pass, as is shown on Fig. 4. This leads to a trade off between nitridation of the gate stacks, which is needed to improve the thermal stability of the stacks and the reliability lifetime.

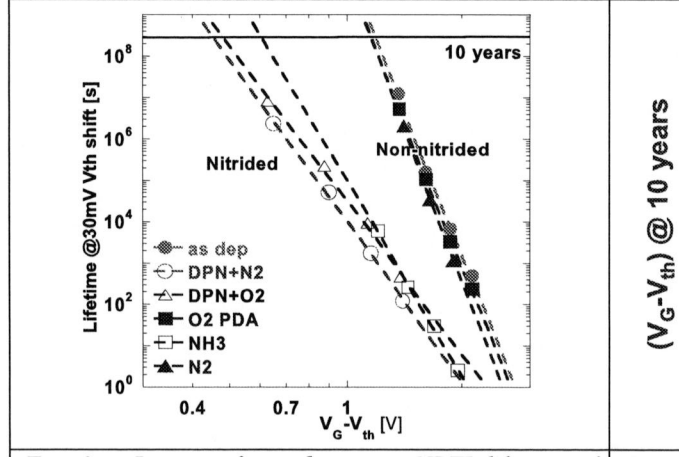 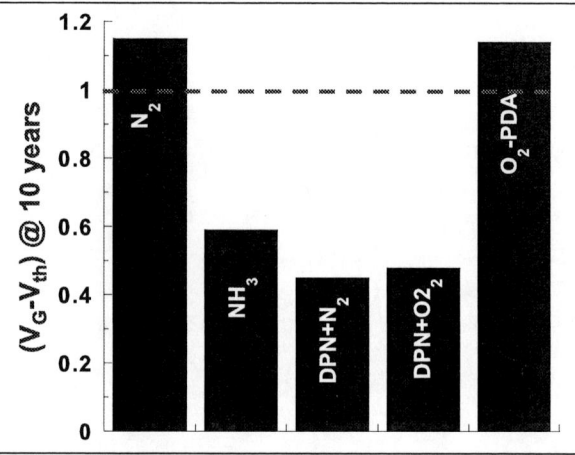

| Fig. 3 – Impact of nitridation on NBTI lifetime of SiOx(N)/HfSiO(N)/TaN stacks with various nitridation conditions | Fig. 4 – Extrapolated (VG-Vth) for 10 years NBTI lifetime for various nitridation conditions of Hf-silicate stacks with TaN meta l gate. |

4. Time-dependent Dielectric Breakdown

It is shown that similar to the case of SiON dielectrics, the degradation process of Hf-silicate layers is dominated by the formation of single localized conductions paths and that the total time-to-failure consists of two parts: the time-to-creation of a conductive path and the time until the wearout of this path has reached a critical threshold [7]. The degradation rate of the progressive wearout is strongly voltage dependent. For thin (1.38 nm EOT) HfSiON we found that the time-to-creation even at operating condition is very small, but the overall TDDB reliability is guaranteed because of the large time-to-wearout [7].

We also demonstrated that even a 0.9 nm EOT ALD HfO_2/TaN gate stack can be intrinsically reliable for TDDB under CVS. During degradation, single traps and two-trap clusters are formed in the HfO_2, the latter giving rise to a considerable SILC. The two-trap clusters wear out with time leading to hard breakdown, but at operating conditions this takes longer than the required lifetime. The main reliability issue for thin HfO2 stacks is therefore the SILC, which has to be taken into account by circuit designers.

Correct low voltage extrapolations of time-to-hard breakdown t_{HBD} are done in 2 steps. 1) Determine time-to-soft breakdown t_{SBD} vs. V_G and use the conventional area and percentile scaling laws. 2) Add the leakage path wear out time $t_{wearout}$. The area scaling on t_{SBD} is, however, irrelevant at low V_G because $t_{wearout} \gg t_{SBD}$, and the reliability can be determined directly by extrapolating t_{HBD} to low voltage without any area scaling! Because of the slow wear out phase, t_{HBD} for HfO_2 is sufficiently high at low voltage (Fig. 5). The main reliability problem of these sub-1nm EOT high k/metal gate stacks is therefore the strong gate leakage current increase due to multiple SBDs.

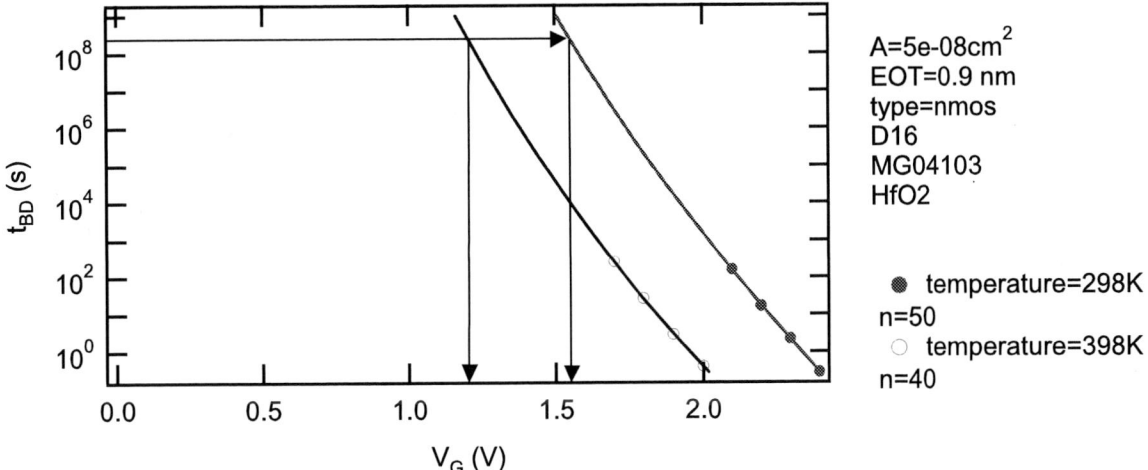

Fig.5: Low voltage extrapolation of tHBD results in sufficient lifetime for 0.9 nm HfO2 both for 298 K and 398 K. A power law extrapolation was used with exponent ~50 at 298 K and ~40 at 398 K. Maximum operating voltage at 10 years is ~1.6 and ~1.2 V respectively

5. Conclusions

This paper reported some of the recent work carried out in understanding the reliability problems in high k/metal gate stacks that will be used in the 45 nm CMOS technology node or beyond. It is shown that high k stacks suffer from transient charging Vt-instability, which necessitates new characterization techniques to be able to measure the defects at its origin. Also NBTI degradation is shown to be potential lifetime limiting mechanisms, especially for sHf-silicate stacks which were nitrided in order to improve their thermal stability. Finally, TDDB lifetime is found to be no major limitation thanks to the progressive nature of the wear-out, both for Hf-silicate and for sub 1-nm HfO_2 gate stacks. The main reliability issue for thin HfO_2 stacks is the SILC-generation, which has to be taken into account by circuit designers.

Acknowledgement

This work was carried out within IMEC Core Partner Industrial Affiliation Program, supported by Intel, Texas Instruments, Matsushita, Samsung, TSMC, Infineon, Philips and ST Microelectronics.

References

[1] W. Tsai, L.-A. Ragnarsson, L. Pantisano, P. Chen, B. Onsia, T. Schram, E. Cartier, A. Kerber, E. Young, M. Caymax, S. De Gendt, and M. Heyns, "Performance comparison of sub 1 nm sputtered TiN/HfO2 nMOS and pMOSFETs," in IEDM Tech. Dig., p. 311, 2003.

[2] A. Kerber, E. Cartier, L. Pantisano, M. Rosmeulen, R. Degraeve, T. Kauerauf, G. Groeseneken, H.E. Maes, U. Schwalke, "Characterization of the VT-instability in

SiO2/HfO2 gate dielectrics", Proceedings of the IRPS-conference (International Reliability Physics Symposium), p. 41, 2003.

[3]   A. Kerber, E. Cartier, L. Pantisano, R. Degraeve, T. Kauerauf, Y. Kim, A. Hou, G. Groeseneken, H.E. Maes, U. Schwalke, "Origin of the Threshold Voltage Instability in SiO2/HfO$_2$ Dual Layer Gate Dielectrics" IEEE Electron Device Letters, vol. 24, p. 87-89, 2003

[4]   C.D. Young et al, IEEE El. Dev. Lett., p. 586, 2005.

[5]   M. Aoulaiche, M. Houssa, R. Degraeve, G. Groeseneken, S. De Gendt and M.M. Heyns, "Contribution of fast and slow states to negative bias temperature instabilities in HfxSi(1-x)ON/TaN based pMOSFETs", Microelectronics Engineering, vol. 80, p. 134-137, 2005.

[6]   M. Aoulaiche, M. Houssa, Thierry Conard, G. Groeseneken, S. De Gendt and M.M. Heyns, "Impact of Nitrogen incorporation in SiOx/HfSiON gate stacks in Negative-Bias-Temperature Instability", Proceedings of the IEEE IRPS-conference (International Reliability Physics Symposium), p. 317-324, 2006.

[7]   T. Kauerauf, R. Degraeve, F. Crupi, G. Groeseneken and H.E. Maes, "Trap generation and progressive wearout in thin HfSiON", Proceedings of the IEEE IRPS-conference (International Reliability Physics Symposium), p. 45, 2005.

[8]   R. Degraeve, T. Kauerauf, M. Cho, M. Zahid, L-Å. Ragnarsson, D.P. Brunco, B. Kaczer, Ph. Roussel, S De Gendt and G. Groeseneken, "Degradation and breakdown of 0.9 nm EOT SiO2/ ALD HfO2/metal gate stacks under positive Constant Voltage Stress", IEDM Tech. Digest, p. 419, 2005.

Nb-Ti/Al/Ni/Au Ohmic Metallic System to AlGaN/GaN

T. Lalinský[1], G. Vanko[1], Ž. Mozolová[1], J. Liday[2], P. Vogrinčič[2], A. Vincze[3], F. Uherek[3], Š. Haščík[1], I. Kostič[4]

[1]Institute of Electrical Engineering, Slovak Academy of Sciences, Dúbravská cesta 9, 841 04 Bratislava, Slovakia
[2]Faculty of Electrical Engineering and Information Technology, Slovak University of Technology, Ilkovičova 3, 812 19 Bratislava, Slovakia
[3]International Laser Center, Ilkovičova 3, 812 19 Bratislava, Slovakia
[4]Institute of Informatics, Slovak Academy of Sciences, Dúbravská cesta 9, 842 39 Bratislava, Slovakia
e-mail: eleklali@savba.sk and gabriel.vanko@savba.sk

In this paper, we report on novel Nb/Ti/Al/Ni/Au metallic system to form ohmic contact to AlGaN/GaN heterostructure. The fabrication and electrical characterization of the Nb-Ti/Al/Ni/Au based ohmic contacts are presented. Auger Electron Spectroscopy (AES) and Secondary Ion Mass Spectroscopy (SIMS) are also used to evaluate the improved ohmic contact formation.

1. Introduction

As reported in the literature [1], Ti/Al/Ni/Au is the most commonly used metallic system to form ohmic source and drain contacts in AlGaN/GaN based HEMTs. The tunneling effect due to nitride formation at the AlGaN interface is generally accepted as the ohmic contact mechanism. When the TiN layer is formed at the Ti/AlGaN interface, a high concentration of nitrogen vacancies is created near the interfaces. This phenomenon causes the AlGaN to be the heavily doped n type, so the tunneling through the thin TiN/AlGaN interfacial potential barrier dominates. The role of AlN interfacial layer in the forming of the ohmic contact is not clarified. In generally, it seems that TiN interfacial layer is more responsible for the ohmic contact formation as compared with the AlN. Recently, Ta/Ti/Al/Ni/Au ohmic contact to AlGaN/GaN has been reported to improve a significantly the both contact resistivity and surface morphology of the conventional contact system [2]. The thick formation of TaN/TiN interfacial nitride phases was appeared to be responsible for the good ohmic contact behavior in Ta/Ti/Al/Ni/Au metal scheme. It is predicted that the TaN formation would cause low ohmic contact resistivity. However, there is no understanding of this mechanism. Thus, further detailed structural studies are necessary for better understanding of this ohmic contact formation. In this paper, we report on novel Nb/Ti/Al/Ni/Au metallic system to form ohmic contact to AlGaN/GaN heterostructure. The fabrication and electrical characterization of the Nb-Ti/Al/Ni/Au based ohmic contacts are presented. Auger Electron Spectroscopy (AES) and Secondary Ion Mass Spectroscopy (SIMS) are used to investigate the contact composition and contact interfacial chemical reactivity. The positive role of the Nb interfacial layer in the ohmic contact formation is discussed.

2. Ohmic contact fabrication

The undoped AlGaN/GaN heterostructure used in this work was grown by metalorganic chemical-vapor deposition on a (0001) sapphire substrate. Fe-doped GaN buffer layer with a

1-4244-0396-0/06/$25.00 ©2006 IEEE

thickness of 3 μm was grown, followed by the growth of 32 nm thick undoped AlGaN barrier layer. The Al composition in AlGaN was set to be 25 %.

To fabricate the ohmic contacts Nb, Ti, Al, Ni contact metals were deposited by electron-beam evaporation and Au by the conventional resistance evaporation through the patterned photoresist mask. The pressure in the cryogen pumping vacuum system during the evaporation was lower than 10-5Pa. A lift-off technique was used to form the circular shaped metallic systems in order to use the circular transfer length method (CTLM) [3] to determine the contact parameters. The patterned contact systems were formed using the rapid thermal annealing (RTA) in flow of pure nitrogen as a forming gas for 35 s.

Three different ohmic contact systems were prepared. The first one (C) corresponded to the conventional ohmic contact based on Ti(20 nm)/Al(100 nm)/Ni(40 nm)/Au(50 nm) metallic system. In the next two ohmic contacts (E, F) a thin interfacial Nb layer of the thickness of 10 nm (E) and 20 nm (F) placed as the first layer in contact with AlGaN barrier layer was additionally incorporated. Contact resistivity measurements at different alloying temperatures (RTA for 35 s) were performed for all contact systems to find optimal conditions of contact formation. Surface morphology of the contact systems after optimal contact formation related to the minimal contact resistance was investigated in first stage. It is shown in Fig. 1.

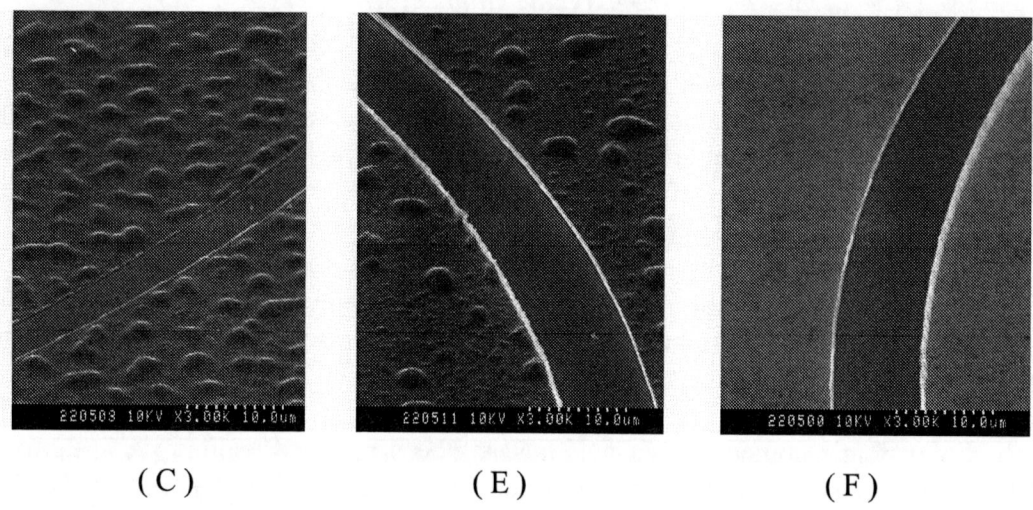

(C)   (E)   (F)

Fig.1. Surface morphology of ohmic contact systems (C, E, F) after optimal contact formation

As compared, it can be seen that surface morphology of the conventional contact system (C) is significantly improved by adding thin Nb interfacial layer (E, F). The surface roughness seems to be practically eliminated adding 20 nm-thick Nb interfacial layer, as shown in Fig. 1 (F).

3. Ohmic contact electrical characterization

Electrical characterization of the fabricated ohmic contact systems was performed using the above mentioned CTLM method. We have proposed the CTLM structures of the inner ring radii 200 μm and gaps 5, 10, 15, 25, 35, and 45 μm. The ohmic contact parameters such as contact transfer length L_T and contact resistivity R_C were extracted by fitting the measured total resistance R between the contacts. The sheet resistance underneath the contacts was assumed to be equal to the sheet resistance R_{SH} between the contacts. No correction was made for non-zero sheet resistance of the contact metallization systems used.

Fig. 2 shows a typical behavior of the extracted R_C values as a function of annealing temperature for all ohmic contact systems analyzed. The corresponding behavior of the extracted L_T values is shown in Fig. 3.

As shown in Fig. 2 and Fig. 3 the presence of Nb in the interface leads to the significant lowering in the both contact resistivity R_C and transfer length L_T.

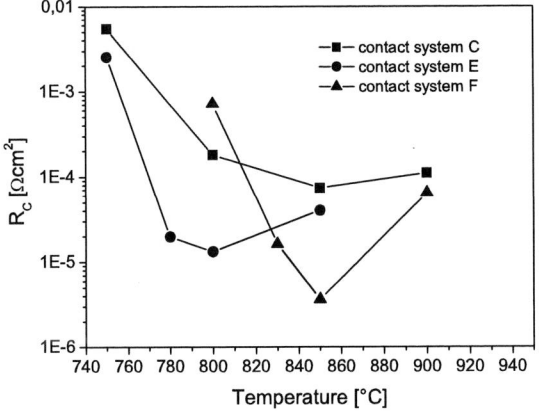

Fig.2. Contact resistivity R_C as a function of temperature annealing

Fig.3. Transfer length L_T as a function of temperature annealing

So, Nb based ohmic contact systems at optimal alloying temperatures seem to be superior to that of conventional Ti/Al/Ni/Au in respect of both surface morphology and contact resistivity evaluation.

4. Contact interfacial composition and reactivity

AES and SIMS depth profiling methods were used to study the contact interfacial composition and reactivity.

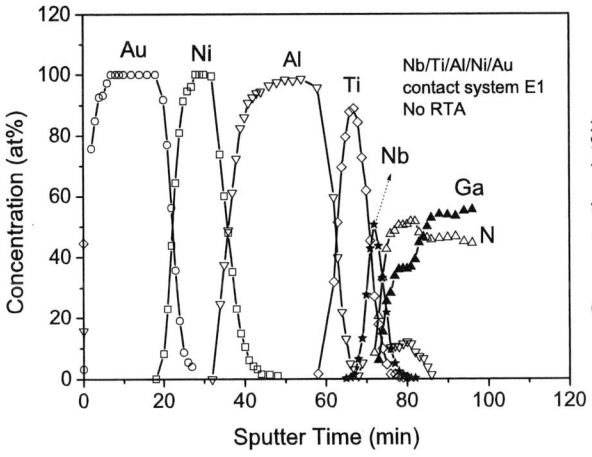

Fig.4. AES depth profile of non-annealed contact system E

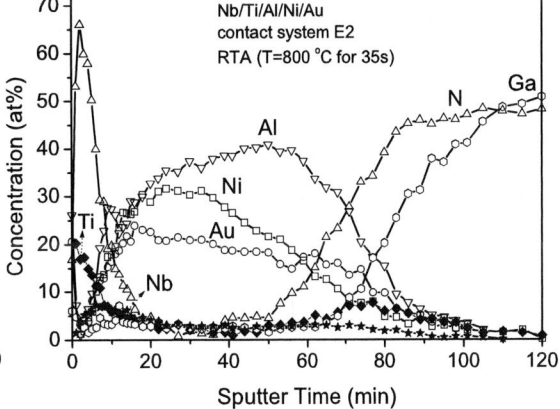

Fig.5. AES depth profile of contact system E after optimal annealing (T=800 $^{\circ}$C for 35s)

Fig.4 shows AES depth profile of contact system E after deposition and patterning without any annealing. As expected, single layer sequence is observed with the first 10 nm-thick Nb layer on the interface. If the contact system is alloyed at optimal annealing conditions (T=800 °C for 35s) however, a significant intermixing of the metallic layers is found as shown in Fig.5. Au and Ni penetrate and diffuse through the Ti/Al layers into the AlGaN interfacial layer. Likewise, diffusion of Ti an Nb from the interface up to the surface of the contact system is recorded. It is accompanied also by out-diffusion of Nitrogen and Gallium from the substrate. To study chemical reactivity of the contact metallic layers directly on the interface, SIMS depth profiling of the same contact systems was performed. The emphasize was given to detect metal nitride phases created on the interface after annealing because they should be responsible for the ohmic contact behavior.

In analogy, Fig.6 and Fig.7 show obtained SIMS depth profiles of the non-annealed and annealed contact system E.

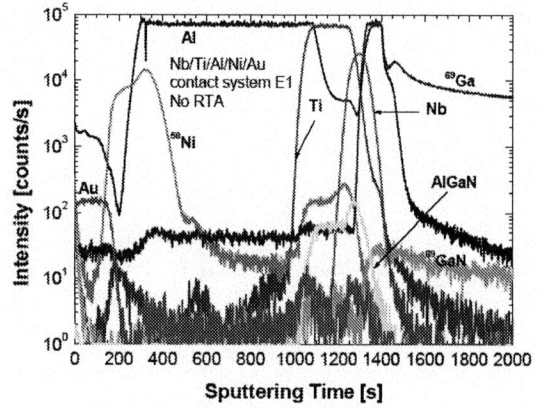
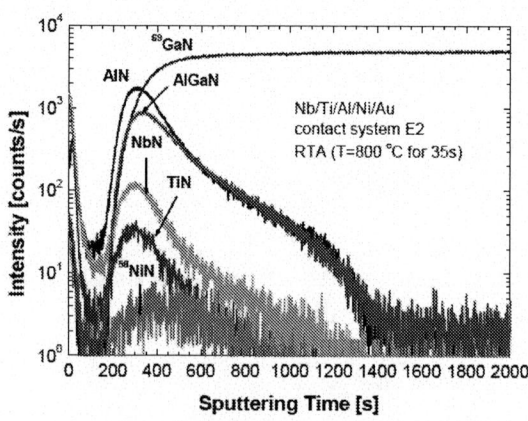

Fig.6. SIMS depth profile of non-annealed contact system E

Fig.7. SIMS depth profile of formed metal nitride phases after optimal annealing

The single layer structure of non-annealed contact system E (Fig.6) was also identify in analogy with AES (Fig.4). SIMS depth profile of annealed contact E (Fig.7) clearly reveals the formation of nitrides such as AlN, TiN, NbN and NiN on the interface. However, the influence of the NiN phase seems to be neglected. We predict the dominant influence of interfacial NbN phase in forming of the ohmic contact with a low and thin potential barrier on the interface. Interface composition is consistent with that of Ta/Ti/Al/Ni/Au contact [2].

Acknowledgement

This work has been supported in part by VEGA Projects 2/6097/26 and 1/3076/06, bilateral projects SK-CZ-09506 and SK-FR-01906 and also by Project APVV-51-040605.

References

[1]   N. Chaturvedi, U. Zeimer, J. Würfl and G. Tränkle, *Semicond. Sci. Technol.* **21,**175, 2006.

[2]   Ki Hong Kim, Chang Min Jeon, Sang Ho Oh, Jong-Lam Lee, Chan Gyung Park, Jung Hee Lee, Kyu Seok Lee, Yang Mo Koo, *J. Vac. Sci. Technol.* **B23**, 322, 2005.

[3]   B. Jacobs, M.C.J.C.M. Kramer, E.J. Geluk, F. Karouta, *Journal of Crystal Growth* **241**, 15, 2002.

Electrical and optical behaviour of nanocrystalline CdS/InP heterojunction p-n diodes

V. Rakovics, Zs. J. Horváth, B. Pődör

(The contribution was not been delivered in time)

High Purity p-type InP Grown by LPE with Rare-Earth Admixtures

J. Grym, O. Procházková, J. Zavadil and K. Žďánský

Institute of Radio Engineering and Electronics, Academy of Sciences of the Czech Republic
Chaberská 57, 182 51 Praha 8, Czech Republic
e-mail: grym@ure.cas.cz

Specific features of rare-earth-oxides (PrO$_x$, TbO$_x$, Tm$_2$O$_3$, Gd$_2$O$_3$, and Eu$_2$O$_3$) and Gd are employed to improve physical properties of InP layers. The InP layers were grown by liquid phase epitaxy (LPE) on (100)-oriented single crystal InP substrates with rare-earth element (RE) addition to the growth melt. The surface morphology and defect density was monitored by optical and scanning electron microscopy, the evaluation of electrical properties was gained from C-V characteristics and Hall measurements, and the low-temperature photoluminescence spectroscopy was used to study the optical properties. Significant improvement of all studied layer parameters with increasing amount of RE in the melt was observed right to critical value of RE concentration. The residual impurity concentration was reduced by up to three orders of magnitude, photoluminescence spectra were markedly narrowed and fine spectral features were resolved. The conductivity changed from n- to p-type when certain limit of RE concentration in the melt was exceeded for majority of studied REs.

1. Introduction

In recent years, there have been high expectations connected with rare-earth (RE) elements as promising sources of a sharp and temperature independent radiation when incorporated into the active layer of a light-emitting semiconductor device. The fortunate coincidence between the Er^{3+} emission band around 1535 nm and the principal low-loss window in the absorption spectrum of aluminosilicate optical fiber has been the main driving force behind recent work on RE-doped materials [1]. Regarding the liquid phase epitaxy (LPE), in most cases, however, REs do not enter the substitutional or interstitial sites of the lattice due to their low solubility in crystalline semiconductors [2].

Another strong motivation to study REs is connected with their high chemical reactivity and reduction capability. The admixture of REs into the molten solution for growing A^{III}B^{V} layers can reduce the background donor concentration by several orders of magnitude. REs form stable compounds with typical residual impurities (e.g. S, Si, Se, C, O, etc.). These compounds are insoluble in the growth solution and precipitate out as a slag atop the melt. Therefore, the impurities are prevented to be incorporated into the grown layers [3].

Recently, we have performed a systematic study of the impact of REs on the properties of InP layers [4-6]. Among the studied REs, Yb and Ce were proved to be incorporated into the InP lattice [7]. In this contribution we compare the influence of Gd and RE oxides (PrO$_x$, TbO$_x$, Tm$_2$O$_3$, Gd$_2$O$_3$, and Eu$_2$O$_3$) with the previously studied REs.

2. Experimental

InP epitaxial layers were prepared by the supercooling method on (100)-oriented semi-insulating InP:Fe and n-type InP:Sn substrates with a certain addition of REs to the growth melt (for the details of the growth technology see [4]). Scanning electron microscopy and Nomarski-contrast optical microscopy were employed to study the substrate-layer interface, the layer thickness, and the surface morphology. Hall effect and capacitance-voltage characteristics were measured to determine the conductivity type and the impurity concentration. Photoluminescence spectra were taken at various temperatures and levels of excitation power by He-Ne and Ar lasers.

3. Results and Discussion

The surface morphology of most layers grown with a small addition of REs was desirably smooth and mirror-like with minimum of surface droplets. The appropriate concentration of RE admixture can decrease the concentration of defects by up to an order of magnitude.

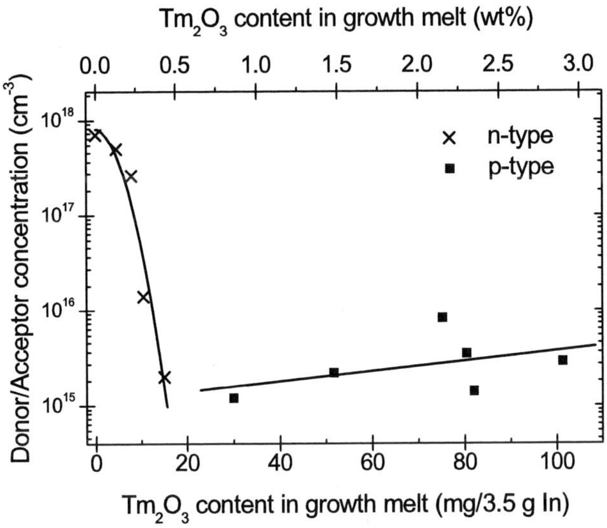

The expected gettering effect was observed for all REs. Introducing REs to the growth solution, simultaneous gettering of shallow impurities takes place. Appropriate RE concentration results in preparation of InP layers with shallow impurity concentration decreased by up to three orders of magnitude. Donor impurities are preferentially gettered due to the high affinity of REs towards Si and group VI elements. The preferential gettering leads to conductivity conversion from n- to p-type when increasing the RE concentration in the growth melt. Further increase of RE concentration results in moderately elevated acceptor concentrations (fig. 1). This elevation is caused by either the incorporation of new impurities to the melt with the REs or increased p-type activity of amphoteric impurities.

Figure 1: Dependence of the donor/acdeptor concentration on Tm_2O_3 content in the growth melt

sample	RE content (mg)	RE content (wt%)	CT	N_D or N_A (cm^{-3})	sample	RE content (mg)	RE content (wt%)	CT	N_D or N_A (cm^{-3})
Gd-10	0.3	0.008	n	2.0E+15	GdO-1	1.9	0.094	n	2.6E+17
Gd-16	0.4	0.011	p	3.5E+15	GdO-2	3.1	0.153	n	1.7E+17
Gd-12	0.5	0.014	p	7.7E+15	GdO-3	15.1	0.741	n	8.0E+16
Gd-11	1.0	0.028	p	6.3E+15	GdO-4	30.0	1.462	n	9.3E+16
Gd-14	2.1	0.059	p	1.2E+16	GdO-5	60.0	2.882	n	1.4E+16
Gd-15	5.0	0.141	p	2.2E+16	GdO-8*	30.2	0.846	n	6.8E+15
PrO-8	5.0	0.250	n	5.7E+15	TbO-0	0	0	n	6.8E+17
PrO-2	15.8	0.790	n	2.2E+15	TbO-1	25.0	1.250	n	6.6E+17
PrO-1	24.5	1.225	p	2.8E+15	TbO-2	50.8	2.540	n	2.3E+17
PrO-6	30.0	1.500	p	2.0E+15	TbO-3	97.6	4.880	n	1.3E+16
PrO-5	59.5	2.975	p	3.8E+15	TbO-6*	49.3	2.465	n	7.8E+15

Table 1: Donor/acceptor concentrations of the samples prepared with different REs admixtures

The data obtained by C-V measurements are summarized in table 1 (asterisk means different conditions of the sample preparation – the RE oxide was dissolved in the In melt at elevated temperature of 800 °C). All layers with TbO_x and Gd_2O_3 addition maintain the n-type conductivity even at relatively high concentrations reaching its solubility limit in the growth melt. Europium oxide appears to have a very high purification effect. The choice of substrate seems to play an important role. Two n-type layers of very low donor concentration were grown on InP:Fe substrate:

- sample EuO-6 (0.39 wt%, n-type) with $N_D = 2.5 \times 10^{14}$ cm^{-3}
- sample EuO-15 (1.13 wt%, n-type) with $N_D = 1.2 \times 10^{14}$ cm^{-3}

The corresponding layers prepared on InP:Sn substrate are of both n-and p-type:

- sample EuO-7 (0.49 wt%, n-type) with $N_D = 1.1 \times 10^{15}$ cm^{-3}
- sample EuO-19 (1.49 wt%, p-type) with $N_A = 2.6 \times 10^{15}$ cm^{-3}

The PL spectra show fine features with narrow peaks supporting the results of C-V measurements. The NBE part of the spectra clearly demonstrates differences between n- and p-type layers (fig. 3).

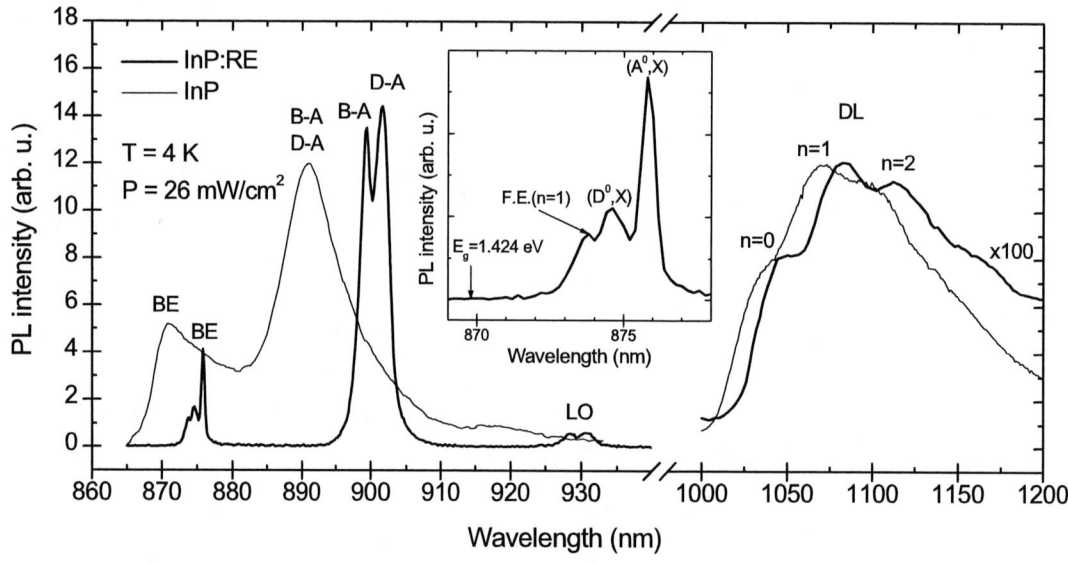

Figure 2: PL spectra of conventionally grown InP layer and that prepared with optimum RE addition

The temperature dependence of PL spectra was studied for various levels of excitation power in order to distinguish and identify fine spectral features. The major manifestation of the RE admixture is a pronounced narrowing of PL spectra and corresponding appearance of fine features, characteristic of pure materials, low in defects. Typical PL spectrum comparing layers grown with and without RE admixture is shown in fig. 2. The observed radiative transitions in studied InP samples could be grouped into three categories: band-edge (BE) transitions at about 1.418 eV (875 nm), shallow impurity related transitions at 1.38 eV (900 nm), and deep-level transitions at 1.14 eV (1090 nm). The peak LO is the phonon replica of the shallow impurities related peak and corresponds to the known value of 43 meV for LO

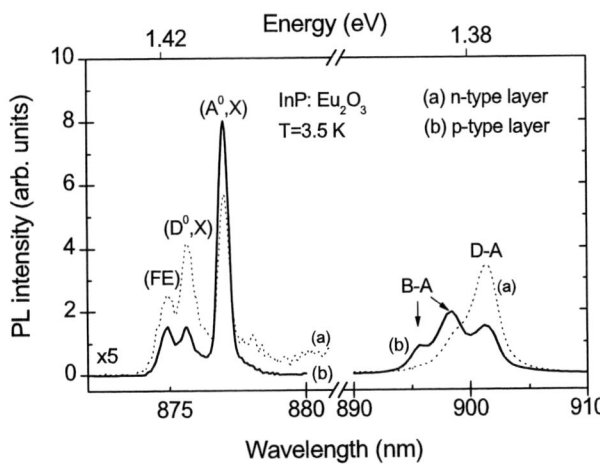

Figure 3: NBE part of the PL spectra of both n- and p-type InP prepared with Eu_2O_3 admixture

phonon. The high energy band (BE) exhibits superlinear behaviour with increasing excitation power and results from the decay of excitons. Fine structure of excitonic transitions is usually well resolved in the case of high purity samples. Three peaks can be resolved in the band of exciton recombination: free exciton (F.E.) at 873.5 nm (1.419 eV), shallow donor bound exciton (D^oX) at 874.2 nm (1.418 eV), and shallow acceptor bound exciton (A^oX) at 875.6 nm (1.416 eV). The shallow impurities related peak is an unresolved convolution of band-acceptor (B-A) and donor-acceptor pair (D-A) transitions in the case of zero admixtures, while separate peaks are well resolved at low excitation power on samples prepared with REs.

The long-wavelength part is usually dominated by Mn related band consisting of three partly resolved peaks at 1.184 eV (n=0), 1.145 eV (n=1) and 1.107 eV (n=2), which are interpreted as a zero phonon line, and one, and two phonon replicas, respectively (fig. 2). This characteristic band, observed in majority of samples, whether rare-earth treated or not, is attributed to recombination of free or loosely bound electrons with holes bound to the Mn acceptor occupying an In site [8].

4. Conclusions

The gettering effect of REs was observed for all studied samples with variable purifying efficiency of individual RE species. The preferential donor gettering leads to conductivity conversion from n- to p-type for majority of the studied REs. Due to the lower reactivity of RE oxides, the concentration at which the conductivity conversion occurs is shifted towards higher values in comparison with elemental REs. RE oxides show some promising advantages over the elemental REs. The process of their introduction into the melt and its subsequent homogenization must be solved to result in repetitive preparation of high quality layers.

Acknowledgement

The support of the Czech Science Foundation project No. 102/06/0153 is acknowledged.

References

[1]   A.J. Kenyon, *Progress in Quantum Electronics* **26**, 225, 2002.
[2]   A. Kozanecki, R Groetzschel, *J. Appl. Phys.* **68**, 517, 1990.
[3]   W. Gao, P. R. Berger, *J. Appl. Phys.* **80**, 7094, 1996.
[4]   O. Procházková, J. Zavadil, K. Žďánský, J. Grym, *Phys. Stat. Sol.* **3**, 950, 2002.
[5]   O. Procházková, J. Zavadil, K. Žďánský, J. Grym, *Mat. Sci. Forum* **480-481**, 483, 2005.
[6]   O. Procházková, J. Grym, J. Zavadil, K. Žďánský, *J. Crystal Growth* **275**, e959, 2005.
[7]   J. Zavadil , O. Procházková, K. Žďánský, P. Gladkov, *Phys. Stat. Sol.*, in press.
[8]   L. Eaves, A. W. Smith, M. S. Skolnick, B. Cockayne, *J. Appl. Phys.* **53**, 4955, 1982.

Performance study of bulk semi-insulating InP radiation detectors with different electrode metallizations

B. Zaťko[a*], F. Dubecký[a], O. Procházková[b], and V. Nečas[c]

[a]Institute of Electrical Engineering, Slovak Academy of Sciences,
Dúbravská cesta 9, SK-841 04 Bratislava, Slovakia
[b]Institute of Radio Engineering and Electronics, AS CR,
Chaberská 57, CZ-182 51 Praha, Czech Republic
[c]Faculty of Electrical Engineering and Information Technology,
Slovak University of Technology, Ilkovičova 3, SK-812 19, Slovakia
*email: Bohumir.Zatko@savba.sk

This work describes the study of three different electrode metallizations with the aim to form the Schottky barrier contact. As a substrates bulk LEC SI InP grown by Fe and Fe+Zn co-doping wafers are used. Results of the study show that no one of used metallizations performs expected blocking behaviour. However detectors with Ti/Pt/Au metallization attained a relatively good energy resolution of 7.0 keV (in FWHM) and the charge collection efficiency higher than 83 % for 122 keV γ–photons at lowered temperature of 255 K.

1. Introduction

InP is one of semiconductor compound material usable in preparation of detectors of ionization radiation due to high atomic number of In (49). The band gap width and the mobility of electrons of InP are 1.35 eV and 5400 $cm^2V^{-1}s^{-1}$ at 300 K, respectively. Maximum drift velocity is higher comparing to other materials (GaAs, Si, and CdTe) [1]. Its detection efficiency for γ–radiation is 1.5–2 times lower in comparing with CdTe. As a base detector material, InP can be used in a form of epitaxial semi–conductive or bulk grown semi–insulating (SI). Initial study of SI InP like a detector of X– and γ–rays was stimulated by its application as a Solar neutrino detector. First paper on this topic was published by Lund et al. in 1988 [2]. InP is also applicable as a fast photoconductive semiconductor detector operating in current mode and can be used in various tasks in the diagnostics of laser–induced plasmas, especially for measuring of X–ray pulse emissions from the hot plasma [3]. SI InP:Fe–based particle detectors were studied for detection of α–particles by Valentini et al. [4] and Dubecký et al. [5]. Our previous papers [6, 7] are devoted to the study of detection performances of SI InP radiation detectors based on various materials. We obtained at 220 K the best energy resolution (in FWHM) 7.0 keV and 10.7 keV for 59.5 keV and 122 keV γ–photons, respectively. Owens et al. studied the best our SI InP detector using sophisticated readout electronics [8, 9]. The measured energy resolution was below 5 keV for 100 keV synchrotron radiation. The main limitation in fabrication of detector based on SI InP is connected with difficult making a contact with blocking behaviour. This problem is not overcome yet. Typically, SI InP detector structure shows a quasi-linear current-voltage characteristic with current limited by bulk resistivitance. Due to a circumstance the operation of such detector at room temperature (RT) is restricted by a high leakage current, thus the detector must be cooled.

In this paper study of SI InP:Fe–based particle detectors with the emphasize to investigation of the role of electrode metallization is presented. Detectors fabricated with various active contact areas (0.0314 – 0.442 mm^2) and three types of a "blocking" electrode metallizations (Ti/Pt/Au, ZnAu and Al/Pt) are studied. Current–voltage characteristics at RT and pulse–height spectra of ^{241}Am and ^{57}Co measured by detectors at various temperatures (300 K – 255 K) are demonstrated.

2. Detector fabrication

Two different base LEC (Liquid encapusulated Czochralski) materials of InP (produced by IREE Prague, Czech Republic, labelled #1 and ROMELAB, USA, labelled #2) are utilized in detector fabrication. Material #1 was grown using Fe+Zn co–doping and material #2 was Fe–doped. Resistivities of 3.2×10^7 and 5.6×10^7 Ωcm and the Hall mobilities of 2600 and 2710 cm^{2}V^{-1}s^{-1} of mentioned materials are observed at RT, respectively (Table 1). Wafer fragments were polished down from both sides to the thickness of about 220 μm. Full area ohmic contact was formed by evaporation of AuGeNi eutectic alloy in vacuum onto chemically

Sample label	Resistivity (Ωcm)	Hall mobility (cm^{2}V^{-1}s^{-1})	Electrode system	Contact area (mm^2)	Detector thickness (μm)
#1	3.2×10^7	2600	Ti/Pt/Au	0.126 to 1.766	220
			Pt/Al		
#2	5.6×10^7	2710	AuZn		

Table 1. Fundamental characteristics of InP base materials and detector structure

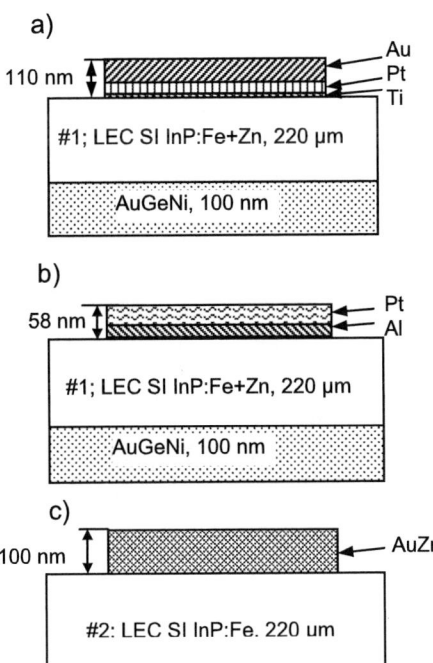

Fig. 1. Cross-section of SI InP detectors.
a) Ti/Pt/Au–SI InP #1–AuGeNi
b) Al/Pt–SI InP #1–AuGeNi, c) AuZn–SI InP #2–AuGeNi.

purified surface [10] and alloyed at 400 ºC (2.5 min) in forming gas. On the opposite side of the wafer circular electrodes with diameters of 0.75, 0.5, 0.3 and 0.2 mm were formed by photolithography masking and lift–off process. Ti/Pt/Au (15/35/60 nm) and Al/Pt (8/50 nm) multilayers were deposited on the surface of material #1 and AuZn (100 nm) eutectic alloy was evaporated on the material #2 (Fig. 1). Closely before evaporation was the surface of materials chemically etched using 1HCl:2H$_2$O solution to remove native oxides. Samples were glued onto PCB holder using silver paste and contacted by wire bonding.

3. Experiment and results

Current–voltage (I–V) characteristics of prepared samples measured at RT are shown in Fig. 2, plotted in log–log scales. Dependencies are linear in low bias voltage region and demonstrate a quasi–ohmic character of the electrical charge carriers transport through the detector structure. Linear part of the I–V dependence (up to ~1 V) with a high dynamic resistance controlled by material resistivity is followed by a slight sub–linear dependence in a range between 10 V and 250 V. Over about 300 V a super–linear part is observed and finally additional linear region with about five times lower dynamic resistance compare with the dependence at in a low current region. Described behavior is less observable when diameter of the contact decreases. Comparison of Fig. 2a and 2b, lower current attain the samples with Ti/Pt/Al contact. The lowest observable

75

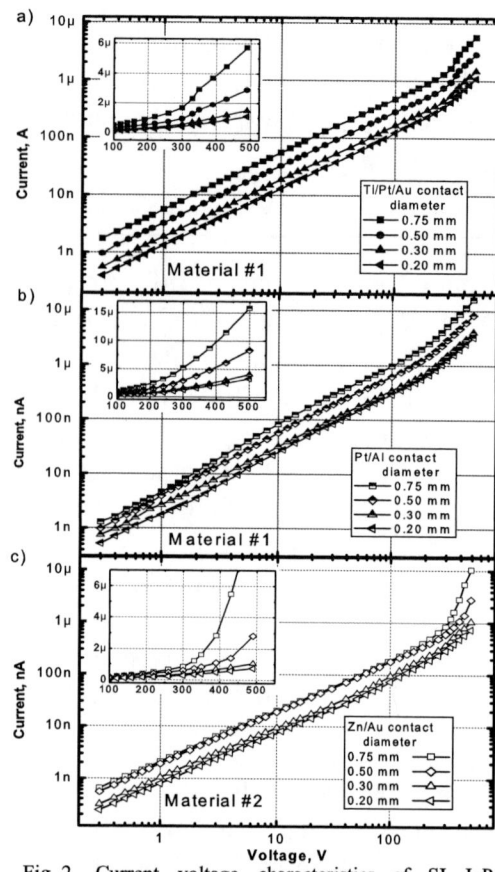

Fig. 2. Current voltage characteristics of SI InP detectors measured at RT. a) Ti/Pt/Au–SI InP #1–AuGeNi, b) Al/Ptl–SI InP #1–AuGeNi, c) AuZn–SI InP #2–AuGeNi.

current revealed samples with AuZn metallization. This reflects the higher resistivity of base material #2. The reduction of current due to decreasing of top contact diameter does not follow precisely to relative area variation, especially for diameter of 0.75 and 0.5 mm. This is probably coupled with surface leakage current.

The required working electric field of SI InP detector is higher than 10^4 V/cm. Bias voltage 220 V or higher should be applied in tested samples. However the flowing current of all samples is too high (> 1 µA) at RT and the cooling of detector is required. Inserted graphs (in linear scale) in the Fig. 2 (a, b, and c) show current at voltages matching to operation region. Current decreases about one order of magnitude if the detector is cooled by 30 °C.

Detectors were connected to the spectrometric readout chain (charge sensitive preamplifier based on CREMAT CR 101D, shaping amplifier ORTEC 572, high voltage supply ORTEC 459, analog-to-digital converter ORTEC 800 with a multichannel analyser M2D). Two stages Peltier cooler was used for detector cooling. Pulse height γ–ray spectra of ^{57}Co (122.1 keV, 136.5 keV) and ^{241}Am (59.5 keV) were measured with the fabricated SI InP detectors at temperatures down to 255 K. Spectra observed at different temperatures are shown in Fig. 3 and Fig.4, respectively. Second peak observed at lower channel represents the escape peak of In. For better visibility the measured data were multiplied to prevent overlapping of displayed curves. Detector Ti/Pt/Au–SI InP #1–AuGeNi reaches the best spectrometric performance at –400 V and temperature 255 K. The attained energy resolution calculated from the spectra is 6.3 keV and 7.0 keV (in FWHM) for 59.5 keV (^{241}Am) 122.1 keV (^{57}Co) γ–photons peak, respectively. Maximum observable CCE is 83 %. Detector Al/Pt–SI InP #1–AuGeNi has worse measured parameters. The CCE reached value about 75 % and energy resolution is hardly qualifiable. Gaussian curve obtained by fitting the peak has 11.9 keV (in FWHM) for 59.5 keV γ–photons at 258 K. The last detector structure AuZn–SI InP #2–AuGeNi did not resolved any peak even at 500 V and temperature of 253 K. Possible explanation of this effect is related to the worse base material characteristics, in particular short charge carriers lifetimes, than used metallization. Noting however, the fact of almost similar electrical transport parameters (resistivity, Hall mobility) of both used SI InP wafers.

3. Conclusions

Radiation detectors based on bulk LEC SI InP with different "blocking" contact metallizations (Ti/Pt/Au, Al/Pt and ZnAu eutectic alloy) were investigated. I–V characteristics show almost linear behaviour with sudden decreasing dynamic resistance at a bias voltage over about 300 V. The pulse height spectra of ^{57}Co and ^{241}Am were measured at

76

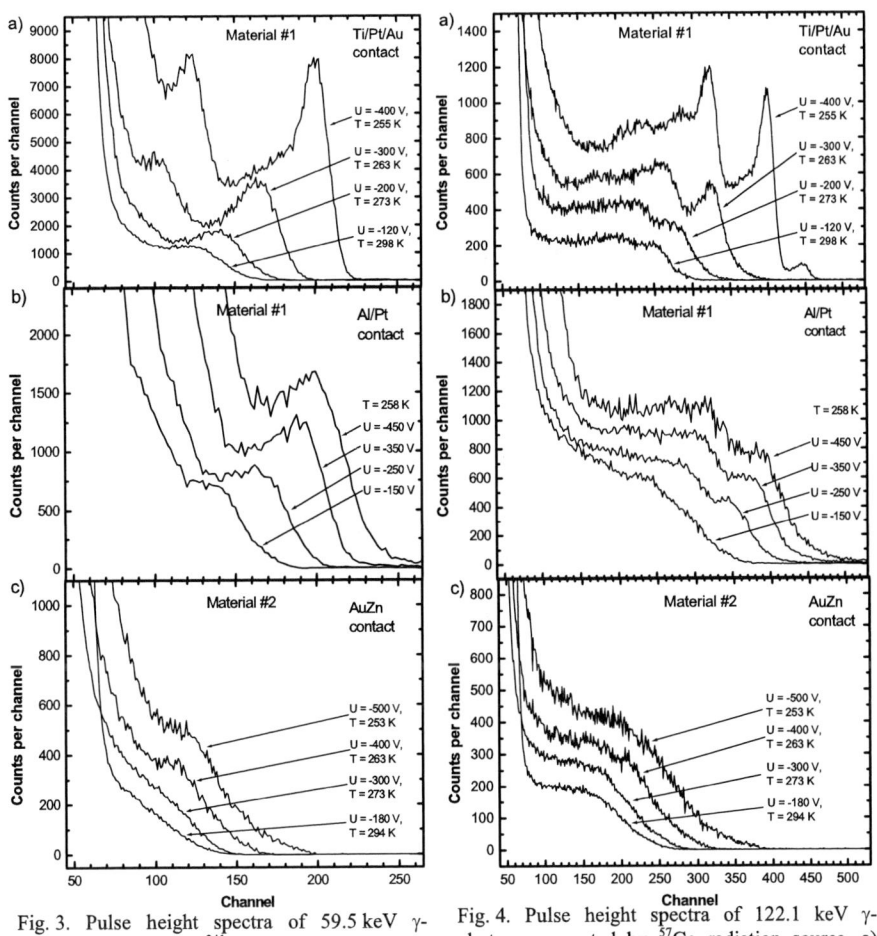

Fig. 3. Pulse height spectra of 59.5 keV γ-photons generated by ^{241}Am radiation source. a) Ti/Pt/Au–SI InP #1–AuGeNi, b) Al/Pt–SI InP #1–AuGeNi, c) AuZn–SI InP #2–AuGeNi.

Fig. 4. Pulse height spectra of 122.1 keV γ-photons generated by ^{57}Co radiation source. a) Ti/Pt/Au–SI InP #1–AuGeNi, b) Al/Pt–SI InP #1–AuGeNi, c) AuZn–SI InP #2–AuGeNi.

different temperatures (253 – 298 K) and applied biases (up to 500 V). Good spectrometric performance was achi-eved with detectors based on the IREE LEC SI InP: Fe+Zn material (#1) with Ti/Pt/Au contact me-tallization. Detector with the contact diameter of 0.5 mm attains an energy resolution of about 6.3 keV and 7.0 keV (in FWHM) for detected 59.5 keV and 122.1 keV γ–photon peaks, res-pectively. The best value of the charge collection effiiency (CCE) 83 % was ob-served at 255 K at a bias voltage of 500

V. Radiation detectors #2 based on ROMELAB LEC SI InP:Fe with AuZn metallization demonstrate bad spectrometric performance even at reduced temperatures and high bias voltage. In such circumstance, the evaluation of FWHM and CCE was rather impossible.

Acknowledgement

This work was partially supported by the Slovak Grant Agency for Science through the grant No. 2/4151/26, Agency for Support of Science and Research No. APVV–99–P06305 and Grant Agency of the Czech Republic, project No. 102/06/0153.

References

[1] Levinshtein M. et al., Handbook Series on Semiconductor Parameters. **Vol. 1,** (1996) p 169.
[2] Lund J.C., et al., Nucl. Instr. Meth. **A 272,** (1988) p 885.
[3] Riesz F. et al., Nucl. Instr. Meth. **A 474,** (2001) p 151.
[4] Valentini A. et al., Nucl. Instr. Meth. **A 373,** (1996) p 47.
[5] Dubecký F. et al., Nucl. Instr. Meth. **A 408,** (1998) p 491.
[6] Zaťko B. et al., In: ASDAM 2000, Piscataway, IEEE, (2000) p 429.
[7] Dubecký F. et al., In: APCOM 2001, Liptovský Mikuláš, Military Academy, (2001) p71.
[8] Owens A. et. al., Nucl. Instr. Meth. **A 487,** (2002) p 435.
[9] Owens A. et. al., Nucl. Instr. Meth. **A 491,** (2002) p 444.
[10] Dubecký F. et al., Nucl. Instr. Meth. **A 531,** (2004) p 181.

GaAs and GaN based SAW chemical sensors: acoustic part design and technology

L. Rufer[1], T. Lalinský[2], D. Grobelný[1], S. Mir[1], G. Vanko[2], Zs. Őszi[2] and Ž. Mozolová[2]
J. Greguš[3]

[1]TIMA Laboratory, 46 Av.Félix Viallet, 38 031 Grenoble, France
[2]Institute of Electrical Engineering, Slovak Academy of Sciences, 841 04 Bratislava,
Slovak Republic
[3]Department of Experimental Physics FMFI, Comenius University, 842 15 Bratislava,
Slovak Republic

e-mail: Libor.Rufer@imag.fr and Tibor Lalinský [eleklali@savba.sk]

In this paper, we present the design considerations and the technology process of SAW (Surface Acoustic Wave) chemical sensors based either on GaAs or GaN structures. These sensors can be used for identifying environmental contaminants and chemical or biological agents in large applications scale; in this study, we aimed at the measurement of low concentrations of gaseous mercury. We describe the design of the acoustic part of the sensor including the structure for the generation and reception of the surface acoustic wave and the chemoselective coating made of gold. We show the technology process that achieves the device operating at the frequency of 250 MHz. Finally we present some preliminary results obtained from the device.

1. Introduction

It is extremely important to monitor, control and reduce the toxic pollutant in the environment. Recently, the European Union (EU) has enacted a new legislation, known as the RoHS Directive (Restriction of the use of certain Hazardous Substances in electrical and electronic equipment), which will restrict the use of six hazardous substances in electronic products made in or imported into the EU [1]. With these environmental aspects in mind, we are developing a chemical sensor targeting gaseous mercury that is one of banned substances specified in the RoHS directive.

Currently, there exist several laboratory techniques for mercury detection. These techniques, based on chromatographic, electrochemical or spectroscopic methods are sensitive but not adapted for in situ measurement. The need of portable sensing device can be satisfied with an integrated solution. Such a device can be fabricated on semiconducting substrate as silicon, gallium arsenide, or gallium nitride, and integrated with electronics. Moreover, this approach can evolve due to the reduction in size to the sensor arrays.

In the case of an integrated device, a surface acoustic wave can be employed as a sensing mechanism. It was demonstrated in the past that a SAW can be generated and detected by means of an interdigital transducer (IDT) deposited on the piezoelectric substrate [2]. One promising strategy leading to portable mercury vapor sensor is based on the SAW-based sensor consisted of two IDTs and a sensing film of gold located in between [3]. Previous works employing SAW-based sensing of mercury were using quartz, gallium arsenide or silicon with a piezoresistive layer deposited on its top surface. In our design, we use a gallium nitride structure that is promising both for better sensing properties and for fast electrical response.

1-4244-0396-0/06/$25.00 ©2006 IEEE

2. Architecture of the Acoustical Part and its Modelling

The schematic view of a two-port SAW device used as a gas-sensitive part of the chemical sensor is shown in Figure 1. The generation of acoustic waves on the substrate is carried out by means of Interdigital Transducers (IDTs). An IDT is a fingerlike periodic pattern of parallel in-plane metallic electrodes, used to build up the capacitance associated with the electric field that penetrates into the material. An IDT allows the conversion of high frequency electric fields in acoustic waves and vice-versa. By changing the length, width, position, and thickness of the electrodes as well as the number of electrodes and the pattern shape of the IDT, the performance of the transducer can be optimized [4]. In our case, the sensor has been designed for the working frequency around 250 MHz. This frequency is the result of the demands put on the substrate size and sensor sensitivity. The dimensions of the acoustic part of the sensor shown in Figure 1 are thus *7.4 x 2* mm^2; the sensitive coating covers the area of *3 x 1.5* mm^2.

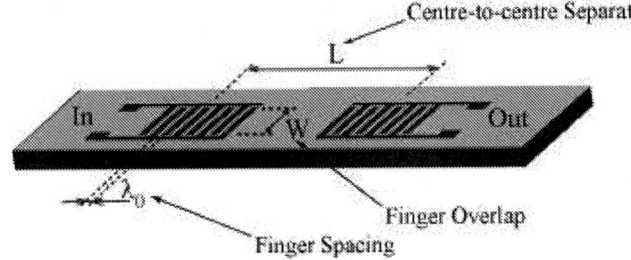

Figure 1: Schematic view of a two-port SAW

There are several ways of describing the function and modelling an interdigital transducer. As we aimed at analyzing the complete model of the acoustical part along with the electronic circuitry necessary for the operation of the sensor, we have adopted an approach using equivalent circuits. Such an approach facilitates the application of any typical software for the simulation of electric circuits. Among the variety of models used for piezoelectric devices (e.g. Mason, KLM, Redwood, Ballato), we have chosen the crossed-field model derived from the Mason equivalent circuit employed for modelling acoustic bulk wave piezoelectric devices. In this model, the electric field distribution under the electrodes of an excited IDT is approximated as being normal to the piezoelectric surface, similarly as between the electrodes of a plate capacitor.

In the adaptation of the Mason equivalent circuit to SAW transducer design, each IDT is represented by a three-port network with two acoustic ports and one electrical port. The approximation by crossed-field model can be considered to hold for electrodes with low values of film thickness to wavelength ratio (h/λ<<1%). In the complete model of the sensor acoustic part shown in Figure 2, the input and output IDTs are described by their corresponding admittance matrices. With the fixed load G_0 on one mechanical port, these matrices can be transformed to the cascade ones. Similarly, the chemically sensitive layer can be described by another cascade matrix and the final cascade matrix of the acoustic part is obtained by the multiplication of each component.

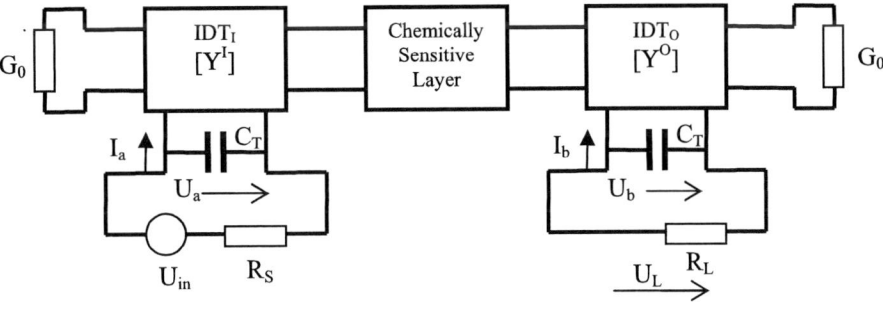

Figure 2: Complete model of the acoustic part of the sensor.

3. IDT Technology

We have designed $2\,\mu m$ – thick undoped GaAs and GaN buffer layers preferentially grown by MOCVD on SI GaAs and Sapphire substrates, respectively to form the basic SAW propagation media of the chemical sensors. The both buffer layers are designed also to form AlGaAs/GaAs and AlGaN/GaN heterostructures to define two-dimensional electron gas (2 DEG) of HEMT structures to be monolithically integrated with the SAW chemical sensors. A schematic cross-section through the SAW chemical sensors is shown in Figure 3.

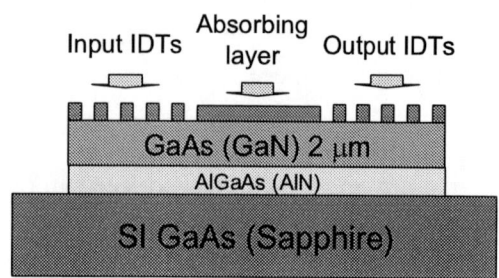

Figure 3: A schematic cross-section through SAW chemical sensors.

It consists of input and output IDTs as acoustic part of the sensor and thin chemical absorbing layer coated between them. To identify the mercury vapours, absorbing layer such as Au was patterned. IDTs consist of a set of closely spaced metal electrodes deposited on the GaAs or GaN buffer layers of HEMT structures.

The metallic system based on Ti/Al ($20\,nm/150$ nm) was used to define interdigital electrodes of IDT. To fabricate the IDTs, Ti and Al metals were deposited by electron-beam evaporation and Au layer was deposited by the conventional resistance evaporation through the patterned photoresist mask. The pressure in the cryogen pumping vacuum system during the evaporation was lower than 10^{-5} Pa. A lift-off technique was used to form the eighty pairs of fingers of the IDTs. The width and spacing between the interdigital fingers were designed to be $5\,\mu m$. To define precisely the acoustic wave propagation part of the sensors a deep "MESA" etching of GaAs and GaN buffer layers was performed afterwards. It followed the topology of IDT acoustic part from both sides. Both ion beam and reactive ion etching methods were applied to etch $2\,\mu m$-thick buffer layers through the photoresist mask up to the substrates (SI GaAs, Sapphire). Scanning electron microscopy (SEM) images of patterned fingers of IDT are shown in Figure 4.

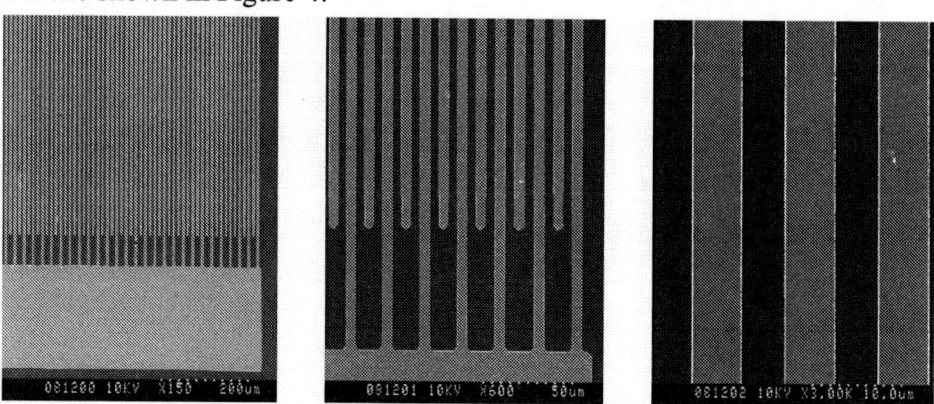

Figure 4: SEM images of patterned fingers of IDT.

4. Interdigital Transducer Characterization

The basic dc electrical characterization of both GaAs and GaN based interdigital finger structures was performed at room temperature. The typical I-V characteristics are shown in Figure 5. Because the interdigital structures in principle represent two reverse-biased Schottky contacts, a symmetrical behaviour in the I-V characteristics is observed.

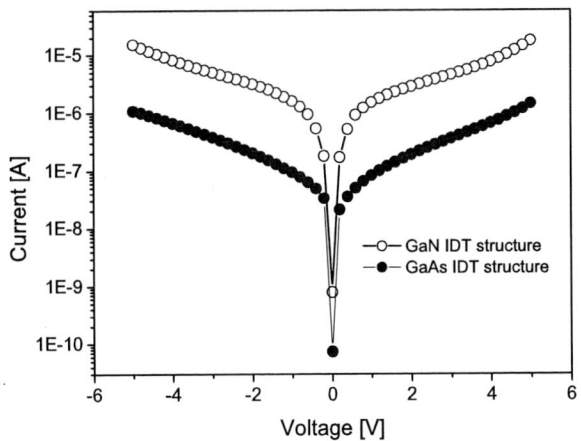

Figure 5: I-V characteristics of IDT structures.

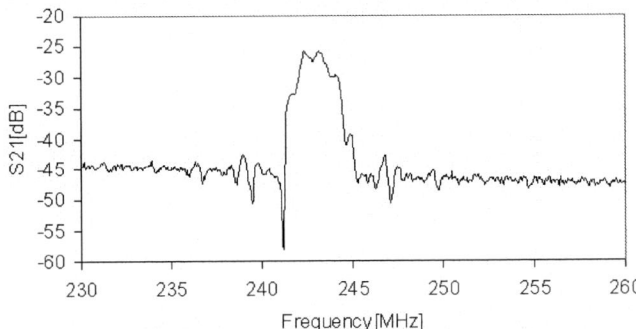

Figure 6: Measured curve of S21 parameter.

The saturation current in the reverse I-V characteristics of GaN based IDT structure was found to be more than one order higher as comparing with that of GaAs based IDT structure. This could be explained by higher excess leakage current induced by high density of surface traps of GaN. A comprehensive I-V electrical characterization at various temperatures is in progress to extract dominant transport mechanism.

The measurement of the transfer characteristics of the GaN based sensor was performed in the test setup using Network Analyzer HP 8753C. The module of the S22 parameter obtained from the measurement is shown in Figure 6. The obtained centre frequency of *243* MHz and bandwidth of *4* MHz are in good agreement with the simulation results.

Acknowledgement

This work is partially funded by the following projects: Stefanik project of French Ministry of Foreign Affaires and French Ministry of Education, Slovak-French project SK-FR-01906, VEGA project 2/6097/26, APVT project 20-026104 and Erasmus mobility program. The authors would like to thank D. Rauly from the IMEP Laboratory, Grenoble for the assistance with the sensor testing.

References

[1] Directive 2002/95/EC of the European Parliament and of the Council of 27 January 2003 on the restriction of the use of certain hazardous substances in electrical and electronic equipment, Official Journal of the European Union L 37/19, 13.2.2003.

[2] R. M. White, Surface Elastic Waves, *Proc. of the IEEE*, Vol. 58, No. 8, 1970, pp. 1238-76.

[3] J. J. Caron, R. B. Haskell, P. Benoit, and J. F. Vetelino, A surface acoustic wave mercury vapour sensor, *IEEE Trans. on Ultrasonics, Ferroelectrics, and Frequency Control*, Vol. 45, No. 5, Sep. 1998, pp. 1393-1398.

[4] A. V. Mamishev, K. Sundara-Rajan, F. Yang, Y. Du, M. Zahn, "Interdigital Sensors and Transducers", *Proc. of the IEEE*, Vol. 92. No. 5, 2004, pp. 808-845.

Physical properties of transparent conductive oxides prepared by RF reactive sputtering

V. Tvarozek[1], I. Novotny[1], P. Sutta[2], S. Flickyngerova[1], L. Harmatha[1],
E. Vavrinsky[1], M. Nigrovicova[1], J. Mullerova[3]

[1] Department of Microelectronics, Slovak University of Technology, Ilkovicova 3, 812 19 Bratislava, Slovakia
[2] Department of Materials & Technology, New Technology Center, West Bohemian University, Univerzitni 8, 306 14 Plzen, Czech Republic
[3] Department of Engineering Fundamentals, Faculty of Electrical Engineering, University of Zilina, kpt. J. Nalepku 1390, 031 01 Liptovsky Mikulas, Slovakia

Transparent conductive oxide thin films (indium tin oxide, zinc oxide and Al doped zinc oxide) were prepared by RF diode sputtering. Selected properties of these films (structure, resistivity and transparency) were investigated with the aim to use them in thin film solar cells.

1. Introduction

Transparent conductive oxide (TCO) coatings have applications in many electro-optical devices [1]. For hydrogenated amorphous silicon (a-Si:H) thin film solar cells, TCO materials used as front contact has a decisive influence on device performance. The TCO layer for solar cell use has to meet many requirements: high optical transparency in the wavelength region from about 350 nm to 900 nm ($> 80\%$), low electrical sheet resistance ($< 15\ \Omega/square$), stability during handling and deposition of the subsequent layers and during use, a textured (rough) surface to enhance optical absorption of red and near-infrared light, and low-resistance electrical contact to the amorphous silicon p-layer. There is an inherent trade-off between how transparent and how conducting TCO layer is, which is represented by the figure of merit: $F = T/R_s$, where T is transmittance (%) and R_s is sheet resistance (Ω). The higher the figure of merit, there is the better the performance of the transparent conducting thin film.

2. Experimental

In this work three types of TCO thin films, Sn doped indium tin oxide (ITO), zinc oxide ZnO and Al doped zinc oxide Al:ZnO (AZO) were investigated. The planar diode RF sputtering unit 2400/8L (Perkin-Elmer) has been used for preparation all thin films. The electrical, structural and optical properties of ITO and AZO are presented. We related surface roughnesses of thin film interfaces with their optical properties. Therefore we measured surface morphology of ITO and ZnO thin films by atomic force microscopy (AFM).

3. Results and discussion

ITO is a n-type degenerate semiconductor with an optical bandgap of about 3.75 eV [2, 3]. The conductivity comes from the oxygen vacancies and the presence of a Sn dopant which has a higher valence than In. Polycrystalline ITO films with preferential orientation (222) were prepared by RF reactive diode sputtering ceramic target in Ar or $Ar+O_2$ working gases (Fig. 1). Sputtering of hot-pressed ITO target in pure Ar produced films with suitable

1-4244-0396-0/06/$25.00 ©2006 IEEE

properties without post-deposition annealing (Fig. 2). These films exhibited sheet resistance $R_S = 7\,\Omega$; transmitance $T(0.55\mu m) = 93\,\%$, and figure of merit $F = 12.2$.

Zinc oxide (ZnO) is a unique material that exhibits semiconducting, piezoelectric, pyroelectric and optoelectronic multiple properties [4]. ZnO is a transparent semiconductor with a direct bandgap of 3.37 eV at room temperature and large excitation binding energy (60 meV), and exhibits near-UV emission and absorption, as well as high transparency and natural n-type in conductivity. Non-stoichiometric ZnO_x films were prepared by RF diode sputtering from ZnO target in pure Ar gas and exhibited disordered grain structure with orientations of (100), (002) and (101). The resistivities of the sputter ZnO_x films were slightly dependent on the RF power, $\rho = 0.4 \div 1.0\,\Omega cm$ and they were determined by the donor carrier concentrations (oxygen vacancies, ZnO intersticials) which varied from

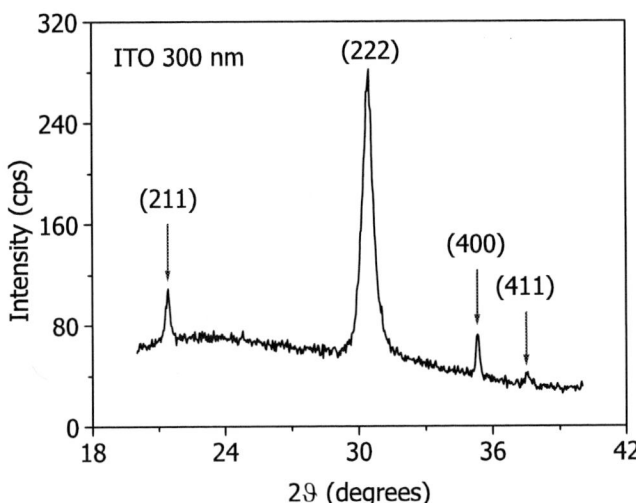

Fig. 1. Polycrystalline ITO film with preferential orientation (222)

$n = 1.4 \times 10^{19}\,cm^{-3}$ to $7 \times 10^{18}\,cm^{-3}$. Reactive sputtering in $Ar+O_2$ working gas (up to 75 % of O_2) at substrate temperatures $T_s = 20\,°C$ and $300\,°C$ improved the structure progressively from more disordered fibrous ZnO grains to columnar crystallites preferentially oriented along c-axis normally to the substrate (<002> direction) (Fig. 3). The X-ray intensity of the reflecting basal (002) planes was compared with the reflecting peaks of (100) planes (Fig. 4). The angular distribution of crystallites along the c-axis oriented normally to the substrate was determined by measurements of the standard deviation angle in the rocking curves $\pm\sigma < 5\,\%$ as a function of O_2 content. Transparency in VIS region was 85 % with adsorption edge in near UV region 3.33 eV which is close to the bandgap of bulk ZnO.

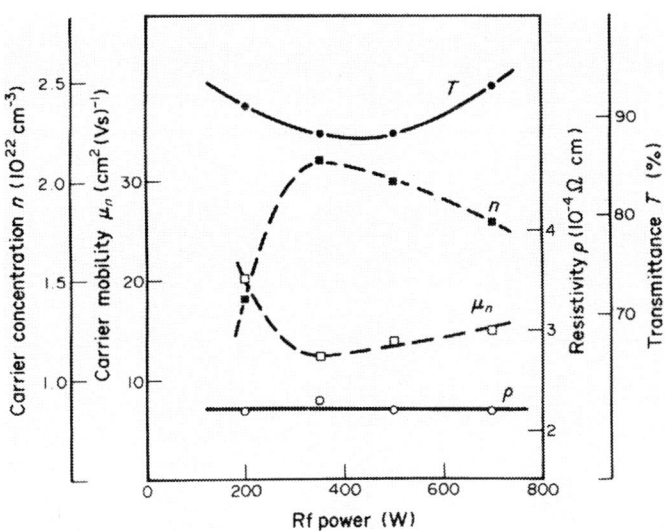

Fig. 2. Variations of ITO thin film properties

Microroughness of ZnO surfaces were measured by AFM, maximal value (10 nm) was reached at high target voltage ($V_{Target} = -900V$) by -175 V bias sputtering. The use of lower voltage ($V_{Target} = -300V$) caused smoother surface (1.2 nm) of the thin film. More rough surface of the thin film was due to atomic scale heating of the film during deposition.

Undoped ZnO_x thin films are not stable, especially at high temperatures. Doping the zinc oxide can reduce this disadvantage. Additionally, doping by replacing Zn^{2+} ions with elements of higher valency such as Al^{3+}, leads to an increase in the conductivity of the AZO thin films [5]. The multilayer structures of the total thickness 250 nm were prepared by

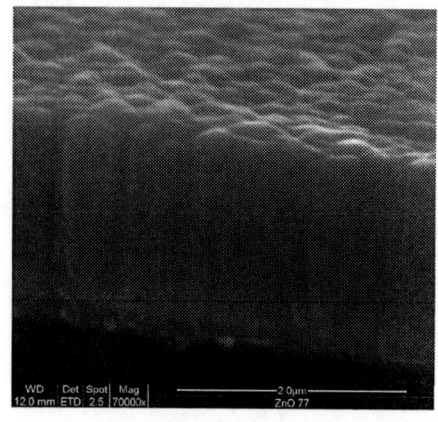

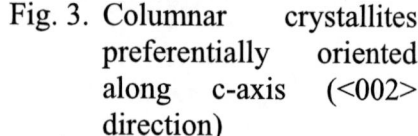

Fig. 3. Columnar crystallites preferentially oriented along c-axis (<002> direction)

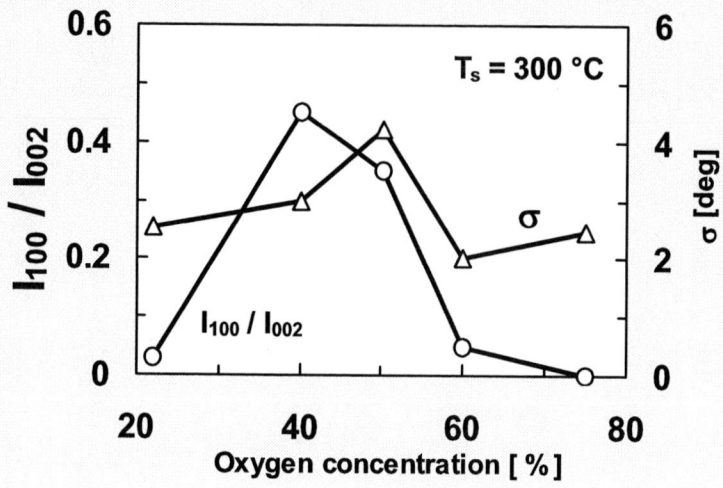

Fig. 4. X-ray intensity and the rocking curves as a function of O_2 content

sequent reactive sputtering of ZnO film (50 nm) and sputtering of Al film (5 or 10 nm) in Ar or Ar + 5 % O_2 gases. After deposition the samples were heat-treated by two methods: rapid thermal annealing in vacuum (at 500 or 800°C for 180 s) or long-term annealing in the furnace in air (at 800 °C for 18 hours). Thin films deposited onto no-heated substrate (room temperature) show high initial resistances (~ 10^{11} Ω), but after annealing, either RTA at 500 °C or in furnace at 800 °C, the resistances were rapidly falling down to values of $10^5 \div 10^7$ Ω. This decrease is particularly caused by diffusion of Al atoms into ZnO creating the new carriers. RTA at 800 °C in vacuum leads to the oxygen desorption which results to AZO with high conductivity. After high-temperature vacuum annealing all kind of samples lowered the sheet resistance value to the order of 10^2 Ω.

The AZO thin films sputtered from ZnO:5% Al_2O_3 target in pure Ar showed changed of properties with RF power (Fig.5). Sheet resistance is decreased with power but it is still high (1.8 kΩ at 800 W). Integral transparency T_s (in the range of 520 and 600 nm) was the hightest at power of 400 W. Therefore the figure of merit was to low, e.g. F = 0.03 at 600 W.

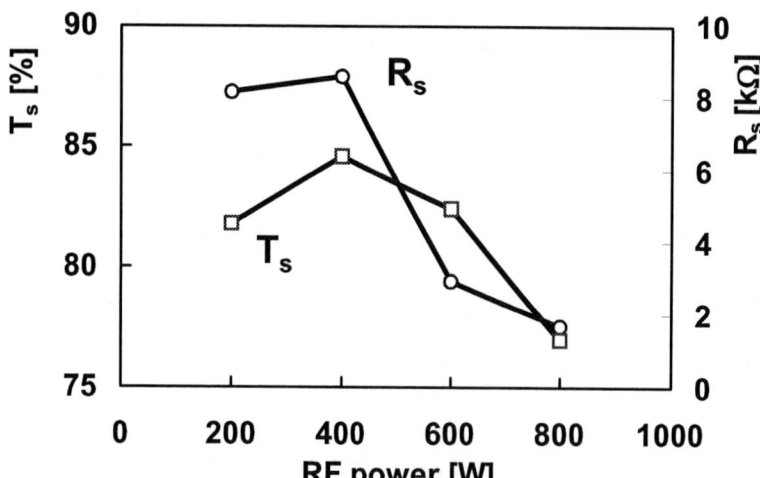

Fig. 5. Integral transparency and sheet resistance of AZO films sputtered from ZnO:5% Al_2O_3 target in pure Ar

4. Conclusions

Main results obtainet by RF diode sputtering of TCO thin films we can summarize:

ITO thin films deposited from ITO target need no reactive sputter gas mixture and no post-deposition heat-treatment for reaching of desired properties. This simplification of technology is economically attractive. Reactive sputtering ZnO produced thin films with good piezoelectric properties and high resistivity. The are not applicable as TCO films without aditional doping. For the degrease of sheet resistance of AZO films will be necessary to optimize all three sputtering parameters (content of O_2, RF power and bias voltage).

Acknowledgement

Presented work was supported by the MSMT Czech Republik Grant in frame of project 1M06031.

References

1. H. Ohta, and H. Hosono, *Materials Today,* **7**, **42**, 2004.
2. A. Salehi, *Thin Solid Films,* **324**, 214, 1998.
3. S.K. Choi, and J.I. Lee, *Journal of Vacuum Science and Technology A,* **19**, 2043, 2001.
4. K.A. Abdullin, A.B. Aimagambetov, N.B. Beisenkhanov, A.T. Issova, B.N. Mukashev, and S.Z. Tokmoldin, Materials Science and Engineering B, **109**, 241, 2004.
5. M.A.Martinez, J.Herrero, and M.T. Gutierrez, Solar Energy Materials and Solar Cells, **45**, 75, 1997.

Translinear Subthreshold MOS Filter for the Wireless Sensors Applications

Adam Boura, Miroslav Husak

Department of Microelectronics, Faculty of Electrical Engineering, Czech Technical University in Prague,
Technická 2, 166 27 Prague 6, Czech Republic
e-mail: bouraa@feld.cvut.cz, tel.: +420 224 352 356, fax: +420 224 310 792

The work describes a design of the filter using log-domain CMOS technology. Filter only consists of capacitances and CMOS transistors in the week inversion mode and they are treated in the subthreshold conduction region. Low-pass, band-pass and high-pass filter is considered in the design. The transfer function of the filter is determined by the feedback around its input and output. My design is based on demand to have a universal building block which is able to filter the frequency band from DC to radio frequencies. This building block is represented by the log-domain differentiator. The frequency band and the transfer function of the filter is determined only aid few devices connected to this building block. It is intended to use these current-mode filters in the design of the wireless sensor systems. In these applications the low-pass filters are usually used as an anti-aliasing filter before the signal is sampled. Also the radio frequency part of the circuit uses the filters. The program CADENCE and 0.18 µm CMOS technology were used for simulations.

1. Introduction

Wireless applications have special demands on the low power consumption and also on the precision and stability of its features. Both properties can be achieved using IC working in the translinear current mode. Current mode allows very low supply voltage and the exponential dependency of the current on the gate voltage permits application of the translinear principle. Originally, the translinear principle is the bipolar junction transistors domain. If the MOS transistors are treated in the week inversion mode and in the subthreshold conduction region the translinear principle is applicable also here. The subthreshold conduction region secures an extremely low power consumption of the circuit. My article describes the log-domain, current mode CMOS filter which is suitable as a universal filtering block for IC design. The filter is represented by the biquadratic section using log-domain differentiator and the transfer function of the filter is dedicated by the feedback loops and connection between each section. LP, BP and HP filter is considered in the design.

1. Multiplication cell

Figure1 shows the basic translinear cell, which works as a multiplier. Transistors NM1-MN4 form the translinear loop. For the drain currents can be derived the equation 1 [2].

$$I_{out} = I_{DMN4} = \frac{I_{DMN1} \cdot I_{DMN2}}{I_{DMN3}} \qquad (1)$$

This dependency is valid only in the range, where the transistors are in the subthreshold conduction region.

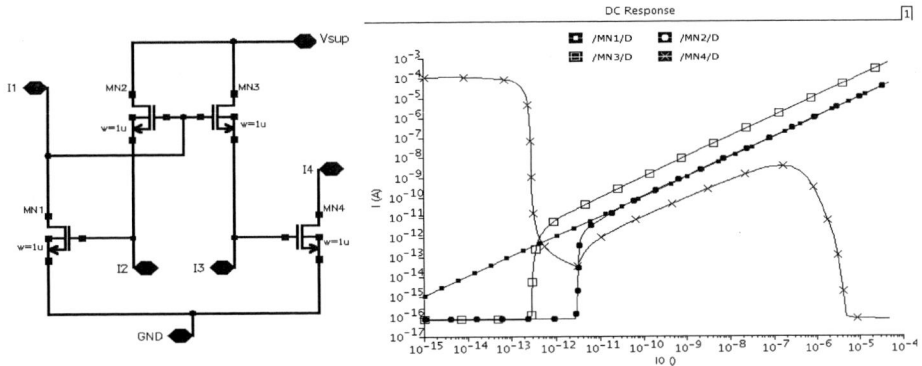

Fig. 1: Log-domain multiplier [2] and a simulation of its function

This range depends on the channel width. On the right side of the figure 1, there is an DC analyse of the cell, where the $I_{MN1}=I_{MN2}=I_0$ and $I_{MN3}=10 \cdot I_0$ (see equation 1) and for channel width 1μm. It is evident that the multiplication cell works correctly for biasing current I_0 in the range of 10^{-11}-10^{-8} A.

2. Current mode differentiator

If the capacitor is connected to the gate of the transistor MN1, the drain current of this transistor depends exponentially on the capacitors voltage (equation 2). Using equation 1 and 2 and considering the topology of the circuit on the figure 2, the transfer function of the circuit can be derived (equation 3 [2]). Output current is present on the positive nod of the Vsens2.

$$I_{MN1} = I_0 \cdot e^{\frac{V_C}{n \cdot V_T}} \qquad (2)$$

$$\frac{i_{out}}{i_{in}}(p) = p \cdot \frac{n \cdot V_T C}{I_0} \qquad (3)$$

This transfer function is typical for lossless differentiator. Time constant of this differentiator is determined by the biasing current (I_1, I_2, I_3) and by the capacity C (figure 2).

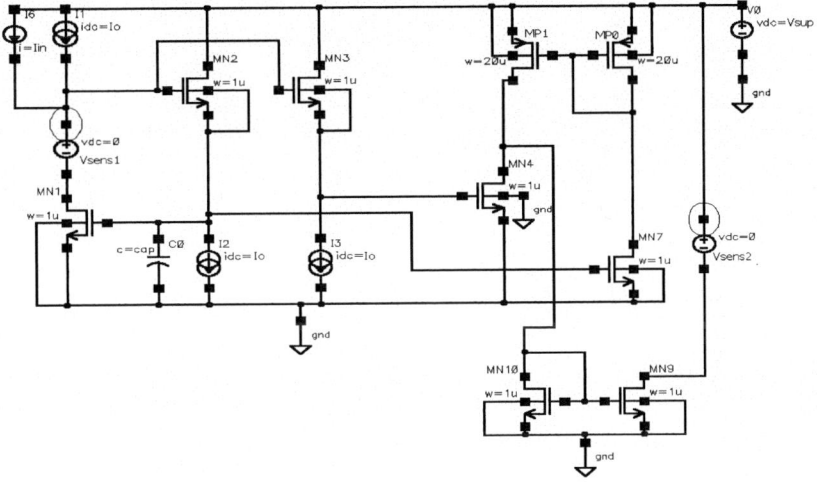

Fig. 2: Scheme of the cmos log-domain differentiator

In the equation 3, the V_T is the temperature voltage and n is the technological factor.
Figure 3 shows the result of the simulation of this differentiator for different biasing currents and for different capacities.

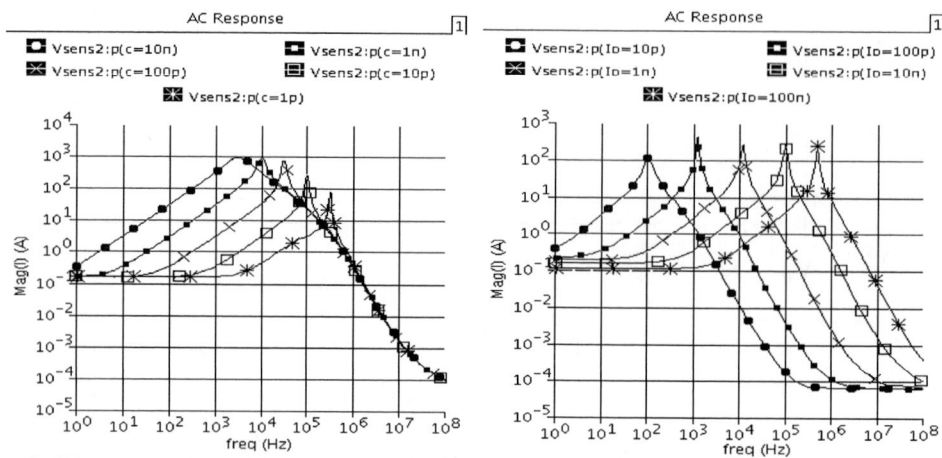

Fig. 3: Simulation result of the differentiator for different capacitors @ Io=10nA (left side) and different currents Io @ c=10pF (right side)

This simulation shows the useful frequency band, where the differentiator works correctly. For example the differentiator with capacitor 10pF and biasing current 1nA can be used in the frequency band 10^2-10^4 Hz.

3. Biquadratic section

Incorporating two differentiators into the feedback and feed-forward loops system, the circuit starts to work as a filter with desired transfer characteristic. Figure 4 shows the block diagram of this system which forms the biquadratic section.

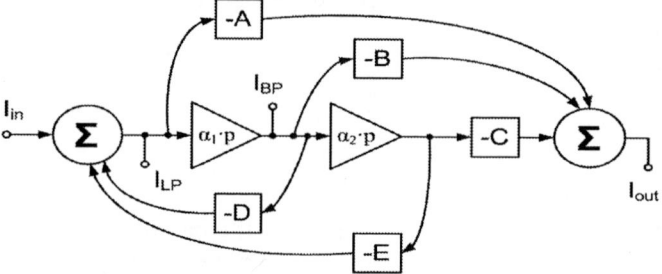

Fig. 4: Block scheme of the general biquadratic section

Transfer function of this system is determined by the equations 4 and 5.

$$\frac{I_{OUT}}{I_{IN}}(p) = \frac{-\left(A + B\alpha_1 p + C\alpha_1\alpha_2 p^2\right)}{1 + D\alpha_1 p + E\alpha_1\alpha_2 p^2}; \frac{I_{OUT\_BP}}{I_{IN}}(p) = \frac{\alpha_1 p}{1 + \alpha_1 p + \alpha_1\alpha_2 p^2} \tag{4}$$

$$\frac{I_{OUT\_LP}}{I_{IN}}(p) = \frac{1}{1 + \alpha_1 p + \alpha_1\alpha_2 p^2} \tag{5}$$

Summing block and interconnections between each block are formed by the current mirrors with scaled ratio of channel width.

5. Result of the simulations

The low-pass, band-pass and high-pass filter were simulated in order to test the influence of the biasing current and capacity. For simulations the parameters from equations 4 and 5 were set to A=B=0, C=-1, D=E=1. For simulations the ideal current sources in the differentiators (figure 2) were substituted by the current mirror. Simulation results are presented on the figures 6 – 8.

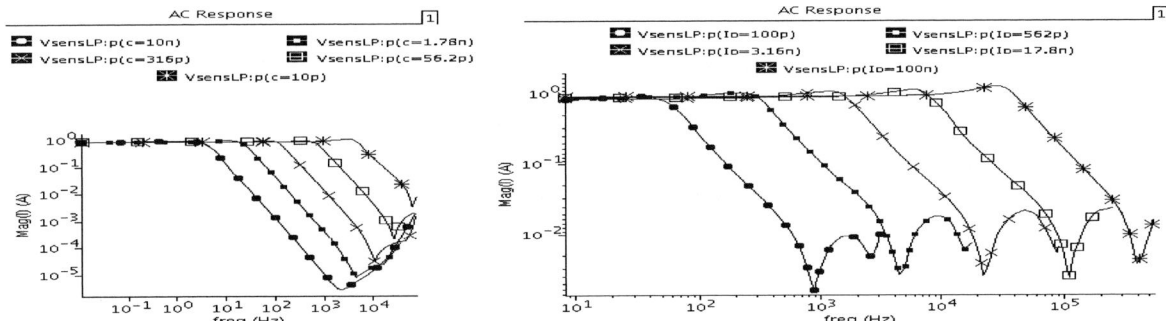

Fig. 5: Simulation result of the biquad, LP output, for different capacitors @ Io=10nA (left side) and different currents Io @ c=10pF (right side)

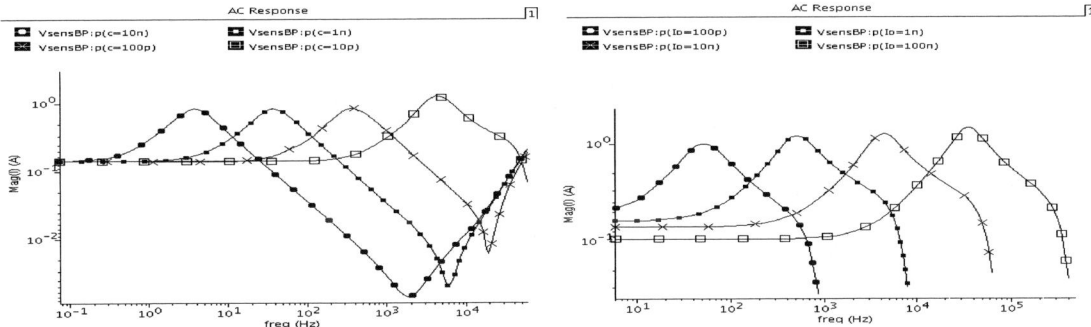

Fig. 6: Simulation result of the biquad, BP output, for different capacitors @ Io=10nA (left side) and different currents Io @ c=10pF (right side)

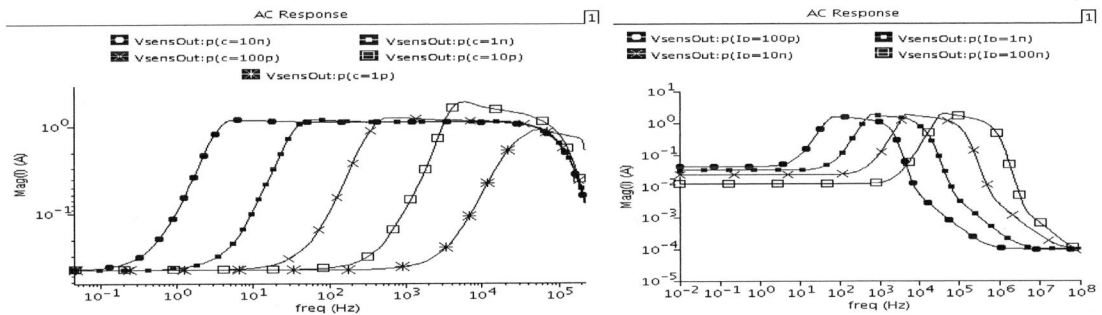

Fig. 7: Simulation result of the biquad, HP output, for different capacitors @ Io=10nA (left side) and different currents Io @ c=10pF (right side)

6   Conclusions

Work described the universal filtering block, which is suitable for wireless sensor IC design. Each filter works correctly only in the several frequency bands, so for the real application it is essential cascade the filters to cut the undesired frequency band.

References

[1]   Kaushik, R., Sharad, P.: *Low-Power CMOS VLSI circuit design*, New York, John Wiley & Sons, Inc., 2000, ISBN 0-471-11488-X

[2]   Ngarmnil, J., Yodprasit, U.: *Log-domain Differentiator for BP-HP Filters*, URL: http://www.ee.mut.ac.th/research/1998/eecon21/A_EECON21_02E.pdf, 24.5.2006

[3]   Mulder, J., Wowrd, A., Serdijn, W., Roermund, A.: *General Current-Mode Analysis Method for Translinear Filters*, IEEE TRANSACTIONS ON CIRCUITS AND SYSTEMS-I: FUNDAMENTAL THEORY AND APPLICATIONS, Vol. 44, No. 3, March 1997

Direction Sensitivity Matrix with PLL Temperature Sensor

M. Husak, A. Boura, J. Jakovenko

Department of Microelectronics, Faculty of Electrical Engineering,
Czech Technical University in Prague, Technicka 2, CZ – 166 27 Prague 6, Czech Republic
husak@feld.cvut.cz, tel. +420-2-2435 2267

In the paper there is presented temperature transducer with PLL signal processing of sensor data. The principle of temperature sensor is based on the MOS structure behaviour in the temperature range. The strong and weak inversion modes are used. There is used matrix of temperature sensors in the design of air flow sensor. Various arrangements of temperature structure are designed. The new circuits design is used.

1. Introduction

The latest knowledge of microelectronic technology is used at development of new types of sensors for measurement of physical and biochemical quantities. For design of a direction sensitive matrix, the anemometric principle may be used. Its operation is based on cooling of temperature sensors of different type. Sensitivity of the sensors is derived from temperature equilibrium in steady state, when electrical energy delivered to the sensor is the same as heat energy lost for cooling of the sensor. The energy loss depends on velocity of flow of the cooling medium. Power delivered for the sensor heating is used for the flow velocity measurement. Sensors using this principle are "hot-wire" sensors (probe is not suitable for integration), thermistors (high sensitivity, not suitable for integration). Another group of sensors using the anemometric principle are p-n junctions (low sensitivity for air velocity), MOS channel (suitable for integration and miniaturization), metal or integrated thermocouples [1] and other types of integrated temperature sensors.

The working principle was introduced in [2]. The analog part contains sensitive temperature sensors S_1 thru S_4, circuits for signal processing, and differential amplifier - Fig.1. The differential output signals are sent to the digital computational part. It serves for computation of the direction of airflow using goniometric functions. According to sensor arrangement, it is possible to measure flow direction up to 360° using two couples of temperature sensors in perpendicular arrangement.

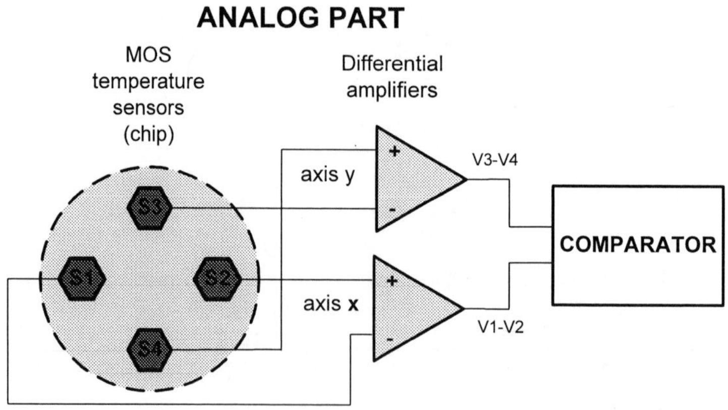

Fig. 1: Block diagram of the sensor part for measurement of velocity and flow direction

1-4244-0396-0/06/$25.00 ©2006 IEEE

2. Temperature MOS sensor design

Simulations of temperature dependence of electric parameters of MOS integrated transistor structures have been performed during the design of the temperature matrix. The basic principle arrangement for modelling of temperature properties of the structure is shown in Fig. 2. The structure served to identify the region of a strong and week inversion of the transistor and to quantify the sensitivity of the drain current on temperature in those regions for one transistor. Reached results are presented in Fig. 4.

The temperature sensors S1 thru S4 are designed as the matrices of NMOS and PMOS transistors. The operating points are set in the area of strong inversion for NMOS transistor M1 and week inversion for the PMOS transistor M2. The temperature dependence of mobility of the charge carrier is dominant in strong inversion. In the strong inversion the temperature coefficient of the drain current is negative, in the week inversion region is positive.

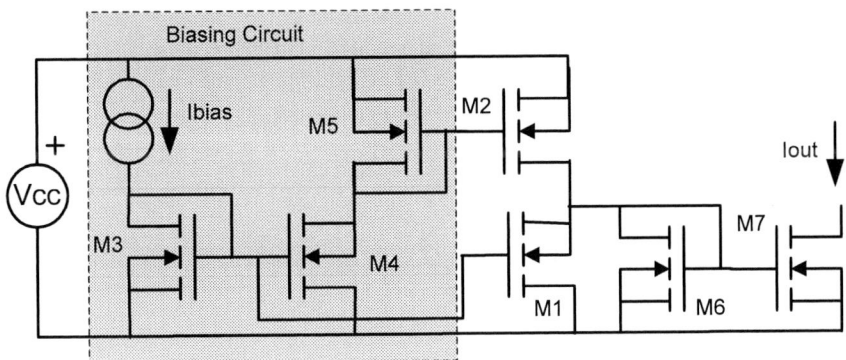

Fig. 2: One MOS temperature sensor

Serial connection of the transistors with different temperature dependence of the drain current improves the sensitivity of the structure. Transistors M3, M4 and M5 serve as a biasing circuit for the sensing transistors M1 and M2. In the ideal case the drain current of M1 and M2 is same but the different temperature coefficients of the transistors causes divergence of this current which is mirrored by M6 and M7 to the output and consequently transformed to the voltage. There are 4 temperature matrices on the chip. They constitute of 2 pairs with mutually orthogonal geometrical layout. Differential amplifiers process the output signals. The differential output signals are sent to the digital computational part. It serves for computation of the direction of airflow using goniometric functions.

3. Processing of a temperature sensor signal

Principle of operation of PLL circuit for processing of a temperature sensor signal is based on comparison of the voltage sensor signal with a triangular waveform signal in a comparator. The triangular waveform signal has a constant and temperature independent frequency. At the output of the comparator is a PWM signal. Width of the signal pulses is a function of the temperature. Constant frequency of the PWM signal is N-times multiplied in the PLL circuit. At the output of the voltage-controlled oscillator VCO there is an N-time multiple of the PWM signal and its duty cycle is 50%. Output signals are processed by a counter. Number of pulses counted by the counter in the course of the PWM pulse duration is proportional to the temperature measured. Block diagram of the transducer designed is in Fig.3. The output signals of the sensors operating in the weak and strong inversion mode must be amplified to a level corresponding to the level of the signal at the output of the generator.

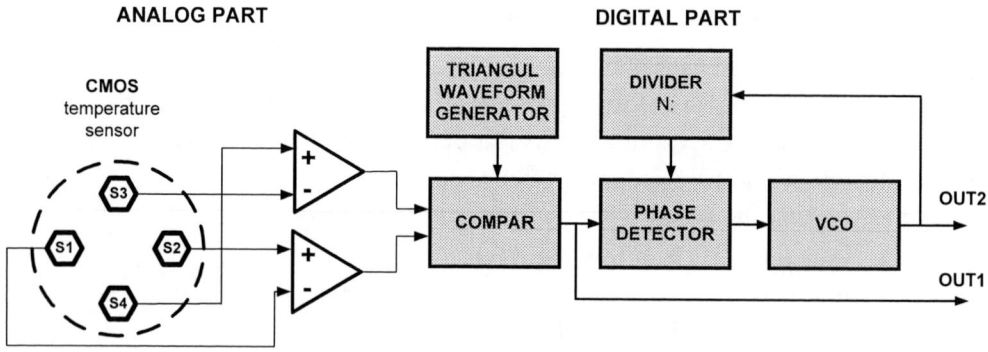

Fig. 3: Temperature transducer with PLL sensor signals processing

4. Reached results

There are evident the week and strong inversion regions for one NMOS transistor in Fig. 4. The same simulation was presented for PMOS structure. There exists optimal biasing current, where the sensitivity of the MOS temperature sensor is maximal but these current changes with the temperature. On the Fig. 5 there is presented the dependence of the sensitivity of the temperature sensor on the biasing current at 60 °C. Optimal current in that case is about 29 µA.

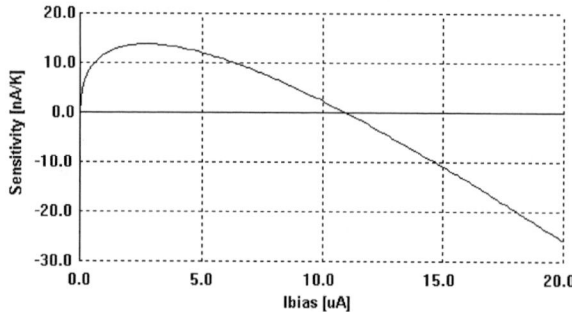

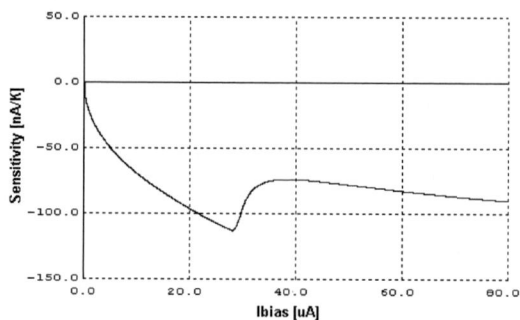

Fig. 4: Temperature sensitivity of the drain current of the NMOS transistor versus biasing drain current

Fig. 5: Temperature sensitivity of the output current from the temperature sensor versus biasing current at 60 °C

Transfer characteristic have been primarily measured of sensor part with output signal V_2-V_1 and V_4-V_3. Temperature gradient is function of flow velocity. Based on the results, it is possible to derive a model for simulation of output voltage from the differential amplifier in dependence on flow velocity [3].

Transfer characteristic of the transducer have been measured for constant temperature 25 °C. Input voltage signal simulating the output signal of the CMOS temperature sensors was applied to the comparator. The output signal of the comparator is a pulse signal and is measured by a counter. Temperature dependence of the transducer output signals U_{OUT1} and U_{OUT2} was measured for the transducer temperatures in the range from 25 °C to 125 °C.

Transfer characteristics of sensor operating in the strong inversion mode are in Fig. 6. The measurements were performed for two gains A_{INV}. The characteristics are nonlinear, increasing temperature causes the higher nonlinearity. This is caused by nonlinearities of the sensor circuit and by the gate voltage temperature dependence. Transfer characteristic of sensor operating in the weak inversion mode are in Fig. 7. Signal from the sensor is applied to the comparator input.

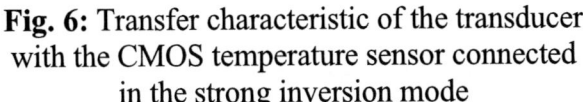

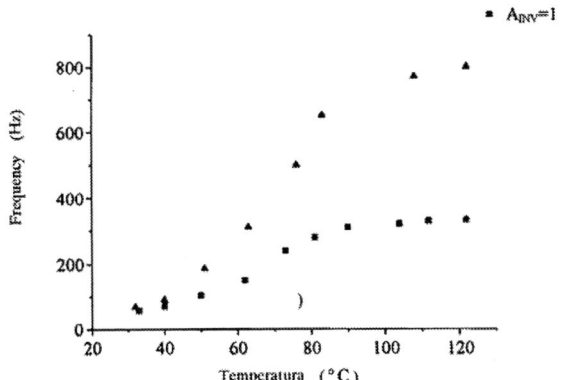

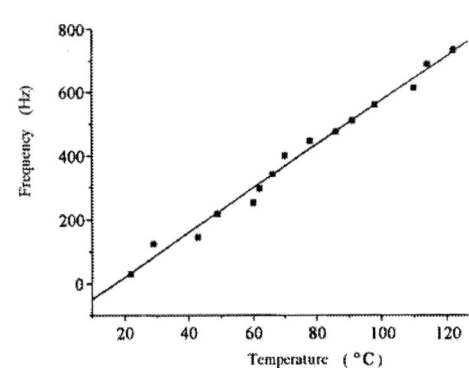

Fig. 6: Transfer characteristic of the transducer with the CMOS temperature sensor connected in the strong inversion mode

Fig. 7: Transfer characteristic of the transducer with the CMOS temperature sensor connected in the weak inversion mode

5. Conclusion

Anemometric system with integrated temperature probe has been designed. Each sensor contains MOS transistors which work in the area of strong and week inversion. Axes of 2 sensor pairs are perpendicular. From the emerging temperature gradients of both sensor pairs, the airflow velocity and direction are computed.

The temperature dependence of the CMOS gates of IC in the weak and strong inversion operation mode have been measured. The transfer characteristics of the CMOS temperature sensors and PLL transducer have been measured. Temperature dependence of the transducer output signals U_{OUT1} and U_{OUT2} in the temperature range from 25 °C to 85 °C is linear, the difference is about 0.5%. The real circuit model of intelligent structure has been realized and tested. The accuracy and reproducibility of the flow velocity measurement has been evaluated to be 4 per cent and has been the same as accuracy and reproducibility of measurement of airflow direction. Measured results correspond with simulated results very well.

Acknowledgement

This research has been supported by the Czech Science Foundation project No. 102/06/1624 "Micro- and nano-sensor structures and systems with embedded intelligence" and partially by the research program No. MSM6840770015 "Research of Methods and Systems for Measurement of Physical Quantities and Measured Data Processing " of the CTU in Prague sponsored by the Ministry of Education, Youth and Sports of the Czech Republic.

References

[1] K.A.A. Makinwa - J.Huijsing: A wind sensor with integrated chopper amplifier. Delft University - DIMES, www.stw.nl/sesens/proc-2001.

[2] Husak,M. – Jakovenko,J. – Kulha,P. – Janicek,V.: Design of Temperature Matrix with Direction Sensitivity. The Nanotechnology Conference and Trade Show (NanoTech 2005), Anaheim, May 8-12.2005. Technical proceedings vol. 3., pp. 407 – 410. ISBN 0-9767985-2-2.

[3] Husak,M.: Microsystem Modeling – from Macromodels to Microsystem Design. Proc. of NanoTech2003, vol. 1, February 23-27, 2003, San Francisco, California, USA, pp. 272 – 275.

Differential blood analysis by thin film interdigitated arrays of electrodes

R. Víglaský *, M. Nigrovičová *, V. Tvarožek *, M. Weis **

* Department of Microelectronics, Slovak University of Technology
Ilkovicova 3, SK-812 19 Bratislava, Slovakia
** Institute of Medical Physics and Biophysics, Comenius University, Bratislava, Slovakia
e-mail: *radovan.viglasky@stuba.sk*

We examined the use of impedance and optical methods for monitoring of human blood sedimentation what is an important factor of health status of a patient. We found that the monitoring of aggregation phase of sedimentation give fast answer on the state of blood. For impedance measurements the planar microsensors with thin film interdigitated arrays of electrodes (IDAE) have been developed.

1. Introduction

The blood sedimentation phenomenon is utilized in clinical diagnostics since 1894 [1]. Blood itself consists of 15 parts (e.g. blood platelets, enzymes, minerals etc) and we are only interested in concentration of erythrocytes and lymphocytes in blood plasma (Fig.1). The composition of the complex fluid has a fundamental physiological significance [2]. The interaction forces between erythrocytes in normal human blood cause that they coalesce and form linear and branched aggregates - rouleaux [3]. Thus, the decrease of the quantity of certain size aggregates in a small time interval is proportional to the total number of these particular size aggregates, and consequently is an exponential function of time (Fig.2).

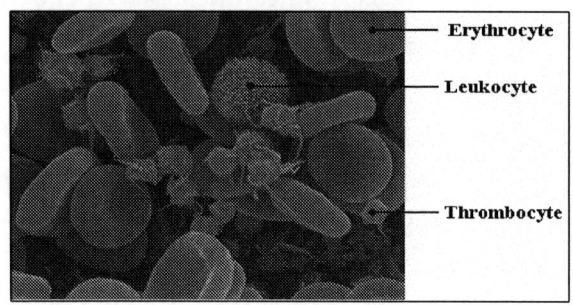

Fig. 1: Composition of human blood.

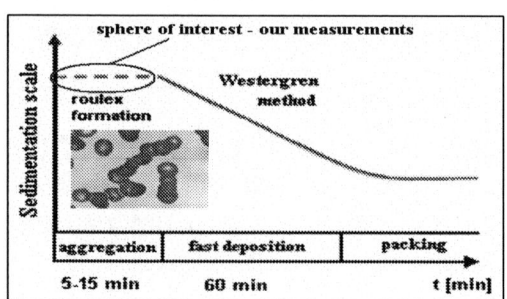

Fig. 2: Three phase sedimentation of blood

The speed of rouleaux agregation creation is influenced by changes of impedance characteristics as well as parameters measured by optical methods. The intensity of the light transmitted by a blood sample increases the result of the rouleau formation. The attractive interaction between erythrocytes provides a necessary condition for a phase separation of the bio-colloidal system [4].

We examined the use of impedance and optical methods for monitoring of human blood sedimentation what is an important factor of health status of a patient. For impedance measurements the planar microsensors with thin film interdigitated arrays of electrodes (IDAE) have been developed.

2. Experimental results

For both our measurements (optical and impedance) we used healthy and cancer blood obtained by a venous puncture. To avoid coagulation the blood was treated with citrate. Samples of uniformly distributed erythrocytes were obtained by gentle mixing before experiment. For measuring of blood it is important that the samples are as fresh as possible.

2.1 Optical measurements

For optical measurements the blood was placed between two glass plates. We compared measurements of blood samples by using one or two glass plates. Two glass plates seem to be better because thickness of blood layer was constant and measurement did not depend on the size of blood drop. The intensity of the light transmitted through a blood sample is increasing due to the rouleaux formation.

Condition for optical measurement:

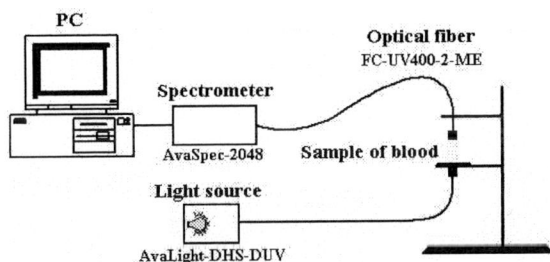

Range of wave-length: 400nm - 1000nm
Maximum light intensity: 4000 count
Time of one measurement: 1 second
Full measurement time: 10 minutes
Temperature: 23 – 24 °C
Sample of blood: 4 healthy and 6 cancer samples

Fig. 3: *Experimental set-up for optical measurements and measuring set-up.*

Full measurement was automatic (Fig.3), PC direct set up and data acquisition from spectrometer (AvaSpec - 2048). For light line we used optical fiber with 400 micrometer diameter for visible radiation (350nm – 1000nm) and light source was calibrated light source.

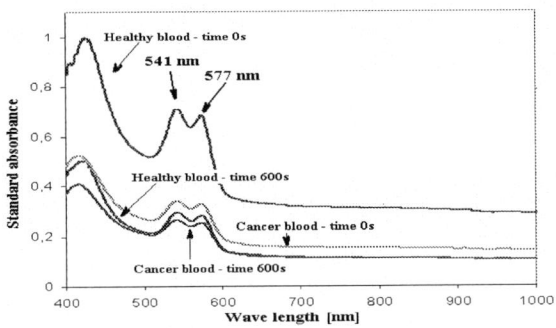

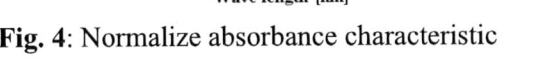

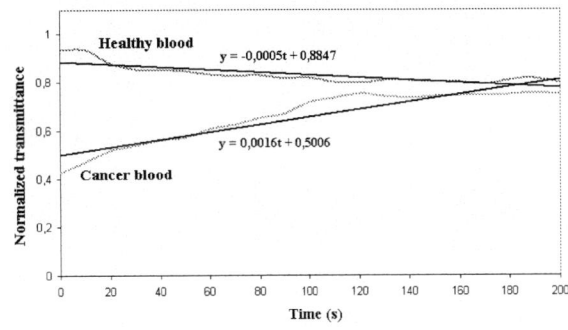

Fig. 4: Normalize absorbance characteristic

Fig. 5: Difference transmittance between healthy and cancer blood with 541nm wave-length.

Cancer blood had faster sedimentation than healthy so changes between healthy and cancer blood in long time should have been more signification. It is not necessary measure full wave-length range because if we used correctly definition wave length measurement will be fast and optical resolution increase. It is better used only wave lengths which place on local maximum (Fig. 4). Healthy blood absorbance declines faster than cancer because in the cancer blood is to more destroyed erythrocytes and rouleaux form so this happened why absorbance slow declines. In healthy blood are erythrocytes uniform diffuse on full capacity and in the first moment have healthy blood has maximal absorbance and when time increase absorbance go down linear.

2.2 Impedance measurement

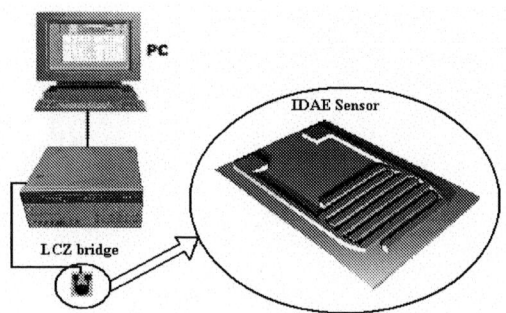

Condition for electrical measurement Fig. 6:
Frequency range: 10 kHz – 1 MHz
Maximum AC amplitude: 20 mV
Time of one measurement: 20 second
Full measurement cycle: 20 minutes
Temperature: 23 – 24 °C
Measured value: Absolute impedance, phase
Sensors: 100/100 µm, 200/200 µm and 400/400 µm (finger/gap width)

Fig. 6: Measurement setup and detail of IDAE chip with dimensions 400x400 µm.

In impedance experiments we used 3 types of IDA Pt microelectrodes on ceramic (Al_2O_3) with symmetric configuration (finger / gap dimensions of 100µm / 100µm, 200µm / 200µm, 400µm / 400µm). We measured dependences of absolute value of impedance and phase on frequency in the range of 10 kHz – 1 MHz. We found that the absolute value impedance for healthy blood was in the range between 245 Ω (for 10 kHz) and 195 Ω (1 MHz). For cancer blood was range of 205-155 Ω (10 kHz-1 MHz). We observed the difference between healthy and cancer blood.

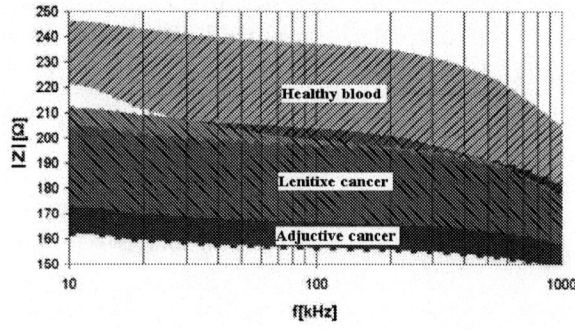

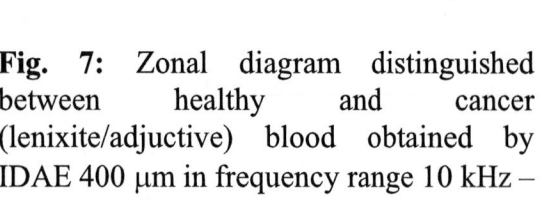

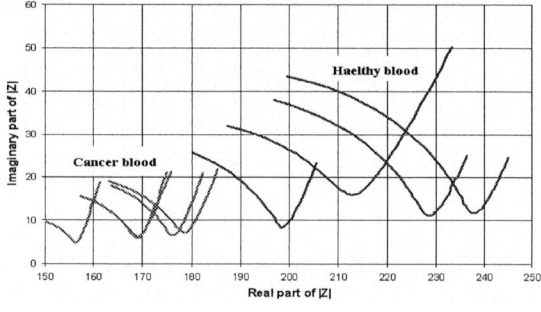

Fig. 7: Zonal diagram distinguished between healthy and cancer (lenixite/adjuctive) blood obtained by IDAE 400 µm in frequency range 10 kHz – 1 MHz.

Fig. 8: Cole-cole graphs show difference between healthy and cancer blood.

Zonal diagram (Fig. 7) was constructed from results of more than 70 blood samples of 45 patients. Frequency range up to 1 MHz has proven itself to be unnecessarily wide because the highest impedance differences occur at low frequencies and with increasing frequency the difference between healthy and cancer blood declines.

On Figure 8 some selected measure ments of healthy and cancer blood samples, are shown where it is used Cole-cole graph approach of real and imaginary impedance part and this way more accurate identification of healthy and cancer blood samples is possible.

3. Discussion

We found out than for optical measurement is better to use two glass plate because blood thickness between two glasses is constant and it is not depend of blood drop. The first measurement was realized with full range of wave spectrum but it is not necessary. We are able to recognize health state of blood even if used only two wave-lengths (541nm and 577nm). The whole optical measurement is very fast when we scan one wave length, therefore it is not a problem to evaluate changes of optical characteristic within first tenths of seconds. We are interested in this short period of time because aggregation phase of blood takes place during it.

All later measurements were performed on 400/400 sensors because these are more suitable for diagnostics of blood impedance characteristic. A model of IDAE sensor has been used with help of ANSYS program and therefore it will be possible to simulate and optimize its geometry for increasing sensivity of given sensor. We can distinguish healthy and cancer blood samples with probability higher than 90 percent for lower frequencies than 30 kHz.

4. Conclusion

In our preliminary measurements we found out that optical or impedance measurements are suitable for early detection (screening) of some human illnesses (for example cancer). These methods are more acceptable for patients, because we need only small amount of blood. A final diagnosis has to be specified by classical medical laboratory methods. Our main goal is to design a sensor system capable to measure optical and impedance characteristic simultaneously within one measurement cycle. The same sensor would be able to mix the given sample before measurement.

Acknowledgement

Presented work was supported by the VEGA Grant 1/3098/06 of the Slovak Grant Agency and partly by the PPP DAAD project 5/2005.

References

[1] E. Biernacki, *Z. F. Physiolog. Chem. 19* , 179 (1894)

[2] B. A . Schottelius, D. D. Schottelius, *Textbook of Physiology*, The C.V . Mosby Company, Saint Louis (1978)

[3] R.W. Samsel, A .S. Perelson, *Biophys.* 45, 805 (1984)

[4] V .L. Voeikov, *Uspekhi Fiziol. Nauk.* 55 (1998)

Influence of the doping material on the benzene detection

P.Ivanov[1*], F.Blanco[2], I. Gràcia[1], N. Sabaté[1], X.Vilanova[2], X.Correig[2], L. Fonseca[1],
E. Figueras[1], J. Santander[1], R. Rubio[1], C. Cané[1]

[1] Gas Sensors Group, Centro Nacional de Microelectrònica, CNM-CSIC, Bellaterra, Spain
[2] DEEEA, Universitat Rovira i Virgili Tarragona, Spain
*Peter Ivanov, National Centre of Microelectronics CNM – CSIC, Campus UAB, 08193
Bellaterra, Spain, Fax (+34) 93 580 14 96 e-mail: peter.ivanov@cnm.es

In this paper we describe the fabrication of low-cost micro-hotplate benzene sensors by using screen-printing technology. Sensitive layers of tin and tungsten oxide (pure and doped with 1% in weight of Au, Pt and Pd) were deposited on silicon micromachined substrates with low thermal inertia. Gas measurements were performed with the fabricated sensors and the results confirm the viability of the techniques introduced to obtain micromachined sensors suitable for battery-powered gas/vapour monitors.

1. Introduction

Benzene is an organic compound that causes serious concerns in environmental health due to its toxicity and carcinogenic properties, even at low concentrations. In particular, benzene is known to be a human carcinogen for all routes of exposure and a risk factor for the occurrence lymphomas. The individual exposure limit per day (time weighted average) to benzene for directly exposed workers fixed by a directive the January 2003 directive of the European Communities is 1 ppm. The Air Quality Directive, which is to be implemented in 2010, has set a much lower limit for benzene exposure: 1.6 ppb (5 $\mu g/m^3$) [1].

The most widely used conventional method for the detection of volatile organic compounds (VOC) is gas chromatography (GC) combined with mass spectrometry (MS). Some authors have developed conducting polymer films to detect toluene, benzene, xylenes and other aromatic hydrocarbons. Ni et al. have studied different materials (polymer films) to coat piezoelectric quartz crystal sensor arrays in order to selectively detect organic vapours. Calvo-Muñoz et al. reported on the use of inorganic (tetramethoxysilane) and hybrid (methyl-ltrimethoxy-silane) polymers prepared by the sol–gel method to optically detect benzene vapours [2]. The usefulness of metal oxide gas sensors for VOC detection has also been established.

2. Process of fabrication

The 4-element integrated micro-hotplates were fabricated on double-side polished p-type (100) Si substrates, 300 μm thick (4–40 Ω cm). The structure of the device basically consists of the gas sensing layer, the electrodes, a thermal insulating membrane and a polysilicon heater.

2.1 Micro-hotplate fabrication

The technological process needed to fabricate the sensors had the following steps: (1) Deposition of the membrane layer. The dielectric membranes consisted of a 0.3 μm thick Si_3N_4 layer grown by LPCVD. Each chip had four membranes, the size of which was 1 × 1 mm. (2) Deposition and patterning of a $POCl_3$-doped polysilicon heating meander. The temperature coefficient of resistivity (TCR) of polysilicon depends on the doping level. If the doping level is low, polysilicon has a positive TCR (e.g. 6.79×10^{-4} for our devices). The heater was also used as a temperature sensor. (3) Deposition of a 0.8 μm thick SiO_2 layer to insulate the heater from the electrodes and the sensing film. (4) Opening of contacts for the heater bonding pads to be accessible. (5) Deposition of interdigited 0.3 μm thick Pt electrodes, patterned by lift-off. A thin layer (20 nm) of Ti is deposited prior to Pt to promote electrode adhesion. (6) Patterning of the backside etch mask. (7) Deposition of the sensing layer onto the electrode area. (8) Backside silicon etching with KOH at 70 °C (40% wt.) to create the thermally-insulated membranes. (9) Wire bonding and packaging.

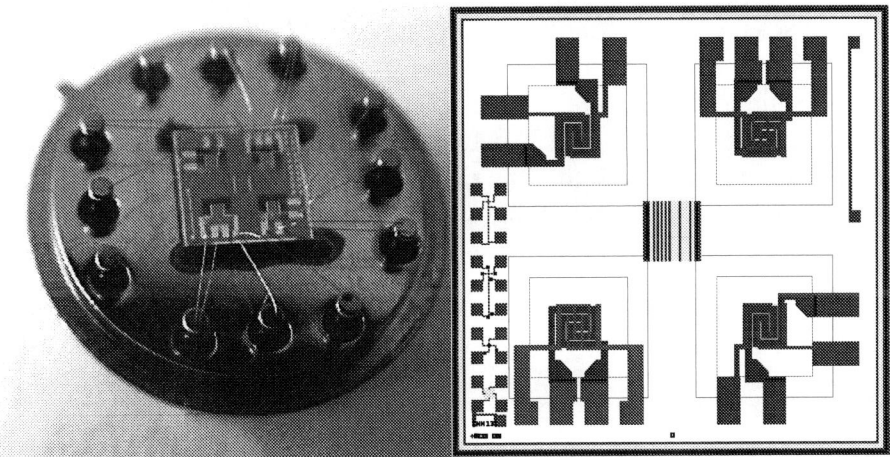

Figure 1. Four-element micro-hotplate mounted on standard TO-8 support (left); electrode, heater and membrane configuration (right).

Figure 1 shows micrographs of SnO_2/WO_3 coated membranes where the 4 membranes are clearly visible (left) and the electrodes, heater and membrane design (right). Each chip was mounted on a TO-8 package. Aluminium wires of 25 μm in diameter were used for standard ultrasonic wire bonding. For preventing the membranes to break due to air expansion in the cavity below the membranes when the device is heated, the chips were not glued directly to the surface of the metallic package but kept elevated by using two lateral silicon spacers.

2.2   Active film deposition

The sensing layers consisted of 5 μm thick SnO_2 or WO_3 nanoparticle films. The thickness of the screen-printed layers can be controlled between 2 and 20 μm by adjusting paste viscosity, mesh aperture, mesh tension and distance between the mesh and the substrate. Commercially available pure tin and tungsten oxide nanopowders were used with the objective of obtaining a good reproducibility in the deposition. In order to obtain high precision in the doping with noble metals and a homogeneous distribution of the doping particles, chlorides of Pt, Pd and Au were used. Each chloride was dissolved in a mixture of HCl and HNO_3. Then, the solution was neutralised with HN_4. Knowing the concentration of dissolved chloride in the neutralised solution allowed for calculating the volume of this solution needed to dope the metal oxide powder. The objective was to dope the metal oxides with 1% in weight of a given noble metal catalyst. Subsequently, the solution containing the metal oxide powder and the doping agent was evaporated and the resulting powders were annealed at 350° for 20 min. Finally, the annealed powders were milled in a mortar and then used for the preparation of a printable paste.

3. Results and discussion

In this chapter we will discuss the morphology and composition of the screen-printed sensing layers. The effect of the noble metal doping on the active layer structure will be also presented. Finally the gas measurement system will be presented and the results obtained from the sensors characterization will be examined.

3.1 Structural studies

The morphology and composition of the screen-printed sensing layers was investigated by scanning electron microscopy (SEM) and energy-dispersive X-ray spectroscopy (EDX). The specimens were coated previously with a thin (around 20 nm) gold layer, which was sputtered on top of the samples to avoid charging effects. Specimens were observed at accelerating voltages 20 kV using a JSM 6400 field emission scanning electron microscope. SEM analyses showed that the different doping did not affect the structure of the active layer. EDX analyses determined that the amount of noble metal catalysts was near 1% in weight.

3.2 Gas measurements

The sensor microarrays were tested in a small-volume (15 ml) sensor chamber using an automated continuous flow measurement system. Ambient temperature was kept constant (30 °C) during the measurement phase and relative humidity was set to 10% R.H. Pure dry air was used both as carrier and also as reference gas (sensor baseline was recorded in pure dry air).

Gas flow was set at 200 sccm. The system enables measuring the gases at different concentrations and with different humidity backgrounds (using a heated bubbling system). Figure 2 shows a schematic view of the measurement setup.

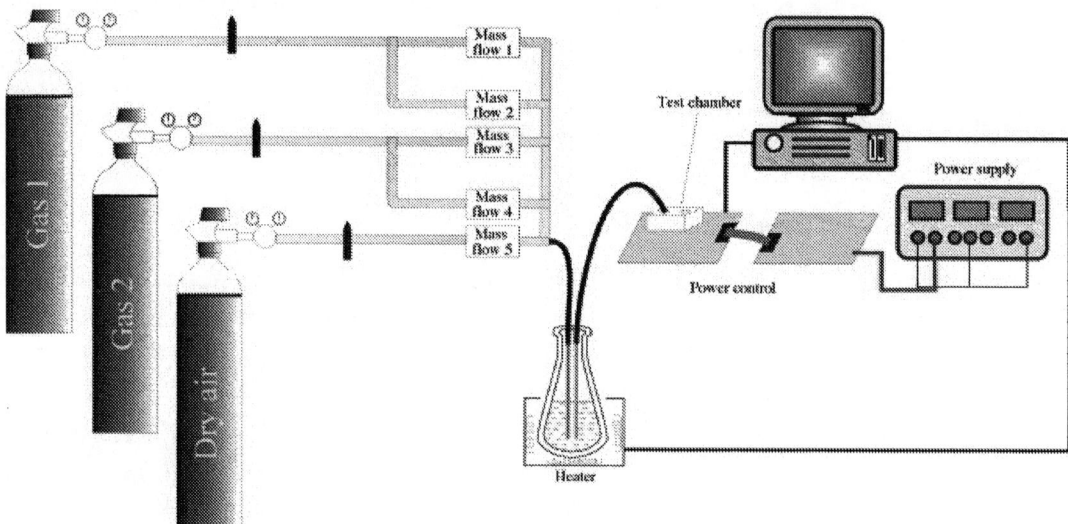

Figure 2. Measurement system.

On figure 3 we can observe a typical dynamic response of three different SnO_2 doped sensors to 10 ppm of benzene. The figure also shows that the sensors regain their baseline resistance when the chamber is cleaned with air.

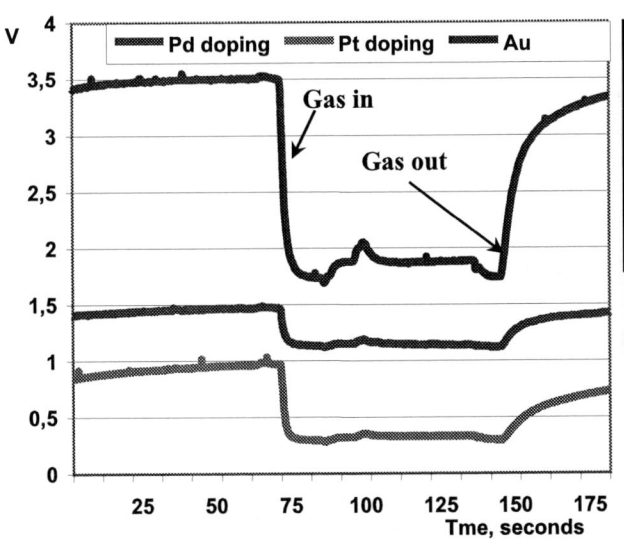

Sensor \ Temp.	250°C	350°C	450°C
SnO₂ pure	1.22	1.11	1.09
SnO₂ + 1% Au	1.46	1.287	1.26
SnO₂ + 1% Pt	1.9	1.75	1.71
SnO₂ + 1% Pd	**3.72**	2.98	2.42

Sensor \ Temp.	250°C	350°C	450°C
WO₃ pure	1.04	1.02	1.02
WO₃ + 1% Au	1.08	1.05	1.03
WO₃ + 1% Pt	1.13	1.09	1.05
WO₃ + 1% Pd	**1,19**	1.17	1,16

Figure 3. Typical response of three doped SnO₂ sensors to 10 ppm at 250 °C.

Table 1. Measurement results.

The measurement results for the SnO_2 and WO_3 gas sensors are summarized in table 1. Every measurement was repeated 5 times in order to obtain a valid result. From the data in table 1 it could observed that the SnO_2 sensors showed better result than the WO_3 ones. To evaluate the influence of the operating temperature, the sensors were operated at three different temperatures 250, 350 and 450°C. From the results obtained it could be clearly seen, that increasing the operating temperature reduces the sensor response. Finally, the effect of the noble metal doping was evaluated. Analyzing the results we can conclude, that the optimal doping material for benzene detection is the Pd with sensitivity almost 2 times higher compared with the Pt.

4. Conclusions

SnO_2 and WO_3 based micro-hotplate sensor arrays have been fabricated via screen-printing. The deposition method employed enables the deposition of a different sensitive layer on each membrane of the integrated microarray. A study on the effects of noble-metal loading on the response of tin and tungsten oxides to benzene was conducted. From the results of this study we can conclude that the SnO_2 must be chosen as material if we want benzene detection. On the other hand the SnO_2 doped with 1% Pd showed best results among the doped sensors. Due to the low power consumption, the micromachined sensors presented in this study are suitable for battery-powered gas/vapour monitors.

Acknowledgement

Part of this work has been financially supported by the Spanish Ministry of Education and Science - project CROMINA (TEC2004-06854-C03-01).

References

[1] Directive 2000/69/EC of the European Parliament and of the Council of 2000 relating to the limit values for benzene and CO in ambient air, *Official Journal L313* (2000) pp. 12–21.
[2] K. Susuki and T. Takada, Highly sensitive odour sensors using various SnO_2 thick films, *Sens. Actuators B,* 24–25 (1995), pp. 773–776

Design and simulation of humidity micro-sensors structure based on Polymers

Pavel Suchánek, Miroslav Husák
Czech Technical University in Prague
Faculty of Electrical Engineering
Technicka 2, 166 27 Prague 6, Czech Republic
suchanp2@feld.cvut.cz

Abstract

The aim of this paper is to report the design of the simple polymers structure humidity sensors and its mechanical and piezo-rezistive simulation in CoventorWare. Acquired simulation data gets a primary image about sizes bend and surface strain on micro-cantilever which forms sensing part of sensor.

1. Introduction

For precise simulation of micro-electro-mechanic systems (MEMS) is used integrated set of tools in programme CoventorWare. These tools allow user to design, simulate and subsequently specify design of (MEMS) structures.

The primary part of structure sensor is micro-cantilever made of thin silicon wafer which has a rectangular shape. With help of applying process we spread thin layer of the hygroscopic polymer materials such as polyamid, polyimid. This micro-cantilever is fixed on the immobility silicon base (Figure. 1). [2], [3]

The humidity which is contained in surrounding gas reacts with hygroscopic polymer material deposited on the surface of micro-cantilever. In this layer occur reversible mechanical changes (contraction, extension), that create stress in micro-cantilever and force it to fold down. [4]

2. Scanning of bend on micro-cantilever

The first way of scanning bend on micro-cantilever is using resistive strain gauge connected into Wheaston bridge (Figure. 1), which is located in the place with the biggest surface strain because of the biggest change resistivity. There are two ways how to produce the resistive strain gauge. The first way uses litography process - we deposit thin layer of metal on the surface polymer material. Second way uses foto-chemical process - it creates structure of strain gauges made of another conductive polymer material. For evaluation of resistive change in the structure of resistive strain gauge we use Ohm act [1].

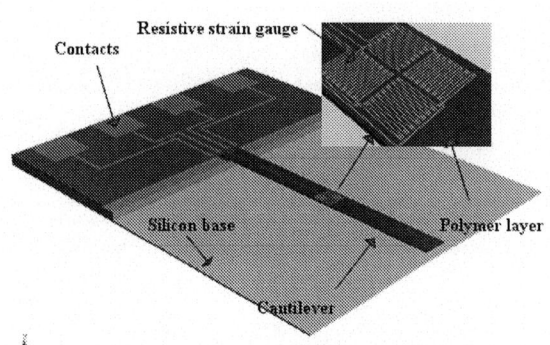

Figure 1. Micro-sensor structure

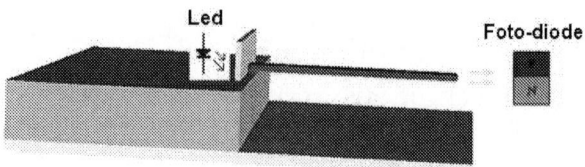

Figure 2. Sensing 2 – using LED and Foto-diode

Another way of scanning bend on micro-cantilever is using an impact location change of ray of light onto foto-diode. The silicon cantilever represents waveguide in this structure (Figure. 2). This structure is using the LED as a source of radiation. This LED generates light in wave-lenght 1,3um. It is for reason of minimal absorption loss in silicon micro-cantilever. Germanium diod is used as a detector of light, which is sensitive on the side of PN crossing.

Another way of scanning bend on micro-cantilever is using electric polarization arising from PVDF (piezoelectric polyvinyliden film) material, which forms micro-cantilever base (Figure. 3). On the surface of micro-cantilever we can deposit thin layer of polymer material with using the one of the most depositing process.

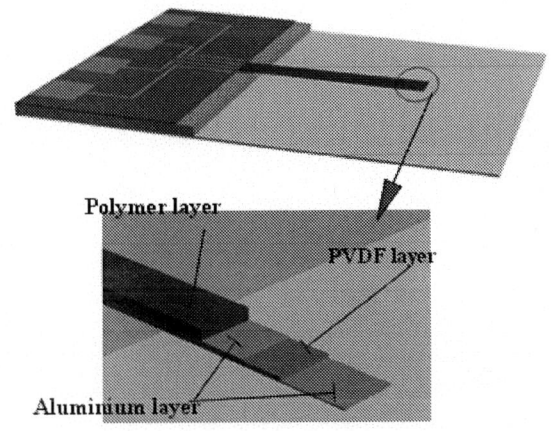

Figure 3. Sensing 3 – using PVDF layer

3. Simulation and results

In the programme CoventorWare it is not possible to simulate bend on micro-cantilever caused by humidity. Instead we use pressure, which effects on the surface of cantilever arising in polymer layer in this programme. To calculate value of this pressure we use base figure for calculation of value of relative humidity and derivative figure for calculating value of saturation pressure.

$$RH = \frac{P_{abs}}{P_{sat}} 100 \qquad (3.1)$$

$$P_{sat} = 610.78 e^{\left(\frac{t}{t+283,3} 17,2694\right)} \qquad (3.2)$$

RH........ relative humidity [%],
P_{abs}...........absolute pressure [Pa],
P_{sat}..........saturation pressure [Pa],
t ....temperature [^{0}C].

After induction of figure (3.2) into figure (3.1) we get figure for absolute pressure. This pressure depends on temperature and humidity. All simulation data were design for changing value of relative humidity with temperature $t = 25^oC$.[5]

From the mechanical simulations we got important information about size of bend (Figure. 4) and surface strain (Figure. 5) on micro-cantilever cause by humidity. [5], [6]

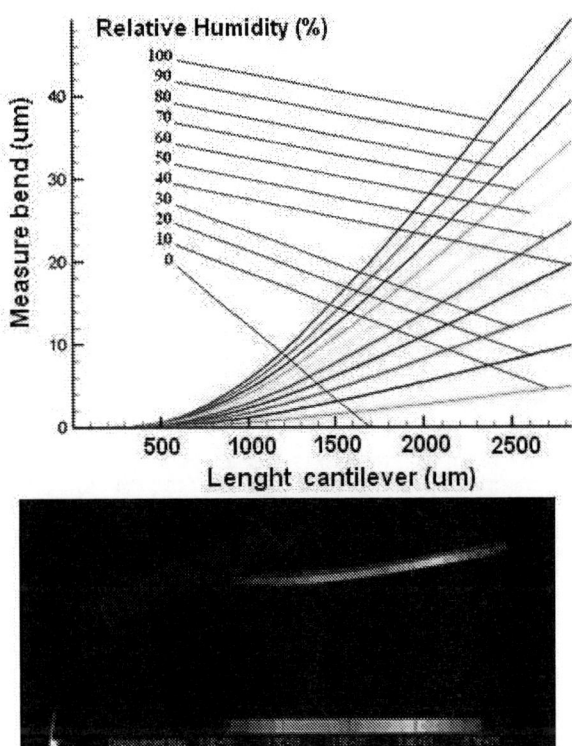

Figure 4. The results of mechanical simulations 1

$$()0000099 \qquad (4.1)$$

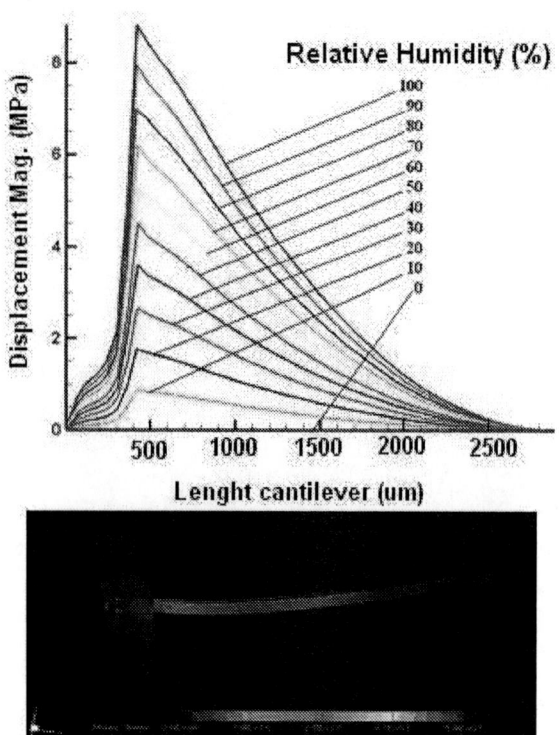

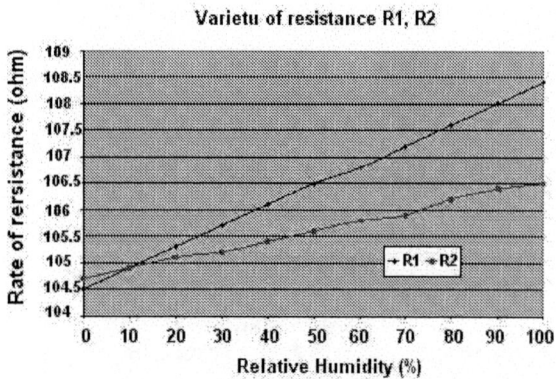

Figure 5. The results of mechanical simulations 2

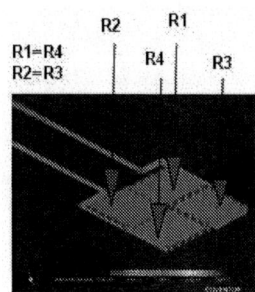

Figure 6.The results of piezo-resistive simulations

From the piezo-resistive simulations we got information about changes of resistivity values in axials and crosswises strain gauges (R1 and R2) placed in Wheaston bridge (Figure. 2). These changes were caused by surface straining evoked by bending of micro-cantilever (Figure. 6). [6]

From measured value of resistive strain gauge it is possible to calculate and picture a course of potential Um, which depends on change relative humidity (Figure. 7).

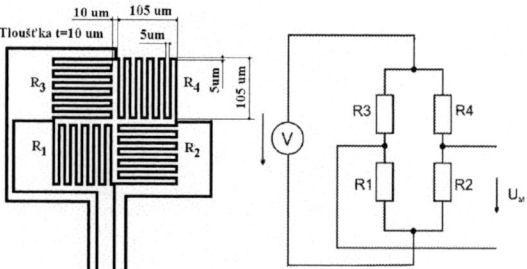

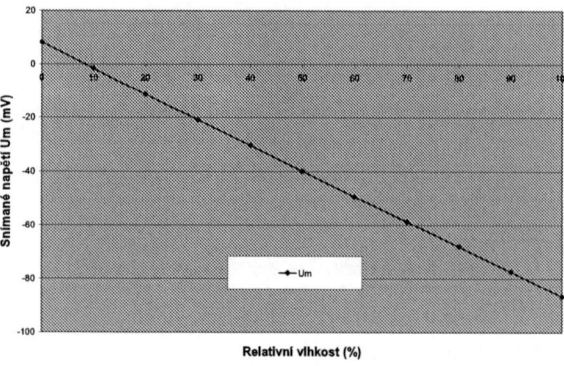

Figure 7. Resistivite strain gauges connection into Wheaston bridge and the result of piezo-rezistive simulations

4. Conclusion

This work is oriented on the design of micro-sensors structure, which changes its geometric dimension caused by humidity. The main aim of this work was the humidity sensor, which has micro-cantilever structure. With this structure we did mechanical and piezo-resistive simulation in programme CoventorWare. This simulation gave us information about size of micro-cantilever bend and size of surface straining on the micro-cantilever surface and resistivity changes in resistive strain gauges.

Another stage of this work will be aimed on designed structures realization and their characterization, which will be compared to simulated data.

5. References

[1] Hygrometrix application note 2005-4 (http://www.hygrometrix.net/documentation/HMX_AN2005.pdf)

[2] J. Brandrup, E.H Immergut "Polymer handbook",1989,

[3] DUPONT "Kapton polyimide film – summary of properties"

[4] J. Fragis, S. Chatzandroulis, D. Papadimitriou, C. Tsamis "Simulation of capacitive type bimorph humidity sensors",2005, Journal of Physics: Conference series 2005, (http://ej.iop.org/links/q35/XN0bovecDP4tncw9ayR7hQ/jpconf5_10_075.pdf)

[5] K. Govardham, Z.C. Alex: Mems based humidity sensor, 2005, (http://www.coventor.com/media/papers/isss2005se04.pdf)

[6] "CoventorWare", User's manual, MEMS simulation software, COVENTOR, Inc.2005

4H-SiC Diode with a RuO$_x$ and a RuWO$_x$ Schottky Contact Irradiated by Fast Electrons

Ľ. Stuchlíková, L. Harmatha, D. Búc, J. Benkovská, B. Hlinka, G. G. Siu*

Department of Microelectronics, Faculty of Electrical Engineering and Information
Technology, Slovak University of Technology
Ilkovičova 3, 812 19 Bratislava, Slovakia
*City University of Hong Kong, Kowloon, Hong Kong
e-mail: lubica.stuchlikova@stuba.sk

The impact of radiation damage on the device performance of 4H-SiC Schottky diodes with a RuO$_2$ and a RuWO$_x$ Schottky contacts, which were irradiated at room temperature with 9 MeV electrons (absorbed dose of 50 gy), is studied. No degradation is observed from measured capacitance-voltage curves. The radiation-induced decrease of the Schottky barrier height is observed from current-voltage measurements. Eight electron deep energy levels are observed before irradiation, and eighteen electron levels are induced after irradiation by a standard Deep Level Transient Spectroscopy.

1. Introduction

Silicon carbide (SiC) is regarded as a promising candidate for high-power and high-frequency devices because of its excellent thermal and electrical properties. More specifically, these properties are: a wide bandgap, high thermal conductivity (better than for example copper at room temperature), high breakdown electric field strength (approximately 10 times that of Si), high saturated drift velocity (higher than GaAs), a high thermal stability and a chemical inertness. Above mentioned suggests its suitability as a material for sensors and high-temperature electronics in the nuclear power industry and satellitebased systems [1]. Moreover, as SiC has a good radiation hardness, it is also expected to be widely applied in electrical devices used in radiation environments such as in space [2].

2. Experiment

The samples used in this experiment were SiC Schottky barrier diodes with a RuO$_2$ and a RuWO$_x$ Schottky contact. The n-type 4H-SiC substrates were fabricated at Linköping University in Sweden [3]. The RuO$_2$ Schottky contacts were prepared by deposition of thin RuO$_2$ films on the 4H-SiC substrates by reactive unbalanced magnetron sputter deposition at 100 W and using an axis target – substrate geometry. The substrate – target distance was 60 mm. The target was a ruthenium disk with a diameter of 50 mm. The sputter environment was an argon/oxygen mixture (16% oxygen) with a total pressure of 0.5 Pa. During deposition, the 4H-SiC substrates were at 500°C and radio-frequency biased which induced a DC offset of -60 V on the substrates [4]. The array of RuO$_2$ contacts with film thickness of 400 nm and lateral sizes of 0.8 mm^2 was formed via metallic contact masks. The electric conductivity of RuO$_2$ film was 60 μΩcm as measured by the four probe method.

The RuWO$_x$ Schottky contacts were prepared by the same manner with two differences: the presence of negatively biased W wire in the vicinity with Ru target and a metallic mask was used to create a round shaped Schottky contact on the surface of SiC with an area of

0.64 mm^2. The W was added to RuO$_2$ with the aim to achieve a stable forming of a RuO$_2$ phase by the thin film growth.

After a detailed examination of the Schottky structures the samples were irradiated 50 hours by fast electrons with energy 9 MeV (absorbed dose of 50 gy) at Centre for Ionizing Radiation, Slovak Institute of Metrology.

The Schottky structures were examined before and after irradiation by fast electrons by several diagnostic techniques. Current-voltage (I-V) measurements in the temperature range of 12.4 to 450 K (using helium and nitrogen cryostats) and capacitance-voltage (C-V) measurements in the temperature range of 83 to 450 K (employing nitrogen cryostat) were first performed. The structures then were investigated by standard Deep Level Transient Spectroscopy (DLTS) methods. In this experiment, wide ranges of reverse voltage biases (from - 10 to -0.2 V), temperatures (from 85 to 450 K) and times of the filling bias pulses (from 0.5 to 50 ms) were employed as the initial measuring conditions.

3. Results and conclusion

The prepared structures were first examined by current-voltage (I-V) method, which confirmed symmetry of the I-V characteristic; the parameters of Schottky diode were defined. The typical ideality factor of investigated Schottky structures RuO$_2$/4H-SiC was $n \sim 1.28$, the current of saturation was $I_S \sim 10$ pA and the Schottky barrier height was $\Phi_b \sim 0.88$ eV. The typical ideality factor of investigated Schottky structures RuWO$_x$/4H-SiC was $n \sim 1.16$, the current of saturation was $I_S \sim 7$ pA and the Schottky barrier height was $\Phi_b \sim 0.7$ eV. After the irradiation the parameters of Schottky diode RuWO$_x$/4H-SiC changed as follows, the ideality factor is rather high $n \sim 1.433$, the current of saturation is noticeable higher $I_S \sim 0.1$ nA and the Schottky barrier height is $\Phi_b \sim 0{,}51$ eV. The radiation-induced decrease of the Schottky barrier height was observed from current-voltage measurements.

The samples were also submitted to high frequency capacitance-voltage measurements, to prove the thermal stability of the structure and a good Schottky contact. Another important parameters of the Schottky structure were defined (Fig. 1); the typical built-in voltage is $U_{bi} \sim 0.963$ V and the bulk concentration is $N_{bulk} \sim 6.26 \times 10^{15}$ cm^{-3} for Schottky structures RuO$_2$/4H-SiC and the built-in voltage $U_{bi} \sim 1.167$ V, the bulk concentration $N_{bulk} \sim 5.79 \times 10^{16}$ cm^{-3} for Schottky structures RuWO$_x$/4H-SiC. No performance degradation of C-V curves was observed after the irradiation by fast electrons.

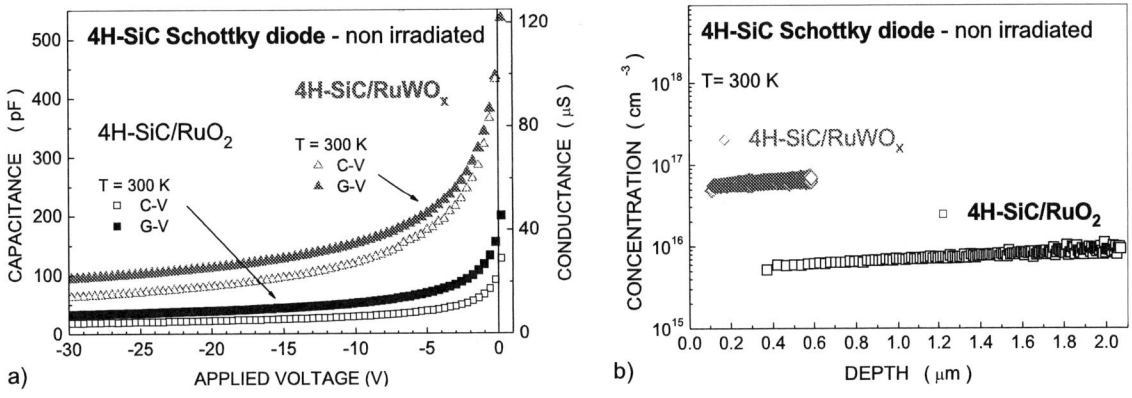

Fig. 1 a) Measured C-V characteristics and measured conductance *vs* voltage of RuO$_2$/4H-SiC and RuWO$_x$/4H-SiC Schottky diode. b) concentration profile for both type of structure

The measured DLTS spectra of both type of investigated Schottky structures RuO_2/4H-SiC and $RuWO_x$/4H-SiC before and after irradiation rather deviate from an exponential dependence which could be caused by the presence of several mutually influenced deep energy levels. The DLTS measurements were performed at the bottom edge of sensitivity (in order of capacitance fF) of the DLTS instrument. We expected DLTS signal much stronger (the order of capacitance up to pF), considering the abundant presence of intrinsic and extrinsic defect centres in SiC crystals (dislocations, micro- and nano-pipes, impurities, ...) which are very well known. The characteristic DLTS spectra of the RuO_2/4H-SiC and $RuWO_x$/4H-SiC Schottky barrier diodes after irradiation by fast electrons are shown in Fig. 2a, b. Five deep levels (ET1 – ET5) were identified in the silicon carbide bandgap from the characteristic DLTS spectra of the RuO_2/4H-SiC Schottky barrier diodes before irradiation and seven deep levels (ET6 – ET12) were identified these structures after irradiation by fast electrons. Three deep energy levels (E1 – E3) were identified from measured DLTS spectra for not irradiated the $RuWO_x$/4H-SiC Schottky barrier diodes and eleven deep levels (E4 – E13) from measured spectra on these structures after irradiation. Their activation energies ΔE_T and capture cross-sections σ_T calculated from an Arrhenius diagram and reference traps in literature are printed in Tab. 1.

Tab. 1. Summary of experimental data from DLTS measurements

Schottky Structure	Trap	ΔE_n (eV)	$\sigma_{n,p}$ (cm^2)	Ref. traps in literature	Schottky Structure	Trap	ΔE_n (eV)	$\sigma_{n,p}$ (cm^2)	Ref. traps in literature
RuO_2/ 4H-SiC	ET1	0.267	2.9×10^{-23}	-	$RuWO_x$/ 4H-SiC	E1	0.364	1.6×10^{-20}	P2 [6] (6H-SiC)
	ET2	0.452	1.8×10^{-19}	-		E2	0.380	1.2×10^{-20}	P2 [6] (6H-Sic)
	ET3	0.563	2×10^{-15}	Z_1 [7]		E3	0.694	8.3×10^{-19}	Z1/Z2 [7]
	ET4	0.581	6.6×10^{-23}	-	$RuWO_x$/ 4H-SiC irradiated	E4	0.124	1.6×10^{-21}	A-centrum [9]
	ET5	0.845	2×10^{-16}	$RD_{1,2}$ [7]		E5	0.133	2.1×10^{-21}	A-centrum [9]
RuO_2/ 4H-SiC irradiated	ET6	0.218	1.3×10^{-19}	V^{-2} [8]		E6	0.103	9.2×10^{-23}	A-centrum [9]
	ET7	0.294	1.8×10^{-17}	E1/E2 [10]		E7	0.132	2.1×10^{-22}	A-centrum [9]
	ET8	0.347	1.5×10^{-21}	E-0.34eV [11]		E8	0.172	1.1×10^{-21}	V^{-2} [8]
	ET9	0.379	3.7×10^{-21}	-		E9	0.336	1.4×10^{-18}	V^{-2} [8]
	ET10	0.788	7.9×10^{-16}	-		E10	0.547	9.4×10^{-18}	-
	ET11	0.627	1.2×10^{-18}	Z1[7],		E11	0.637	1.6×10^{-16}	-
	ET12	0.649	7.1×10^{-19}	E_c-0.64eV[5]		E12	0.569	8.4×10^{-19}	-
						E13	0.330	2×10^{-22}	-
						E14	0.789	2.2×10^{-17}	Z1[7]

We have, however, no straightforward explanation of these peaks. The deep energy level ET3 (0.56 eV) was detected in close vicinity of the known deep level Z1 with ionization energy 0.63 - 0.68 eV. In literature it is presented that level Z1 is not the only residual defect detectable in as-grown n-type layers, but also Z1 centre is acceptor-like and can be generated by either He+ or H+ implantation. It is therefore assumed that the Z1 defect is caused by the same intrinsic defect complex suggested that it may consist of a non-axial C-Si nearest-neighbor divacancy. The deep energy level ET5 (0.85 eV) was detected in close vicinity of deep level RD1/2 with activation energy 0.89 to 0.97 eV. It is an implantation-induced centre. It still has not been identified. The origin of three levels ET1 (0.27 eV), ET2 (0.45 eV) and ET4 (0.58 eV) are not known at this time. They could be induced by intrinsic material defects but therefore, further experimental investigations are needed to determine the origins of these traps precisely.

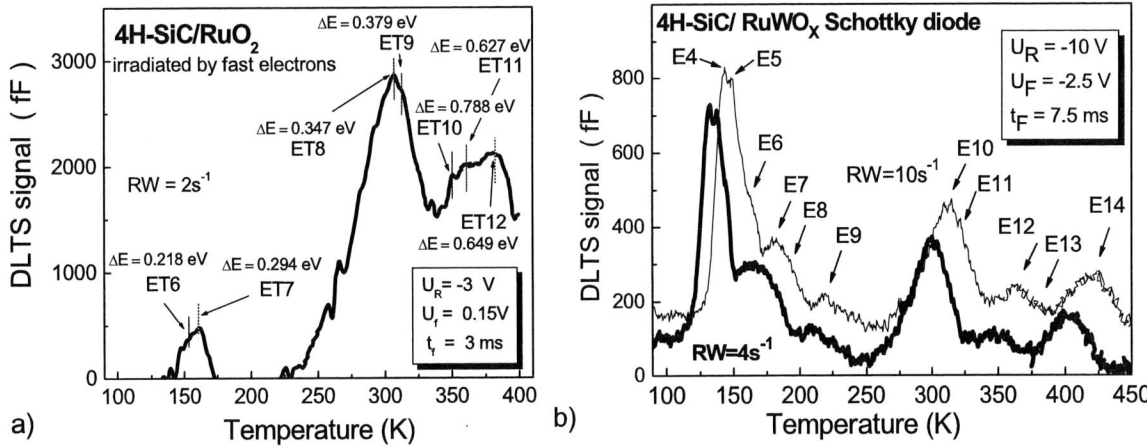

Fig. 2 The best characteristic measured DLTS spectra of the a) RuO_2 /4H-SiC and
b) $RuWO_x$ /4H-SiC Schottky barrier diodes after irradiation by fast electrons

Deep levels E1 (0.364 eV) and E2 (0.38 eV) have similar parameters and it is probable, at least one of these levels correspond with deep level P2 identified in 6H-SiC substrate (0.362 eV). The origin of this deep level could be the presence of wolfram in the structure. The deep level E3 (0.694 eV) has similar parameters as the structural defect in SiC Z1/Z2. Levels E4 and E5 are detected in a close vicinity of temperature level 150 K. It is apparent this peak represents two influencing energy levels. It is very expectant, that these deep levels along with levels E3 – E4 correspond to a radiation defect marked as A-centre (0.08 - 0.18 eV) which refers to an active bond vacancy-oxygen in Si. The deep energy levels E8 and E9 as well as ET3 in RuWOx/4H–SiC Schottky structures were detected in close vicinity of divacancies V^{-2} in Si. The irradiation by fast electrons (interaction of electrons and semiconductor structure) causes temporary or permanent structural or material changes in this structure. New deep energy levels were created after the irradiation of the sample, as it is shown in Tab. 1. These deep energy levels are situated in the range of lower temperatures and we can assume that they correspond to the defects originated as result of irradiation. All measured data are still under investigation also with other available methods.

Acknowledgement

The authors are indebted to J. Dobrovodský of Centre for Ionizing Radiation, Slovak Institute of Metrology for irradiation by fast electrons of Schottky structures. This work was supported by grant of the Slovak Grant Agency Vega 1/3091/06, 1/3095/06 and a grant from the Research Grants Council of the Hong Kong Special Administrative Region, China [CityU 102003].

References

[1] C. Claeys, E. Simoen, *Radiation Effects in Advanced Semiconductor Materials and Devices, Springer*, New York, 2002

[2] J. Kim, et al., *Appl. Phys. Lett.* **84**, 2004, 371.

[3] Tuominen, M., Yakimova, R., Prieur, E., Ellison, A., Tuomi, T., Vehanen, A.,.: *Diamond Relat. Mater.* 6, 1997. p. 1272,

[4] Búc, D., Music, D., Helmersson, U.: *In: Proceedings of the ASDAM' 2000 Conference,* Smolenice, Slovakia, Piscataway: IEEE, 2000, pp. 465-468

[5] *Doyle, J. P., Aboelfotoh, M. O., Svensson, B. G., Schoner, A., Nor, N.:* Diamond and Related Materials 6, *1997, p. 1388*

[6] MITRA, S.: *Solid-State Electronics*, 47, Issue 2, 2003, p. 193

[7] *Dalibor, T., Pensl, G., Matsunami, H., Kimoto, T., Choyke, W. J., Shöner, A., Nördell, N.: Phys. Stat. Sol. (a) 162, 1997, p. 199*

[8] Watkins, G.D.: In: Deep Centers in Semiconductors (Si). Edited by S.T. Pantelides, Gordon and Breach Science Publishers, 1986, pp. 147-183

[9] Fukuoka, N., Yoneyama, M., Honda, M., Atobe, K.: Jpn. J. Appl. Phys., 32, 1993, no. 5A, pp. 2059-2062

[10] Irmscher, K.: *Materials Science and Engineering B*, 91-92, 2002, pp 358-366

[11] Davydov, D. V., Lebedev, A. A., Kozlovski, V. V., Savkina, N. S., Strelchuk, A. M.: *Physica B*, vol 308-310, 2001, pp.641

Rapid thermal annealing and performance of Al₂O₃/GaN metal-oxide-semiconductor structures

K. Čičo[1], J. Kuzmík[1,2], D. Gregušová[1], T. Lalinský[1], A. Georgakilas[3], D. Pogany[2] and K. Fröhlich[1]

[1]Institute of Electrical Engineering SAS, Dubravska cesta 9, 841 04 Bratislava, Slovakia
[2]Institute for Solid State Electronics TU Vienna, Floragasse 7, A-1040 Vienna, Austria
[3]Microelectronic Research Group, FORTH-IESL and University of Crete, Dep. of Physics, P.O. Box 1527, 711 10 Heraklion, Crete, Greece
e-mail: karol.cico@savba.sk

In this paper we investigate growth of Al₂O₃ and performance of Al₂O₃/GaN MOS structures using O₂, Ar and NH₃ pre-treatment of GaN surface. Rapid thermal annealing (RTA) after the growth is also tested. Current-voltage (I-V) and thermal activation energy measurements were used for characterization of MOS and reference Ni/GaN Schottky contact structures. From the I-V characteristics, reduction of the leakage current was observed in MOS structures compared with the Schottky contacts for all types of processing, from three to five orders of magnitude in the reverse bias. Barrier height at the semiconductor-insulator interface (Φ_{S-I}), which is responsible for the reduction of the forward bias leakage was extracted from the thermal I-V measurements. Values of Φ_{S-I} was found to be ~ 1. 6eV for O₂ and ~ 2.5 eV for Ar treated samples and these samples show substantial leakage reduction in both senses. On the other hand I-V performance of the MOS structures with the NH₃ pretreatment resemble Schottky contact diode-like characteristic.

1. Introduction

GaN is a promising semiconductor electronic material for applications in high-temperature, high-speed and high-power electronics due to its basic physical properties such as wide bandgap, high electron saturation velocity and high thermal conductivity. Recently, there has been a growing interest in exploring high electron mobility transistors (HEMTs) based on GaN for high-frequency high-power applications. However, conventional Schottky-barrier transistors suffer from the high gate leakage current and in particular the leakage in the forward direction can cause fast device degradation [1]. Reduction of the gate leakage current and long-term stability of transistors can be achieved by employing an insulated gate metal-oxide-semiconductor (MOS) technology [1-5]. Several dielectric materials of metal-oxide semiconductor (MOS)-type GaN HEMTs, such as SiO₂ [1, 2], ZrO₂ [3], Al₂O₃ [4] and Ga₂O₃ [5] have been reported however, analysis of the MOS functionality is rarely performed. In this paper, we investigate basic electrical properties of Al₂O₃/GaN MOS structures with various pretreatments by current-voltage (*I-V*) and thermal *I-V* measurements.

2. Experiment

For the present study, the MOS and the reference Schottky-based structures were fabricated using N-type GaN (doped to 2×10^{17} cm⁻³) grown on sapphire by MOCVD system. Prior to deposition of the ohmic contacts, the native oxide had been removed from GaN substrate with HCl/H₂O (1/10) solution. The ohmic contacts were fabricated using a Ti/Al/Ni/Au (20nm/100nm/40nm/50nm) system. In the next step, the ohmic contacts were annealed at 800 °C for 60s using rapid thermal annealing (RTA) in a nitrogen atmosphere. Before deposition of Al_2O_3 film, the samples were dipped into HCl/H₂O (1/2) solution for 20s. Several technological parameters were varied for the GaN MOS structure processing. We tested in-situ exposure either to O_2 for 5 min. at 500 °C, or ramp heating up to 600 °C in Ar atmosphere, or ramp heating up to 750°C in NH_3 atmosphere, respectively. Afterwards Al_2O_3 was grown by MOCVD technique at 600 °C using precursors of aluminum acetylacetonate dissolved in toluene. Thickness of the deposited film was 14 nm. We used either Ar or NH_3 as a carrier gas. NH_3 pretreatment was combined with NH_3 as a carrier gas only, while O_2 and Ar pretreatment was combined with Ar carrier gas. Post-deposition ex-situ RTA was performed in N_2 for 40 s at 400 °C before Ni/Au MOS metal was evaporated from the e-gun. As a reference we prepared also a Ni/Au Schottky barrier diode on identical wafer of GaN. Parameters of used pretreatmens are described in Tab. 1. *I-V* measurements of MOS and Schottky structures were performed using a Keithley 2400 and Keithley 238.

3. Results and Discussion

Fig. 1 shows *I-V* characteristics of MOS structures and of the reference Schottky contact. We measured break-down voltage of the Al_2O_3 films and the electrical strength was determined to be about 6.1 MV/cm for all Ar pre-treated samples (Fig. 1 c) and annealed O_2 pre-treated samples (Fig. 1 b). Electrical strength of non-annealed sample with O_2 pretreatment was 3.1 MV/cm. Value of the break down voltage in NH_3 pre-treated samples could not be determined for high values of leakage currents even before the break down.

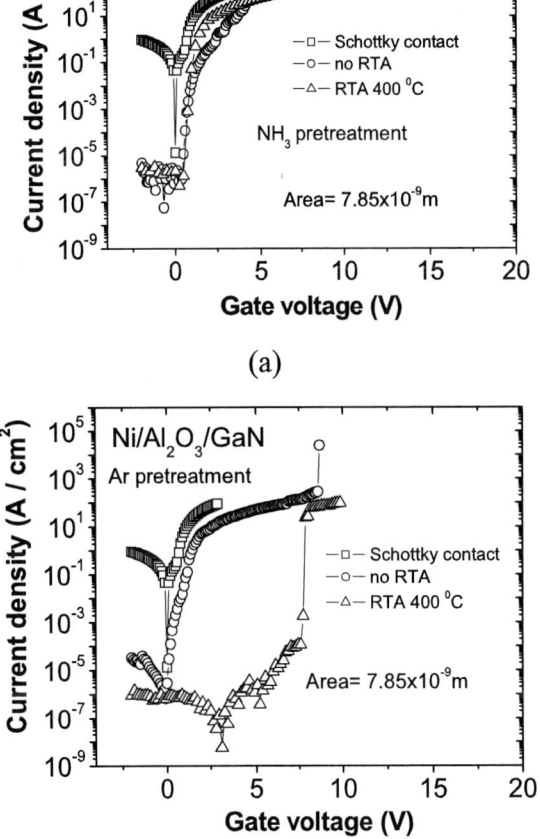

(a)

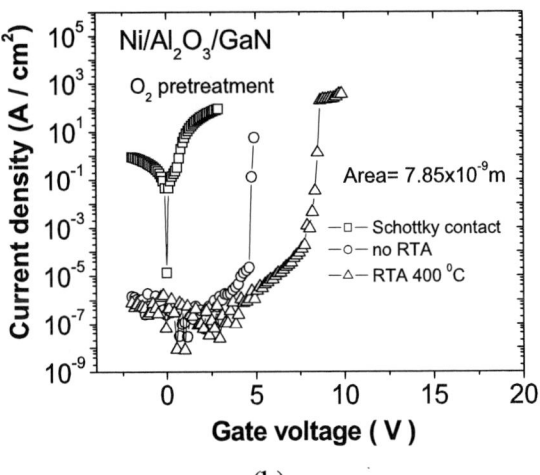

(b)

(c)

Fig. 1 Current -voltage characteristics of MOS structures and a reference Schottky structure for NH_3 (a), O_2 (b) and Ar (c) pre-treated samples

111

Tab. 1 Summary of pretreatment parameters

Pretreatment	Carrier gas	RTA in N_2 (°C)
NH_3 (ramp heating up 750 °C)	NH_3	
O_2 (5min.at 500 °C)	Ar	0, 400
Ar (ramp heating up 600 °C)		

In the reverse direction (negative gate bias), all characteristics of MOS structures show clear reduction of the gate leakage current (from three to five orders of magnitude) compared to the Schottky contact. Similarly in the forward direction (positive gate bias), all O_2-treated samples and annealed Ar-treated samples (Figs. 1b, c) show clear reduction in the leakage current, from seven to eight orders of magnitude compared to Schottky contact. Only NH_3-treated samples show diode-like characteristics, resembling the Schottky contact (Fig. 1a) performance.

We assume that the thermo-emission applies over the semiconductor- insulator (Φ_{S-I}) and the metal-insulator (Φ_{M-I}) barriers, see Fig. 2b, and consequently the barrier heights can be determined from the thermally dependent current $I(T)$ plots as:

$$\Phi_{S-I} = Ak/q \qquad (1)$$

where A is the slope of a linear dependence of $ln(J/T^2)$ on $1/T$.(activation energy plot), k is Boltzmann constant, q is elementary charge and J is the current density over the barrier. We measured thermally dependent $I(T)$ values in the reverse (at -1 V) and in the forward (at 1 V) directions to construct the activation energy plots. Results of all samples annealed at 400 °C with the positive gate bias is shown in Fig.2 a. Using equation (1) extracted barrier height Φ_{S-I} was found to be ~ 1.6 eV for O_2 and ~ 2.5 eV for Ar treated samples. However it seems that no semiconductor-insulator barrier Φ_{S-I}, which would reduce the forward leakage current is created if the NH_3 pretreatment is used. Alternatively, barrier height at the metal-insulator interface Φ_{M-I} (reverse direction) was determined to be ~ 3.5 eV. Values of the barrier heights are in good agreement with theoretical values [6]. We note that $I(T)$ values deviate from the ideal line (Fig. 2 a), what is probably caused by thermally-stimulated emissions of carriers from the interface states during the measurement.

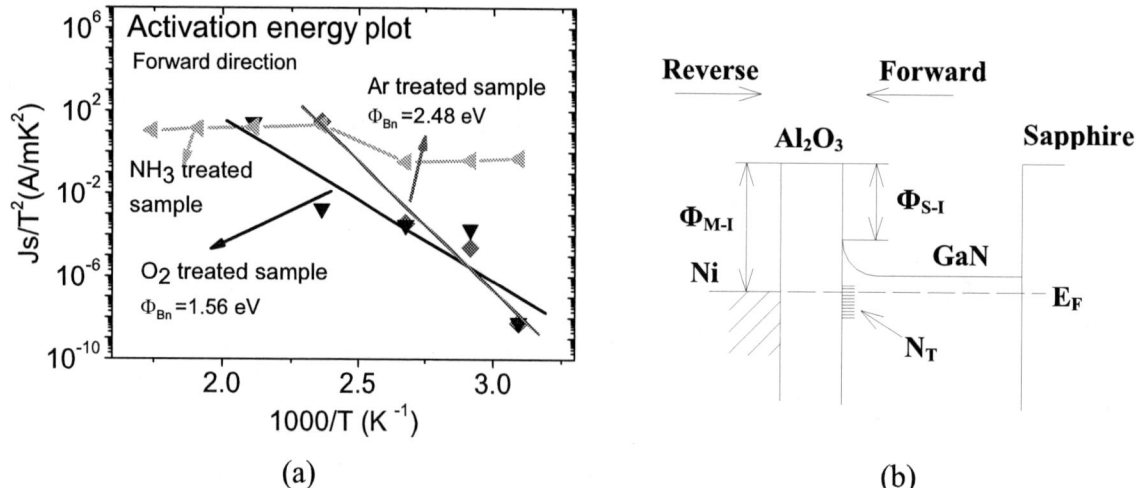

Fig. 2 a) Activation energy plot for the determination of the barrier height of samples annealed at 400 °C with a positive gate bias. b) Energy band diagram of the MOS structure.

Different values of the barrier height at the semiconductor-insulator junction can be explained by considering the theoretical expression for the barrier height value [6]:

$$\Phi_{S-I} = (\chi_S - \Phi^C_S) - (\chi_I - \Phi^C_I) + S(\Phi^C_S - \Phi^C_I) \qquad (2)$$

where χ is the electron affinity, Φ^C is the material charge neutrality level of the interface states (CNL) and S is a barrier pinning factor, $0 < S < 1$. It has been reported elsewhere [4] that GaN pretreatment before the insulator deposition may change the interface states distribution. The interface states subsequently change the CNL value at the GaN surface resulting in a different band alignment of the oxide to GaN.

4.   Conclusions

We have analyzed MOS structures on GaN with Al_2O_3 oxide film. Reduction of the leakage currents from three to five orders of magnitude was observed for all MOS structures compared to the Schottky contact in the reverse direction. From thermal *I-V* characteristics, the extracted barrier height at the oxide-semiconductor interface of samples annealed at 400 °C was found to be ~ 1.6eV for O_2 and ~ 2.5 eV for Ar treated samples. We have shown that pretreatment has an important role for the barrier height formation at the oxide-semiconductor interface because it influences interface states distribution and changes the materials band alignment and consequently determines the junction barrier height.

Acknowledgement

The work was supported by EU ULTRAGAN project, contract No. 6903. We acknowledge Aixtron A.G. for supplying GaN substrate and discussion with M. Ťapajna.

References

[1]   V. Adivarahan et al. *IEEE Electron Dev Lett*, **26**, 535, 2005.
[2]   J. Bernat et al. *Electronics Lett.* **41**, 667, 2005.
[3]   J. Kuzmik et al. *Semicond. Sci. Technol.*, **19**, 1364, 2004.
[4]   T. Hashizume et al. *J. Vac. Sci. Technol.* **B 21**, 1828, 2003
[5]   Ch-T. Lee et al. *Appl. Phys. Lett.*, **82**, 4304, 2003.
[6]   J. Robertson and B. Falabretti *J. Appl. Phys.*, **100**, 014111-1, 2006

Helium Irradiation for Advanced Lifetime Control in Silicon: New Recombination Centers and Their Interaction Stimulated by Isochronal Annealing

Volodymyr Komarnitskyy and Pavel Hazdra

Department of Microelectronics, Czech Technical University in Prague,
Technická 2, CZ-16627, Prague 6, Czech Republic
e-mail: komarv1@fel.cvut.cz and hazdra@fel.cvut.cz

The article presents results of systematic investigation on annealing of radiation defects introduced into n-type oxygen-rich float-zone silicon by single (7 MeV) and double energy (7 and 7.6 MeV) alpha-particle irradiation with fluences form 8.5×10^8 to 1×10^{12} cm^{-2}. Effect of isochronal anneal in the temperature range from 100 to 500°C on introduced defects and their interaction was studied by deep level transient spectroscopy. Shallow donor levels arising during annealing were investigated by C-V profiling. It is shown that formation of thermal donors (TDs) in the temperature range from 375 to 500°C is significantly enhanced by radiation damage (vacancy-oxygen clusters) produced by alpha-particle irradiation.

1. Introduction

Nowadays, irradiation of silicon with high energy swift ions is irreplaceable tool for arbitrary lifetime control in silicon power devices. The irradiation introduces recombination centers locally and turn-off of irradiated device speeds-up without significant deterioration of static parameters. It is usually assumed that irradiation with alpha-particles, in contrast with protons, has no negative doping effects on irradiated device. Application of alphas therefore allows to maintain high blocking capability of irradiated devices and increase their safe operation area. However, irradiation has to be completed by annealing to stabilize introduced defects and remove undesirable defects, generation centers, which increase device leakage [1]. The annealing leads to formation of new centers in silicon band gap and also activates thermal donors (TDs) [2] if annealing temperature exceeds 350°C. The aim of this paper is to show a systematic investigation of defect interaction in silicon irradiated with high energy alphas using isochronal annealing in range from 100 to 500°C. The main emphasis is put on properties and evolution of new recombination centers, as well as enhanced formation of thermal donors.

2. Experimental

Recombination centers were investigated on commercial 100A/1700V planar p$^+$nn$^+$ chip diodes. The diodes were fabricated from the low-doped <100>-oriented float zone n-type silicon using field ring technology. Long thermal diffusion, which was required for production of deep anode (p$^+$) and cathode (n$^+$) emitters, enhanced concentration of oxygen in the samples. Radiation damages were introduced by single- (7 MeV) and double-energy (7 and 7.6 MeV) alpha-particle irradiation. The fluences of alphas ranged from 8.5×10^8 to 1×10^{12} cm^{-2}. Isochronal (30 minutes) annealing in temperature range form 100 to 500°C was applied after irradiation. Deep levels resulted from irradiation and the subsequent isochronal annealing were monitored by deep level transient spectroscopy using DLS-82 E spectrometer. Profiles of shallow levels were characterized using HP 4280 1MHz capacitance analyzer at temperature 85°C to eliminate influence of ionized deep acceptors. High level lifetime was calculated using open circuit voltage decay measurement (OCVD). Leakage current of samples was monitored at 85°C and reverse voltage of 300V.

2. Results and Discussion

Deep levels. Spectra corresponding to radiation defects appearing after irradiation with 7 MeV alphas to a fluence of 8.5×10^8 cm^{-2} and their evolution during isochronal annealing up to 450°C are shown in Fig.1. DLTS spectrum of not annealed sample (n.a.) shows three major peaks that correspond to different deep levels in silicon bandgap which are attributed to the following defects: the vacancy-oxygen pair VO$^{(-/0)}$ at E_C-0.163eV (E1), the double charge state of divacancy $V_2^{(-/=)}$ at E_C-0.252eV (E2) and the single negative state of divacancy $V_2^{(-/0)}$ at E_C-0.436eV (E3) with a contribution the acceptor level of vacancy-phosphorous pair (VP$^{(-/0)}$). The spectrum also contains levels labeled as T1 and T2 which originate from diode fabrication since they were also detected in unirradiated samples. The dependence of the DLTS peak amplitude versus annealing temperature, which was received from spectra shown in Fig.1, is plotted in Fig.2.a. Figure shows that all deep centers are stable up to 150°C when VP pair (E–center) anneals out ($VP \leftrightarrow V+P$). The peak E3* at E_C-0.425 eV (see Fig.1) then can be attributed to pure divacancy $V_2^{(-/0)}$. During annealing of VP centers, part of free vacancies is being captured by interstitial oxygen and participates to formation of VO ($V+O \leftrightarrow VO$). This can explain slight reverse annealing of A-center (see Fig.2a). Annealing at 250°C shifts divacancy related peak E2 and E3* to A1 located at E_C-0.238 eV and A2 at E_C-0.454 eV. This shift is attributed to annealing of divacancy and formation of a new center V_2O with two charge states $V_2O^{(-/=)}$ and $V_2O^{(-/0)}$ lying close to $V_2^{(-/=)}$ and $V_2^{(-/0)}$, respectively [3]. The new center forms as a result of reaction of divacancy with interstitial oxygen ($V_2+O_i \rightarrow V_2O$). This reaction facilitates annealing out of divacancy in oxygen rich silicon. Fig.2a shows that annealing above 350°C removes both VO and V_2O centers.

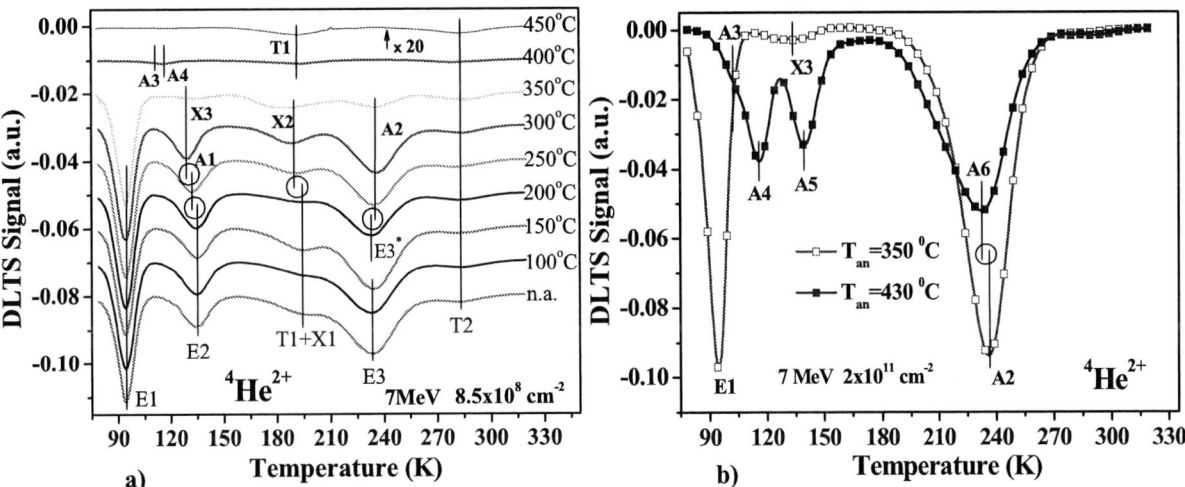

Fig. 1. Majority carrier DLTS spectra of diode irradiated with alphas to a fluence of 8.5×10^8 cm^{-2} measured after irradiation (n.a.) and subsequent isochronal annealing in range from 100 to 450°C (a). Two spectra recorded after annealing at 350 and 430°C for sample irradiated with alphas to fluence of 2×10^{11} cm^{-2} (b). Rate window of 260 s^1 was used in both cases.

115

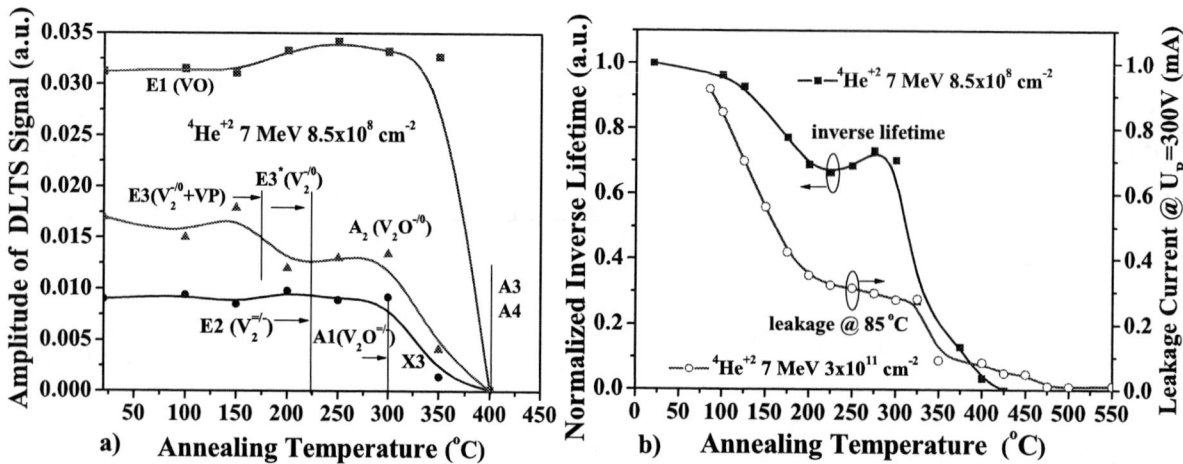

Fig. 2. Amplitude of DLTS signal of dominant deep levels after irradiation with 7 MeV alphas to fluence 8.5×10^8 cm^{-2} and consequent isochronal annealing (a). Proposal defect identification is shown in brackets. Normalized inverse lifetime for sample irradiated to fluence of 8.5×10^8 cm^{-2} and leakage current for samples irradiated to fluence 3×10^{11} cm^{-2} (b).

At higher temperatures, new defects are formed as a result of dissociation or reaction of VO and V$_2$ centers. This is evidenced by appearing of deep levels A3 at E$_C$-0.181 eV and A4 at E$_C$-0.211 eV. According to DLTS measurement, all new centers are localized in the defect peak maximum. Evolution of radiation defects and formation of the new centers is strongly influenced by irradiation fluence. Fig.1.b shows new peaks labeled as A5 (E$_C$-0.248 eV) and A6 (E$_C$-0.44 eV) in DLTS spectrum of sample irradiated to a fluence of 2×10^{11} cm^{-2} and subsequently annealed at 350 and 430°C, respectively. Both levels A5 and A6 are probably related to complexes consisting of multi-vacancies and oxygen V$_n$ and VO$_m$. Influence of defect thermal stability on normalized inverse lifetime and generation current is shown in Fig.2b for samples irradiated with low- and high-fluence of alphas. Decreasing of leakage current in temperature range from 85 to 200°C and corresponding increase of the carrier lifetime can be explained by annealing of unstable VP pairs. Between 200 and 275°C, the inverse lifetime increases due to additional formation of VO centers. Since these centers have negligible influence on carrier generation, the increase of VO concentration does not affect diode leakage. In agreement with results of DLTS measurement, for samples irradiated with low fluences, lifetime recovers after annealing at 425°C. In samples irradiated with higher fluences of alphas, new centers (A5, A6) remain after annealing at 400°C and contribute to increased leakage of diodes. In this case, the recovery of the carrier lifetime is also prolonged.

Shallow levels. At higher annealing temperatures (> 350°C), majority of the vacancy related defects is annealed out if lower irradiation fluences are applied (5×10^9 cm^{-2}). At these temperatures, C-V profiling shows enhanced formation of shallow donors at the depth close to the projected range R$_P$ of alphas. One can conclude that the enhanced formation of thermal donors is caused by an easier transformation of O$_i$ configuration supported by vacancies from annealed vacancy-related defects. At higher radiation fluences, temperatures exceeding 400°C are necessary to anneal more stable vacancy-oxygen complexes and to stimulate TDs formation. Since for higher fluences of alphas, the distribution of vacancy-related defects is shifted closer to the irradiated surface, the TDs distribution develops with temperature in a different way. This is depicted in Fig.3 where evolution of donor doping with temperature is compared for irradiation with 7 MeV (a) and for double irradiation with 7 and 7.6 MeV alphas

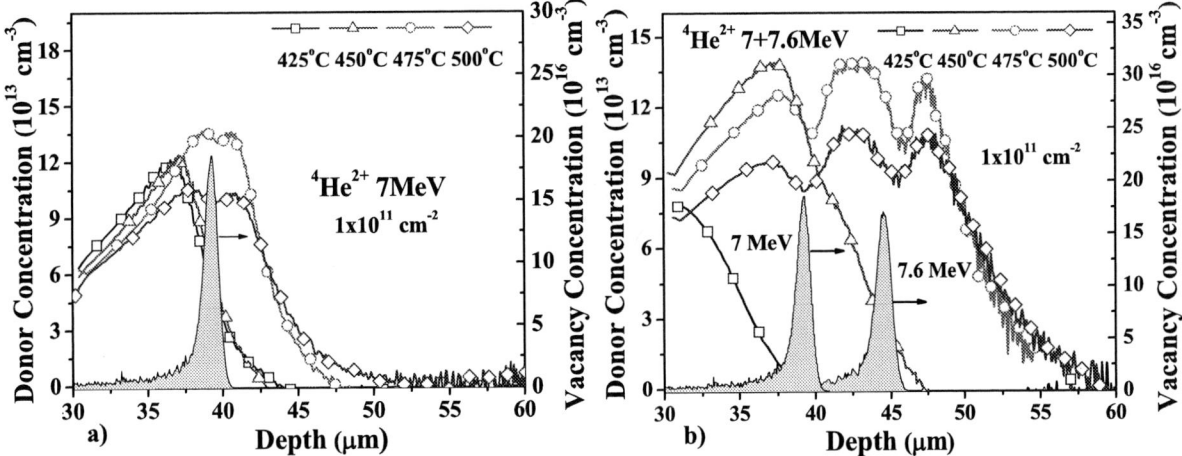

Fig. 3. Evolution of excess donor doping with annealing temperature in FZ silicon irradiated with single 7 MeV (a) and double 7 and 7.6 MeV (b) energy of alphas to fluence of 1×10^{11} cm^{-2}. Distributions of primary vacancies for 7 and 7.6 MeV irradiation are also shown.

(b) to fluence of 1×10^{11} cm^{-2}. Higher annealing temperatures spread-out distribution of TDs well behind projected range of projectiles and make it more complex compared with single irradiation. It is clear from the Fig.3a and Fig.3b that TDs achieved their maximum concentration 1.3×10^{14} cm^{-3} at 475°C. This TDs saturation level was observed in all samples irradiated with high fluences of alphas and a finite concentration O_i is found to be a limiting factor for TDs formation.

Conclusions

New recombination centers were investigated after isochronal annealing of FZ silicon irradiated with alpha-particles (7 and 7.6 MeV) in wide range of fluences. It is shown that annealing leads to formation of variety of new centers with energetic levels in silicon band gap. Their appearing is attributed to interaction of radiation defects created by alphas with intrinsic defects originated from FZ silicon. Results show that radiation damage (increased formation of vacancy-related defects) stimulates thermal donor formation during annealing at temperatures from 375 to 500°C. The evolution of the donors depends on alpha particle's fluence and temperature of subsequent annealing. The maximum enhancement of TD's formation is observed at 475°C. For higher fluences, the TD formation at the defect peak saturates due to the limited amount of interstitial oxygen.

Acknowledgement

This work was supported by the Research Programme MSM 6840770014 and the project of the Ministry of Education, Youth and Sports of the Czech Republic LC06041. Authors also acknowledge ABB Switzerland Ltd, Semiconductors for diode preparation and FZ Rossendorf for sample irradiation.

References

[1] P. Hazdra and V. Komarnitskyy, *Microel. Journ.*, V **37**, 3, 2006, pp. 197-203.

[2] E.P. Neustroev, I.V. Antonova, V.P. Popov, D.V. Kilanov, A. Misuk, *Physica,* **B 293**, 2000, 44-8.

[3] G. Alfieri, E.V. Monakhov, B.S. Avset, B.G. Svenson, *Phys. Rev,* **B68**, 2003, 233202.

Electrical and Memory Properties of Non-volatile Memory Structures with Embedded Si Nanocrystals

Zs. J. Horváth, P. Basa, T. Jászi, A. E. Pap, P. Szöllősi, K. Nagy, and V. Hardy

Hungarian Academy of Sciences,
Research Institute for Technical Physics and Materials Science,
Budapest 114, P. O. Box 49, H-1525 Hungary,
e-mail: horvzsj@mfa.kfki.hu, basa@mfa.kfki.hu, jaszi@mfa.kfki.hu, pap@mfa.kfki.hu,
szollosi@mfa.kfki.hu, nagyk@mfa.kfki.hu, and hardy@mfa.kfki.hu

Memory structures with an embedded sheet of separated Si nanocrystals were prepared by low pressure chemical vapour deposition using a Si_3N_4 control layer and a SiO_2 or a Si_3N_4 tunnel layer. It was obtained that the charging behaviour of structures with SiO_2 tunnel layer was better than with Si_3N_4 tunnel layer.

1. Introduction

Dielectric layers with embedded semiconductor nanocrystals have been widely studied recently in order to overcome difficulties of non-volatile memory devices connected with technology scale-down [1-6]. The main problem with scaling down of floating gate memory transistors is that the required thin tunnel oxide layers of 2-4 nm are not reliable enough [7]. In the case of the presence of any defect, the whole stored charge will leak through the defected area. However, if separated semiconductor nanocrystals (nc-s) are used as charge storage media, the stored information will not be lost.

But, in metal-nitride-oxide-silicon (MNOS) devices, which were the first realized memory structures, the charge is stored in traps located in the Si_3N_4 layer close to the Si_3N_4/SiO_2 interface. In these structures traps are isolated a'priori. However, formation of semiconductor nanocrystals in MNOS structures can enhance their charging and/or retention behaviour.

Both the tunnel and control layers in current memory arrays are usually SiO_2 layers. However, if using Si_3N_4, as a control layer, due to its higher dielectric constant, higher electric field will be developed in the tunnel oxide for the same layer thicknesses and voltage pulses, than for a SiO_2 control layer.

Usually formation of semiconductor nanocrystals requires a high temperature annealing process [1-6]. However, it was obtained recently that Si nc-s can be formed in Si-rich SiN_x layers by different CVD deposition techniques even without postdeposition annealing [8,9]. But, in this case Si nc-s are distributed along the whole layer, while memory structures require a thin nc layer just near the tunnel layer.

In this work our goal has been to create memory structures with a thin Si nc layer by low pressure chemical vapour deposition (LPCVD) using a SiO_2 or a Si_3N_4 tunnel layer and a Si_3N_4 control layer, i.e. to prepare MNOS and MNS (metal-nitride-oxide) structures with an embedded Si nc layer.

2. Experimental

For tunnel layer either a SiO_2 or a Si_3N_4 layer was prepared after cleaning the wafers in 1 wt% HF. SiO_2 layers were prepared using HNO_3 treatments [10,11]. n-type Si wafers were immersed in 68 wt% HNO_3 at the boiling temperature (121 °C) for 60 minutes. This method yielded a SiO_2 layer with a tickness of 2.5 nm, as obtained by cross-sectional transmission electron microscopy (XTEM) [12]. The Si_3N_4 tunnel layer, the Si nc layer and the Si_3N_4 control layer were deposited by LPCVD on n-type Si substrates at 830°C at a pressure of 30 mPa using SiH_2Cl_2 and NH_3. The Si_3N_4 layers were grown at gas flow rates of SiH_2Cl_2 and NH_3 of 21 and 90 sccm, respectively, while the Si nc layer with a gas flow rate of SiH_2Cl_2 of 100 sccm. The duration of deposition for the nc-Si layer was 30 s, which yielded average layer thicknesses of 2 nm in MNS structures (Si_3N_4 tunnel layer) [13]. (XTEM measurements of structures with SiO_2 tunnel layer are in progress.) Reference structures without Si nc layer were also prepared. The duration of deposition for the Si_3N_4 tunnel and control layers was 5 min and 10 min in the case of MNS structures, which yielded average layer thicknesses of 15 nm and 32 nm, respectively, as also obtained by XTEM measurements [13]. Si_3N_4 control layer for MNOS structures was grown during 15 min, which yielded a layer thickness of 37-40 nm, obtained by ellipsometry [14]. For electrical and memory measurements Al capacitors were formed with dimensions of 0.8 mm by 0.8 mm by evaporation. For the backside ohmic contact also Al was used after an appropriate chemical surface treatment using a mixture of H_2SO_4 and H_2O_2 [15].

Current-voltage (I-V), capacitance-voltage (C-V), and memory window measurements were performed at a frequency of 1 MHz at room temperature.

C-V measurements were carried out in dark using step like bias in the range of ±10 V with step of 0.2 V. The capacitance was measured in 1 s after changing the bias. Memory window measurements were performed using voltage pulses with an amplitude of ±20 V and width of 100 ms. To enhance the development of inversion layer, and so to avoid high voltage drop on the deep depletion layer during negative voltage pulses, the structures were illuminated with white light during these measurements.

3.   Results and discussion

The current flow through the structures at high current levels was dominated by the Poole-Frenkel mechanism, which can be expressed as [13]

$$J_{PF} = C_{PF1} \cdot E \cdot \exp\left(C_{PF2} \cdot \sqrt{E}\right) \tag{1}$$

The fit of this expression yielded parameters of $C_{PF1}= 8.0 \times 10^{-26}$ $AV^{-1}m^{-1}$ and $C_{PF2}=1.48 \times 10^{-3}$ $V^{-1/2}m^{1/2}$. However, an exponential excess current was also obtained at low current levels (see Fig. 2), which can be expressed as [13]

$$J = C_{EX1} \cdot \exp\left(C_{EX2} \cdot E\right) \tag{2}$$

with parameters of $C_{EX1}= 1.5 \times 10^{-7}$ Am^{-2} and $C_{EX2}=1.55 \times 10^{-8}$ Vm^{-1}.

Table 1 presents the initial flat-band voltage values obtained in the as-prepared structures. The initial flat-band voltage was higher for MNS structures, than for MNOS ones. It was also higher for structures with Si nc-s, than without Si nc-s. The actual values of the effective charge calculated for the Si nc layer are in the range of $(0.9-1.7) \times 10^{12}$ q/cm^2, which is mainly related to the fixed charge. These values are reasonable.

During C-V measurements a relatively high flat-band voltage shift was obtained for the structures, which is connected with charge injection during the measurement, i.e. with the memory effect. The obtained values are presented in Table 2. A higher shift (4.9 V) was obtained for MNOS structures with the chemical oxide, than for the MNS structures (3.8-4.4 V). The results indicate that the presence of the thin oxide layer makes easier the charge injection. This is probably connected with shorter tunneling distance in MNOS structures due to higher electric field through the oxide layer, and so higher potential drop on it. However, the presence of the Si nc layer affected the charging behaviour of MNS structures only.

Table 1: Initial flat-band voltage values obtained in the as-prepared structures

Tunnel layer	Without Si nc-s	With Si nc-s
Chemical SiO_2	-0.9±0.1 V	-1.3±0.1 V
LPCVD Si_3N_4	-1.3±0.1 V	-1.74 V

Table 2: Flat-band voltage shift obtained by capacitance-voltage measurements for bias range of ±10 V

Tunnel layer	Without Si nc-s	With Si nc-s
Chemical SiO_2	4.9±0.2 V	4.9±0.2 V
LPCVD Si_3N_4	4.4±0.2 V	3.8±0.2 V

Table3: Memory window width obtained by capacitance-voltage measurements for charging pulses with amplitude ±20 V and width of 100 ms

Tunnel layer	Without Si nc-s	With Si nc-s
Chemical SiO_2	10.9±0.2 V	10.7±0.2 V
LPCVD Si_3N_4	10.1±0.2 V	9.2±0.2 V

Memory window measurements yielded similar results, as C-V measurements. They also indicated easier charge injection in MNOS structures, than in MNS structures. The actual values of memory window width for charging pulses with amplitude ±20 V and width of 100 ms are presented in Table 3. (During memory window measurements charging pulses are applied to the sample with alternating sign. The memory window width is the difference between the flat-band voltage values after positive and negative charging pulses.)

4. Conclusion

Memory structures with an embedded sheet of separated Si nanocrystals were prepared by low pressure chemical vapour deposition using a Si_3N_4 control layer and a SiO_2 or a Si_3N_4 tunnel layer. It was obtained that the charging behaviour of structures with SiO_2 tunnel layer was better than with Si_3N_4 tunnel layer. This is probably connected with shorter tunneling distance in MNOS structures due to higher electric field through the oxide layer, and so higher potential drop on it.

Acknowledgement

This work has been partially supported by the European Commission through project called SEMINANO under the contract NMP4-CT-2004-505285, and by the Hungarian Scientific Research Fund under Grant No. T048696.

References

[1] G. Molas, B. De Salvo, D. Mariolle, G. Ghibaudo, A. Toffoli, N. Buffet, and S. Deleonibus, *Solid-State Electron.*, **47**, 1645, 2003.

[2] P. Normand, E. Kapetanakis, P. Dimitrakis, D. Skarlatos, K. Beltsios, D. Tsoukalas, C. Bonafos, G. Ben Assayag, N. Cherkashin, A. Claverie, J.A. Van Den Berg, V. Soncini, A. Agarwal, M. Ameen, M. Perego, and M. Fanciulli, *Nucl. Instr. and Meth. B*, **216**, 228, 2004, and references therein.

[3] S. Lombardo, D. Corso, I. Crupi, C. Gerardi, G. Ammendola, M. Melanotte, B. De Salvo, and L. Perniola, *Microel. Eng.*, **72**, 411, 2004.

[4] A. G. Nassiopoulou, A. Salonidou, A. Olzierski, M. Kokonou, E. Tsoi, P. Normand, and K. Giannakopoulos, in *Semiconductor Nanocrystals; Proc. First Int. Workshop on Semiconductor Nanocrystals SEMINANO2005*, Budapest, Hungary, p.405; http://www.mfa.kfki.hu/conferences/seminano2005/

[5] Zs. J. Horváth, *Current Appl. Phys.*, **6**, 145, 2006, and references therein.

[6] C. Y. Ng, T. P. Chen, D. Sreeduth, Q. Chen, L. Ding, and A. Du, *Thin Solid Films*, **504**, 25, 2006.

[7] P. Cappelletti, *Microelectron. Reliab.*, **38**, 185, 1998.

[8] R. A. Rao, R. F. Steimle, M. Sadd, C. T. Swift, B. Hradsky, S. Straub, T. Merchant, M. Stoker, S. G. H. Anderson, M. Rossow, J. Yater, B. Acred, K. Harber,E. J. Prinz, B. E. White Jr., and R. Muralidhar, *Solid-State Electron.*, **48**, 14633, 2004.

[9] K. S. Cho, N.-M. Park, T.-Y. Kim, K.-H. Kim, G. Y. Sung, and J. H. Shin, *Appl. Phys. Lett.*, **86**, 071909, 2005.

[10] H. Kobayashi, Ashua, O. Maida, M. Takahashi, and H. Iwasa, *J. Appl. Phys.*, **94**, 7328, 2003.

[11] S. Imai, M. Takahashi, K. Matsuba, Ashua, Y. Ishikawa, and H. Kobayashi, *Acta Phys. Slovaca*, **55**, 305, 2005.

[12] L. Dobos, B. Pécz, and L. Tóth, unpublished.

[13] P. Basa, Zs. J. Horváth, T. Jászi, A. E. Pap, L. Dobos, B. Pécz, L. Tóth, and P. Szöllősi, *Physica E*, in press.

[14] P. Petrik, T. Lohner, and M. Fried, unpublished.

[15] Zs. J. Horváth, M. Ádám, I. Szabó, M. Serényi, and Vo Van Tuyen, *Appl. Surf. Sci.*, **190**, 441, 2002.

Leakage current and Physical properties of Tantalum oxide thin films for Micro capacitor integration

Insung- Kim, Jaesung-Song and Bokki-Min

Electric and Magnetic Devices Research Group, KERI,
Sungju-Dong 28-1, South Korea 641-120
e-mail: kimis@keri.re.kr

Tantalum oxide (Ta_2O_5) were grown on the Ti and Ti-O_2 layers as a oxide barrier using RF-sputtering method. Measurement of physical and dielectric properties of the reactive sputtered Ta_2O_5 as two forms of simple MOS structure, states that the amorphous Ta_2O_5 grown on Ti gave high dielectric constant ($\varepsilon_r=30\sim70$) and high leakage current ($10^{-1}\sim10^{-4}$ A/cm^2), whereas relatively low dielectric constant (~10) and low leakage current ($\sim10^{-10}$ A/cm^2) were observed in the amorphous Ta_2O_5 deposited on the Ti-O_2. As a result, the Ta_2O_5/Ti capacitor exhibits three dominant conduction mechanism regimes contributed by the ohmic emission at low electrical field, by the Schottky emission at intermediate field and by the Poole-Frenkel emission at high field. In the case of Ta_2O_5/TiO_2 capacitor, the two conduction mechanisms, the Ohmic and Schottky emissions, governs the leakage current density behavior.

1. Introduction

The capacitance and leakage current of the simple MOS capacitor structure can be lowered to the desired values by diffusion loss of the other dense oxide layer with low dielectric constant between Ta_2O_5 and Si or electrode. In this study, we investigated the electrical properties of Ta_2O_5 MOS capacitor grown onto metal layer Ti and oxide layer TiO_2. As an oxide buffer layer, TiO_2 was chosen in that it possesses low crystallization temperature, relatively low dielectric constant and dense crystal structure. Use of TiO_2 as a leakage current barrier was successful and gave an insight to the capacitor structure that meets the requirements of low capacitance and leakage current density. Ti was used as a buffer layer to avoid oxide SiO_2 formable between Ta_2O_5 and Si wafer. The rapid annealing effect, capacitance-voltage and leakage current density-electric field characteristics of the capacitors were studied.

2. Experimental procedure

The Ta_2O_5 films were deposited in an O_2 atmosphere. Two substrates were employed in this experiment; one is Ti 30 nm coated on Si wafer and the other TiO_2 deposited on Si. Titanium was chosen to improve adhesion and to avoid the possibility of formation of new oxide layer. TiO_2 was employed as a barrier of oxygen diffusion from the Ta_2O_5 to the Si wafer. The Pt bottom electrode beneath the Si wafer was deposited at 100W and 10^{-3}torr. The Ta_2O_5 films were deposited on the Ti and Ti-O_2 coated Si substrates by a RF reactive sputtering. The sputtering gas was 80% Ar and 20% O_2 mixture with a pressure of 10^{-3}torr. Film thickness was estimated to be 50nm using a surface profile meter. The Au top electrode with a thickness of 200nm and a diameter of 300 μm were patterned by a shadow mask process. The RTA treatments (a heating rate of 30°C/min) of capacitor structures Ta_2O_5/Ti/Si

1-4244-0396-0/06/$25.00 ©2006 IEEE   122

and $Ta_2O_5/Ti-O_2/Si$ were carried out at 600, 650, 700, 750 °C on a vacuum of $< 5\times10^{-7}$ torr to reduce the defect of Ta_2O_5, prior to the formation of the electrodes. The current-voltage (I-V) characteristics of the as-deposited and annealed Ta_2O_5 films were measured on the MOS structure with a impedance analyzer. The capacitance-voltage (C-V) characteristic was read at frequency ranging from 100 Hz to 100 MHz.

3. Results and Discussion

To observe the effect of the RTA on the electrical characteristics, the dielectric constant of as-deposited and annealed Ta_2O_5/Ti and Ta_2O_5/TiO_2 MOS structures was measured as a function of annealing temperature as shown in the Fig. 1. Ti-TiO $_2$ (buffer layer) and Ta_2O_5 (dielectric layer) thickness were in the rage of 35~50 nm and 180~220nm respectively.

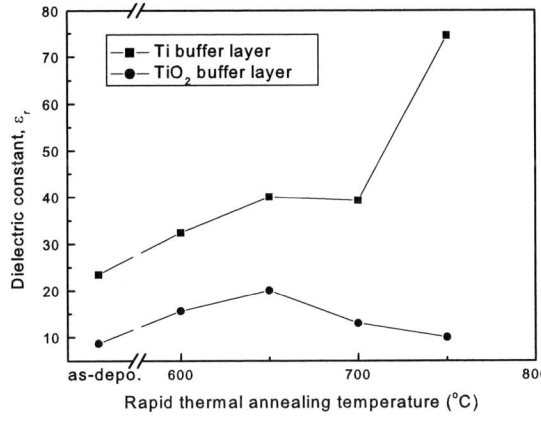
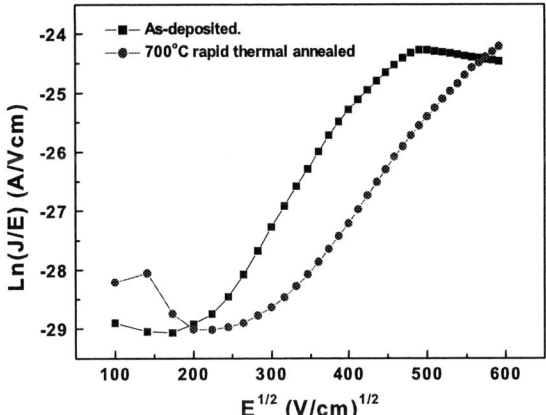

Fig. 1. Dielectric constants of Ta_2O_5 thin films deposited on Si wafer as a function of buffer layer and RTA

Fig 4. J-E characteristics of Ta_2O_5 thin films deposited on Si wafer as a function of RTA

In the Ta_2O_5/Ti structure, the dielectric constant ε_r increases from 25 to 75 with increasing annealing temperature T_a, whereas 650°C annealed Ta_2O_5/TiO_2 films has the maximum (23) in relative dielectric constant ε_r and other temperature annealed TiO_2 capacitor structures exhibit low ε_r (~10). The 70 of a dielectric constant for 700 ℃ annealed films may seems unrealistic. However, dielectric constant of >35 have also been reported in the literature [1].

The high frequency (100 kHz) capacitance-applied voltage characteristics of the as-deposited and 700°C annealed amorphous Ta_2O_5/Ti and Ta_2O_5/TiO_2 MOS capacitors were measured. For both as-deposited and 700 °C annealed Ta_2O_5/Ti films (Fig. 2 (a)), the relatively large residual capacitances were seen at even applied bias voltages up to 2.5 V. As seen in the Fig. 2 (b), the Ta_2O_5/ TiO_2 films showed the distinct flat band bias voltages 1 V at 700 °C RTA and 1.3 V at as-deposited condition. The relatively small capacitance of the annealed Ta_2O_5/TiO_2 is probably due to a low dielectric constant of compositionally gradient Ta-Ti-O layer forming in the MOS structure during high temperature (700 °C) annealing[2]. The different flat band voltage observations of the two capacitors indicate that Ta_2O_5/TiO_2 structure is a preferred one in that capacitor must lose easily the capacitance when given switching bias voltage. Keeping the small cut-off voltage for the Ta_2O_5/TiO_2 capacitor is probably caused by forming Ta-Ti-O inter layer as well as reducing broken bond between Ta and O which diffuse through the new oxide interface from TiO_2 to Ta_2O_5 layers. The leakage current in the Ta_2O_5 films can be explained by the several conduction mechanisms[3] including Ohmic emission, Schottky emission, Poole-Frenkel emission, Fowler-Nordheim tunneling and a space charge limited current. Among them, the current density due to the Ohmic emission[4] is expressed as a equation of J, E, T^2 exp. where, J denotes the current density, T the absolute temperature, k the Boltzmann constant, q the electronic charge, E the electric field, Q_B the barrier height to conduction quantity and ε_r the dielectric constant of the insulator material.

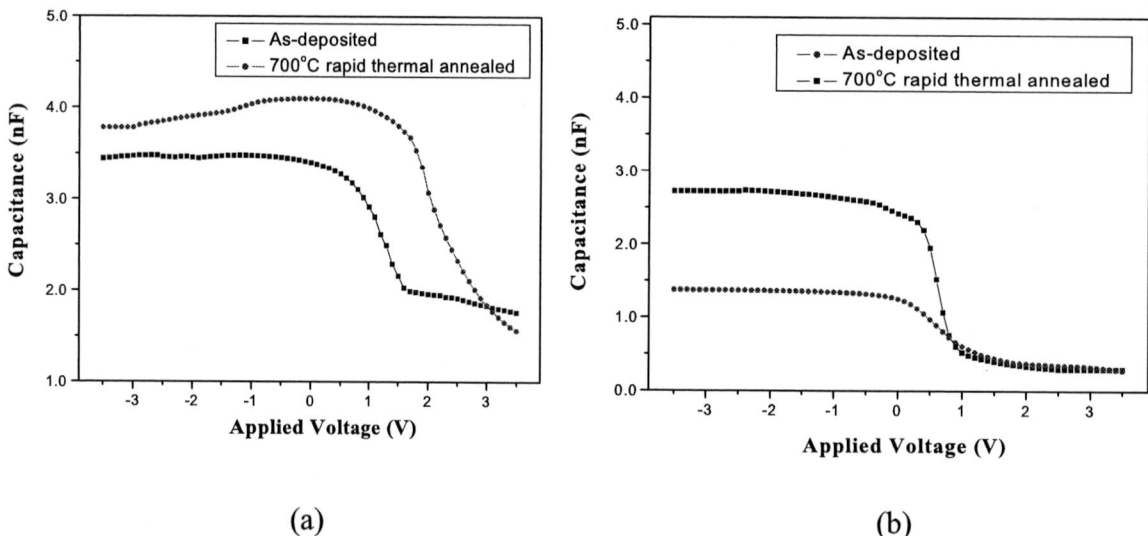

(a)   (b)

Fig 2. Capacitance-applied voltage of as deposited and 700 °C rapid annealed Ta_2O_5 thin films deposited on Si wafer as a function of buffer layer (a) Ti and (b) TiO_2.

The current densities of Ta_2O_5/Ti and Ta_2O_5/TiO_2 films are in the order of 10^{-3} and 10^{-8} A/cm^2 at an applied electric field of 2×10^5 V/cm, as shown in the Fig. 3. In the as-deposited and 700°C annealed Ta_2O_5/Ti, the current density increases linearly with increasing electric fields up to 3.6×10^5 V/cm for RTA and 5.2×10^5 V/cm for as-deposited condition. Then, the Ta_2O_5/Ti film exhibited dramatic increase of leakage current density when imposing the higher electric field. In the case of Ta_2O_5/ TiO_2 films as shown in the figure 3 (b), in the range of low electric field up to 1×10^5 V/cm, current density increases exponentially with applied electric field. In the higher electric fields than 1×10^5 V/cm, current density was observed to increase linearly with electric field for as-deposited and 700°C RTA conditions. This current density behaviours as a function of electric field was different from those of Ta_2O_5/Ti films. The dominant leakage current mechanisms of the two MOS capacitors were determined by plotting the logarithmic current density as a function of electric field. As shown in the Fig. 4, at low field area ($E < 1 \times 10^5$ V/cm) leakage current density increases as a form of electric field times exponential of the field, which represents Ohmic emission mechanism. In the range of intermediate field (10^5 V/cm $< E < 3.6 \times 10^5$ V/cm) a straight line of logarithmic current density to the square root of electric field ($\log(J)$ vs $E^{1/2}$) can be obtained.

The dielectric constant determined from the two slopes ($\varepsilon_r = 28$ for Ta_2O_5/Ti and 8 for Ta_2O_5/TiO_2 at 700°C RTA) are consistent with those (30 for Ta_2O_5/Ti and 10 for Ta_2O_5/TiO_2) observed in the Fig. 1.

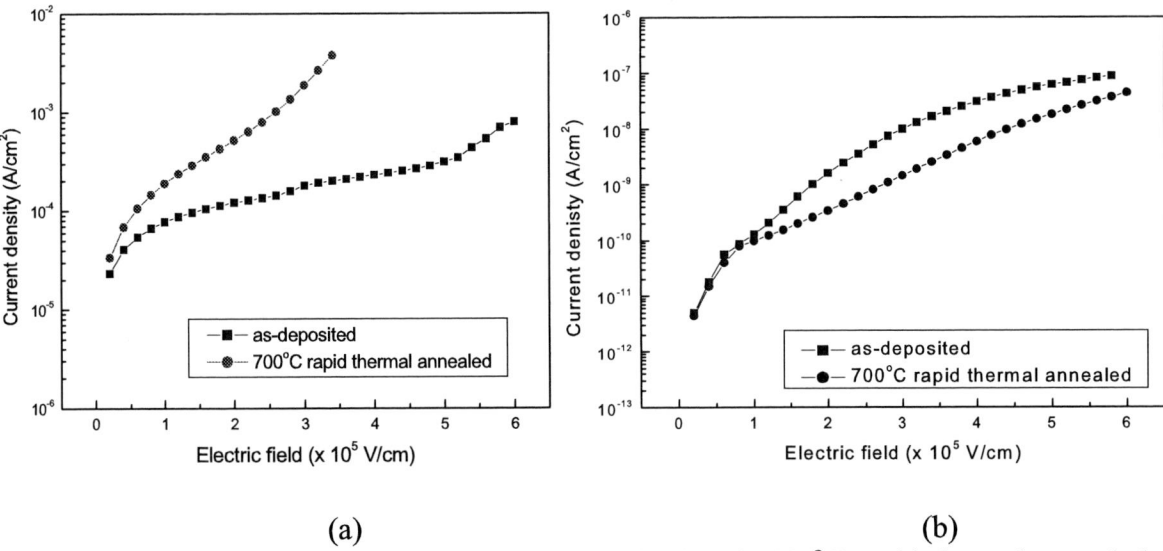

(a)   (b)

Fig 3. Current density vs electric field of as-deposited and 700 °C rapid thermal annealed Ta_2O_5 thin films deposited on Si wafer as a function of buffer layer (a) Ti and (b) TiO_2.

For the Ta_2O_5/TiO_2 films, the current density increase as a function of applied field was found to continuously obtain at given high electric field range ($E > 3.6 \times 10^5$ V/cm) as compared with that in the intermediate filed. By contrast, when imposing the high electric field for the Ta_2O_5/Ti capacitor, the logarithmic current density divided by the electric field has a linear relation as a function of electric field. Therefore, the conduction mechanism in the high field range for Ta_2O_5/Ti was determined to be the Poole-Frenkel (PF) emission [5, 6].

4. Conclusions

Measurement of electrical characteristics of the $Au/Ta_2O_5/Ti/Si/Pt$ and $Au/Ta_2O_5/Ti-O_2/Si/Pt$ capacitor structures indicates that the amorphous Ta_2O_5/Ti shows high dielectric constant (30~70) and high leakage current (10^{-1}~10^{-4} A/cm^2). In the amorphous Ta_2O_5 deposited on the $Ti-O_2$, a relatively low dielectric constant (~10) and low leakage current (~10^{-10} A/cm^2) were observed. The different electrical behaviours for the two Ta_2O_5 are caused by the contribution of $Ti-O_2$ as the other capacitor layer and an oxygen donor. The Ta_2O_5/Ti capacitor exhibits three dominant conduction mechanism regimes contributed by the Ohmic emission at low electrical field, by the Schottky emission at intermediate field and by the Poole-Frenkel emission at high field. In the case of Ta_2O_5/TiO_2 capacitor, the two conduction mechanisms, the Ohmic and Schottky emissions, explained the leakage current density behaviour.

References

[1] X. Wu, S. R. Ross, E. J. Rymaszewski , and T. M. Lu, Mater. Chem. Phys., 38, (1994)
[2] H. Shinriki, M. Nakata and K. Mukai: IEEE Electron Device Lett. **10** (1989) 514.
[3] C. Isobe and M. Saitoh: Appl. Phys. Lett. **56** (1990) 907.
[4] X.M. Wu, S.R. Soss, E.J. Rymaszewski and T.M. Lu: Mater. Chem. Phys. **38** (1994) 297.
[5] S. Ezhilvalavan, M. S. Tsai and T.Y. Tseng: J. Phys. D: Appl. Phys. 33 (2000) 1137.
[6] IS Kim, SJ Jeong, JS Song and PS Shin, "Dielectric Properties of Ta_2O_5 Thin Films Deposited on Buffer Layer" Metal & Materials International, Vo., 8, No 6, (2002)

Leakage characteristics of advanced MOS capacitors with hafnium silicate dielectric and Ru electrode

M. Ťapajna[a,b], K. Hušeková[a], K. Fröhlich[a], E. Dobročka[a], and F. Roozeboom[c]

[a] Institute of Electrical Engineering, Centre of Excellence CENG, SAS, Dúbravská cesta 9, 841 04 Bratislava, Slovakia
[b] Faculty of Electrical Engineering and Information Technology, STU, Ilkovičova 3, 812 19 Bratislava, Slovakia
[c] Philips Research Eindhoven, WAG 02, High Tech Campus 4, 5656 AE Eindhoven, The Netherlands
e-mail: Milan.tapajna@savba.sk

We have studied the leakage characteristics of $Ru/Hf_xSi_{1-x}O_y/Si$ MOS capacitors projected for advanced CMOS gate technology. Prior to Ru gate electrode deposition, the gate dielectrics were annealed by RTA in the temperature range of 700 – 1000 °C in O_2. The influence of RTA has been analyzed by X-ray diffraction, X-ray reflectivity and capacitance-voltage techniques. RTA at temperatures ranging from 800 – 900 °C improves the leakage characteristics. Presumably, the Pool-Frenkel mechanism controls the current flow through the $Hf_xSi_{1-x}O_y$ gate oxide with Ru electrode.

1. Introduction

The conventional polycrystalline Si/SiON gate stack used in high-performance CMOS technology reaches its physical limitations due to polycrystalline Si (poly-Si) depletion effect and the high tunnelling current through SiON dielectric for equivalent SiO_2 thicknesses (EOT) below 1.5 nm [1]. Therefore, the scientific community makes an effort to search the appropriate replacement of poly-Si by a dual-metal gate and SiON by dielectrics with high permittivity (high-κ). In this work, we have studied the current leakage mechanisms in advanced $Ru/Hf_xSi_{1-x}O_y/Si$ MOS gate structures. It was pointed out that the leakage current in thin $Hf_xSi_{1-x}O_y$ dielectrics decreased dramatically after rapid thermal annealing (RTA) in oxygen at 900 °C, which was attributed to the removal of the carbon impurities from the film and improvement of the interfaces [2]. However, a more detailed study on the defects evolution in the $Hf_xSi_{1-x}O_y$ upon RTA in O_2 is still missing. We focused our attention on the influence of the RTA temperature on the leakage characteristics of $Ru/Hf_xSi_{1-x}O_y/Si$ MOS capacitors with slanted dielectric (dielectric with varying thickness). As the leakage characteristics are extremely sensitive to the structural properties of the gate stacks and band alignment of the MOS structure, we have also employed the X-ray diffraction (XRD), X-ray reflectivity (XRR), and capacitance-voltage, (C-V), techniques in our investigation.

2. Experimental

The $Hf_xSi_{1-x}O_y$ dielectrics with thickness of 9 nm were prepared by metal-organic chemical vapor deposition (MOCVD) by a dual source approach by Aixtron. Part of the dielectrics was gradually etched in an HF-based solution in order to prepare the slanted dielectric. The dielectrics underwent RTA in a temperature range from 700 to 1000 °C in oxygen for 10 s. Afterwards, Ru thin films were deposited also using MOCVD at 350 °C on both, slanted dielectrics as well as dielectrics with constant thickness (in following referred as

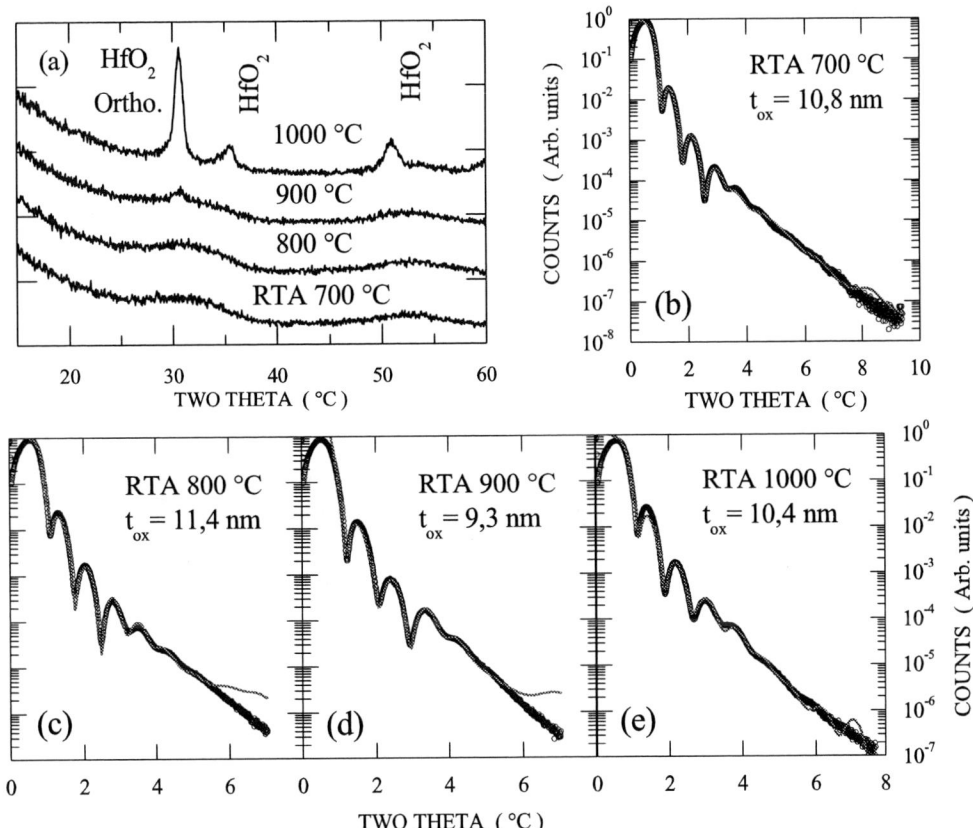

Fig. 1 (a) XRD pattern of uniform $Hf_xSi_{1-x}O_y$ thin layers after RTA performed at various temperatures. Measured (symbols) and fitted (grey line) XRR patterns taken from uniform $Hf_xSi_{1-x}O_y$ thin layers after each RTA temperature together with extracted layer thickness.

uniform dielectrics). MOS capacitors were manufactured by standard optical lithography and Ar ion milling. To form the bottom contact, Al was sputtered on the back-side of the samples. Finely, samples received a forming gas anneal (FGA, H_2+N_2) at 430 °C for 30 min.

XRD was taken in grazing incidence mode (incidence angle of 1.5) on Bruker AXS – D8 Discovery equipment. Cu Kα radiation generated by an X-ray tube with rotating anode operating at 12 kW was used. All measurements were performed in parallel-beam geometry with Goebel mirror in the primary beam. The thickness of the films was extracted from the XRR measurements. Current-voltage, (*I-V*), and *C-V* measurements were carried out using a Keithley 6517A Electromer and an Agilent 4284A LCR-meter, respectively. *C-V* curves were taken at 1 MHz and 100 kHz. The data were corrected by two-frequency correction [3] and analysed using NCSU CV model [4].

3. Results and Discussion

RTA can have a strong influence on the structural parameters of the thin dielectric layers and, thus, the leakage characteristics. Therefore, we have measured the XRD pattern after each annealing [Fig. 1 (a)]. The layers are amorphous or nanocrystalline up to 900 °C. Phase separation appears after RTA performed at 1000 °C. This is clearly visible in the XRD pattern by developing of HfO_2 peaks.

The dielectric thickness was measured by XRR and the measured as well as simulated spectra are depicted in Fig. 1 (b-d). Reasonable fits were obtained after employing the model

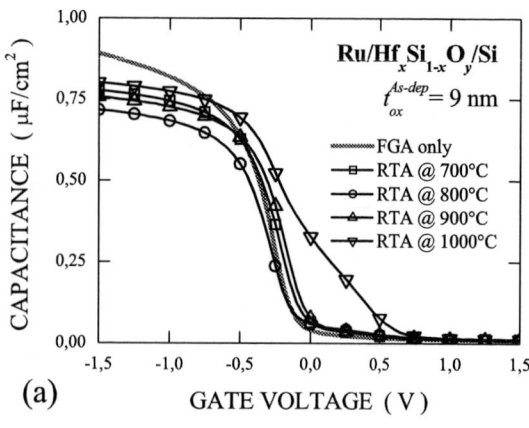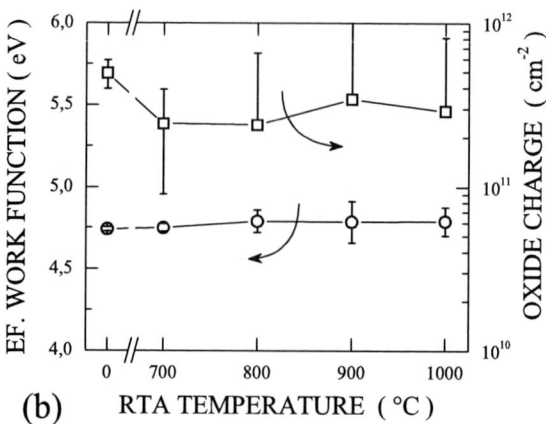

Fig. 2 (a) *C-V* curves of Ru/Hf$_x$Si$_{1-x}$O$_y$/Si MOS capacitor with uniform dielectrics t_{ox}=9 nm after deposition and FGA and after RTA at different temperatures. The $\Phi_{m,eff}$ and Q_{ox}/q were determined using MOS capacitors with slanted dielectrics (b).

of the thin layer, in which the density changes linearly with the thickness. Except the Hf$_x$Si$_{1-x}$O$_y$ layer annealed at 900 °C, there is a considerable increase of dielectric thickness. Such behaviour can be explained by occurrence of two competing effects: (i) growing of the interfacial oxide and (ii) densification of the Hf$_x$Si$_{1-x}$O$_y$ layer.

In order to determine possible change of electrical properties during RTA, we performed *C-V* measurements on the entire set of samples. The curves, depicted in Fig. 2(a) show no considerable shift in flat band voltage, V_{FB}, hence constancy of the effective metal work function, $\Phi_{m,eff}$ and the oxide charge, Q_{ox}/q. More detailed analyses were carried out on the slanted samples, by plotting V_{FB} against EOT. The linear fit to this data allows precise determination of the mentioned entities and the summary is given in Fig. 2(b). One can conclude a negligible change of the $\Phi_{m,eff}$ and thus the band alignment at the metal/oxide interface with RTA temperature.

The *I-V* characteristics were measured on slanted samples in the same position in the direction of etching, hence, approximately the same dielectric thickness before RTA. The results are shown in Fig. 3(a). To analyze these curves, first, we determine the thickness of the high-κ dielectric and the interfacial SiO$_x$ layer. As the EOT changes only slightly during RTA, we have considered the constant κ of the high-κ dielectric, determined from the EOT vs. t_{ox} data on as-deposited samples [6]. Using the XRR thicknesses, t_{XRR}, and the κ value, the high-κ dielectric thickness, $t_{high-\kappa}$, was calculated using the simplification: $t_{XRR}=t_{high-\kappa}$. Then, from the EOT, we were able to determine the EOT of the interfacial layer. We obtained oxide thickness of 3, 4.4, 4 and 5.2 nm for dielectrics annealed at 700, 800, 900 and 1000 °C, respectively. The dielectric thicknesses obtained this way were used to calculate the electric field in the oxide, E_{ox}, in the analysis of the experimental *I-V* characteristics together with the known electron barrier height between the $\Phi_{m,eff}$ and the bottom of the Hf$_x$Si$_{1-x}$O$_y$ conduction band, ϕ_b=2.4 eV [5]. The E_{ox} was calculated after subtracting the surface potential from V_g.

Now we consider the injection of electrons from the metal gate to Si, thus negative gate bias. As t_{ox} of the MOS capacitor with its dielectric annealed at 700 °C is very thin, one should consider the direct tunneling (DT) in this structure. The dashed thick grey line in Fig. 3(a) represents the simulated DT current with t_{ox}=3 nm, ϕ_b=2.4 eV, and m_{ox}=0.26m_e. This curve fits well with the experimental data and m_{ox} is similar to the one reported for MOCVD-grown Hf$_x$Si$_{1-x}$O$_y$ [2]. Leakage currents of MOS capacitors with dielectric annealed at 800 and

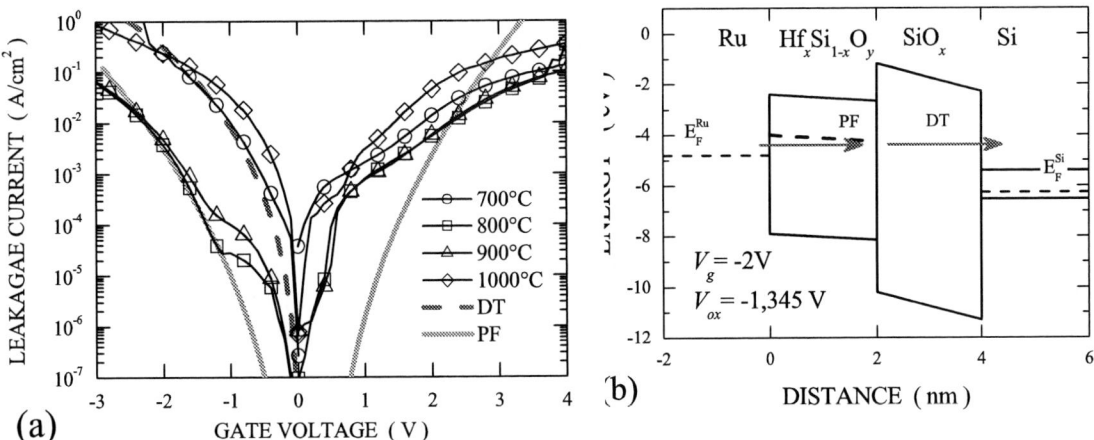

Fig. 3 (a) Leakage characteristics of the Ru/Hf$_x$Si$_{1-x}$O$_y$/Si MOS capacitors measured at the same position of the slanted dielectric as a function of RTA temperature. Grey lines shows the simulation of direct tunneling (dashed line) and Pool-Frenkel (solid line) current mechanisms with particular sample parameters. (b) Illustration of Pool-Frenkel (PF) mechanism in the gate dielectric under gate voltage V_g=-2 V. The energy is referred to vacuum level and "DT" represents the direct tunneling.

900 °C were fitted by Pool-Frenkel bulk limited current mechanism at higher electric fields. Since the voltage drop across SiO$_x$ is much higher than that across high-κ dielectric [Fig. 3(b)], the traps level should be located energetically deep in the high-κ dielectric. Spatially, the traps are probably located at the Hf$_x$Si$_{1-x}$O$_y$/SiO$_x$ interface. After emission from the trap, electrons tunnel through the SiO$_x$ layer to the Si conduction band, but the limiting factor is the thermal emission from the traps. Temperature dependent *I-V* measurements should be done to confirm this suggestion. The leakage current of the MOS capacitor with dielectric annealed at 1000 °C is extremely high due to presence of grain boundaries in the polycrystalline film [see Fig. 1(a)].

We were not able to fit the data under positive bias. We suppose the current is limited by the supply of minority carriers by Si, although the measurements were done upon light.

Acknowledgement

The authors gratefully acknowledge the preparation of Hf$_x$Si$_{1-x}$O$_y$ layers by Aixtron A.G. This work was supported by the projects APVT-51-017004 and VEGA 1/3091/06.

References

[1] International Technology Roadmap for Semiconductors 2005 Ed., http://public.itrs.net/.
[2] M. Lemberger, A. Paskaleva, S. Zürcher, A.J. Bauer, L. Frey, H. Ryssel, *Microelectron. Reliab.* **45**, 819, 2005.
[3] K.J. Yang, Ch. Hu, *IEEE Trans. Electron Dev.* **46** 1500, 1999.
[4] N. Yang, K. Henson, J. Hauser, J. Wortman, *IEEE Trans. Electron. Dev.* **46** 1464, 1999.
[5] K. Fröhlich, M. Ťapajna, K. Hušeková, A. Vincze, J.P. Espinos, in: *Proceedings of the 2006 MRS Spring Meeting*, San Francisco, United States of America, 2006, p. E5.2.
[6] M. Ťapajna, K. Hušeková, J.P. Espinos, L. Harmatha, K. Fröhlich, *Mater. Sci. Semicond. Proc.*, in press.

First results observed with test X-CT system using GaAs radiation detector working in single photon counting regime

F. Dubecký[1], B. Zaťko[1], I. Frollo[2], J. Juraš[2], J. Přibil[2], J. Jakubek[3] and J. Mudroň[4]

[1] Institute of Electrical Engineering, Slovak Academy of Sciences
Dúbravská cesta 9, SK-841 04 Bratislava, Slovakia
[2] Faculty of Electrical Engineering and Information Technology
Slovak University of Technology, Ilkovičova 3, SK-812 19 Bratislava, Slovakia,
[3] Institute of Experimental and Applied Physics, Czech Technical University in Prague
Horská 3a/22, Prague 2, CZ-128 00, Czech Republic
[4] Magic Trading Corporation, a.s.,
Kuzmányho 11, SK-03101 Liptovský Mikuláš, Slovakia
e-mail: elekfdub@savba.sk

The aim of present work is application of novel single photon counting system for detection of gamma-rays in CT using semiconductor radiation detector based on semi-insulating (SI) GaAs compound. A simple positioning test system consisting of two stepper motors with ^{241}Am gamma source of 60 keV photons and single SI GaAs radiation detector operated in single photon counting mode is used for taking CT projections. First, rather preliminary results of CT reconstruction applied to imaging of inner structure of a small phantom using developed reconstruction algorithms is demonstrated.

1. Introduction

The study of inner structure of objects has a beginning in discovery of X-rays. Non-destructive evaluation of various objects is perspective in many fields of applications [1]. Computer tomography presents a dedicated non-destructive technique allowing imaging of the internal structure of an object. Computer tomography is organized by principle of working: X-rays (CT), Positron Emission Tomography (PET), Isotope Imaging (II), ultrasound (US-CT) and Magnetic Resonance Imaging (MRI). In Computed Tomography, CT, the tested object is exposed by X- or gamma-rays at chosen amount of angles. Moreover, every scan is sampled by a sampling interval. Collected data are called projections. There are several methods one can use for image reconstruction (e.g. filtered backprojection, iterative methods, or Fourier reconstruction). Reconstruction methods depend on measurement set-up: beams can be either parallel, equidistal or equiangular aligned. In general, more angles and more samples imply finer resulting images; unfortunately, scanning times and computing requirements increase as well. First computer tomograph used in medicine was invented in 1971. Standard computer tomograph for detection of X-rays uses large volume scintillation detectors. New approach toward better image quality and smaller dimensions of this instrumentation includes semiconductor X-ray detectors working in the progressive single photon counting opening an improvements of resolution in contrast of the imaging system comparing with currently used integration mode of operation.

The aim of our work is application of novel single photon counting system for detection of X-rays in CT using developed single radiation detector based on semi-insulating (SI) GaAs [2]. In this preliminary study we used a source of photons passive γ-radiation source ^{241}Am with photons energy of 60 keV.

2. Experiment

Block scheme of the developed test X-CT system is depicted in Fig. 1. The beam of photons impinges to the investigated object. Passive γ-radiation source [241]Am is used as a source of photons with an energy of 60 keV. Radiation is partially absorbed by the phantom test object and detected by one of monolithic 24 strip detectors named "SAMO" based on bulk SI GaAs illustrated in Fig. 2. In the detector volume a transformation of the absorbed radiation into the electric signal occurs [2]. The detector is working in single photon counting regime with direct conversion of absorbed photons to the electric signal [3]. The number of registered pulses is recorded by computer for every point of x axis. After measurement of all points corresponding to the sample projection data for one angle are observed. Then the sample is rotated by a small angle and following projection data are collected. Position resolution of the test system is 0.25 mm in x axis and minimum rotation angle is 1.8 deg. given by used stepping motor of the testing X-CT mini platform. Investigated object can rotate from 0 to 360 deg. The shifting and rotation of the object ensure automatically two stepping motors controlled by personal computer, output power microcontroller through the application software.

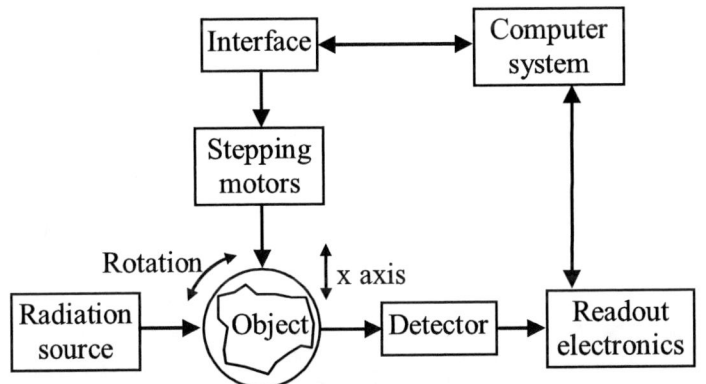

Fig. 1: Block scheme of developed test X-CT system.

Fig. 2. Photo of the monolithic line detector based on SI GaAs.

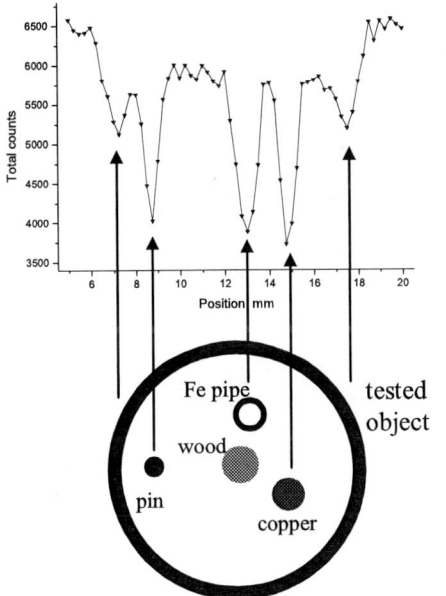

Fig. 3: The tested sample and the corresponding X-ray projection (position in mm).

Fig. 3 shows a data of one projection of tested sample, which consists of steel pin, a little Fe pipe, wood and copper sticks placed into the Teflon ring filled with polystyrene. The dependence of total counts vs. position represents the absorption of γ–rays in each point. Five peaks in graph correspond to object in the used test phantom: Teflon ring, pin, wood with Fe pipe, copper and again to the Teflon ring. The first measurement was done with 8 projections from 0 to 180 deg. and the step of rotation was 22.5 deg. The collected data are processed by a routine written in C++ using iterative reconstruction algorithm [4, 5] observed from 8 projections only at angle step of 22.5 deg. Fig. 4 shows the final image of inner structure – cross section – of the tested phantom sample. As can be seen from the figure only three of four objects are visible: pin, copper and wood with Fe pipe joined into

131

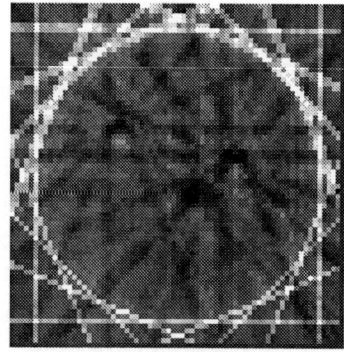

Fig. 4: First tomogram of the tested object observed from 8 projections.

one object. This is because the wood and Fe pipe are too close. The Teflon ring is visible as a darker shadow circle only. Image with better quality needs more projections (our X-CT test system takes as many as 200 projections with used stepping motor).

The same data of 8 projections were used for various 2D tomographic reconstructions performed by both: by the standard filtered back projection algorithm (FBP) and also by iterative algorithm based on the Expectation Maximization (EM) methods [6]. The observed tomograms are shown in Fig. 5. Five peaks instead of six necessary for the right image reconstruction

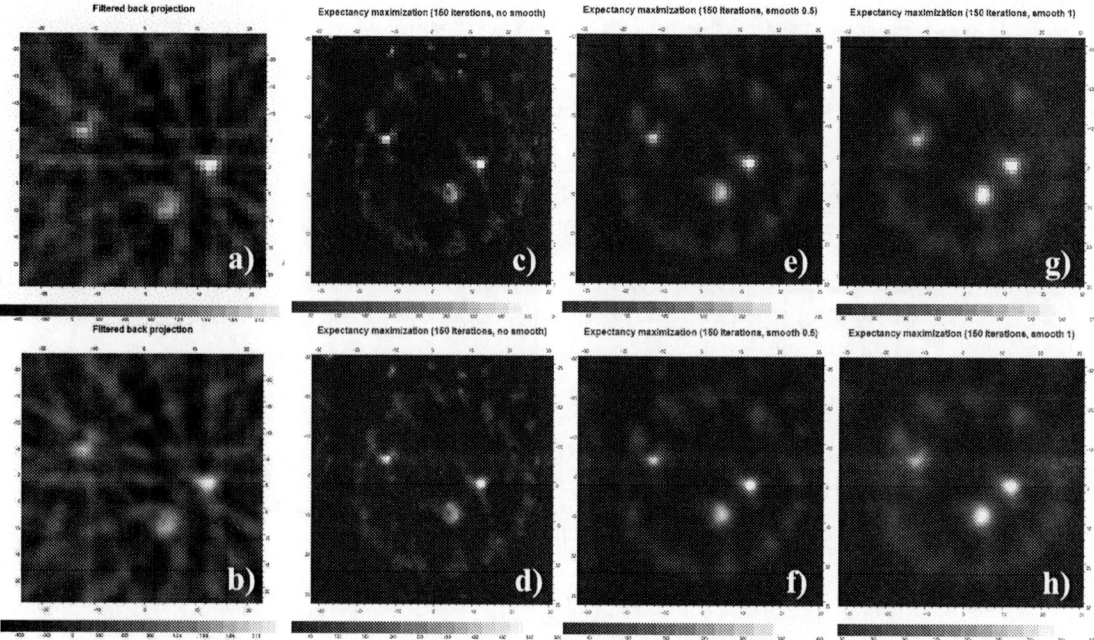

Fig. 5: Tomographic reconstructions (the same object and data as in Fig. 3) observed using different reconstruction algorithms, FBP (a, b) and EM (c - h): without (a, c, e, g) and with smooth correction (b, d, f, h).

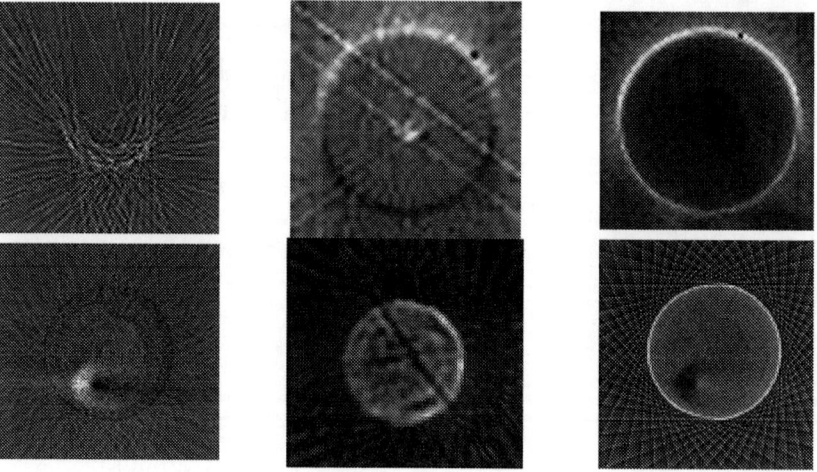

Fig. 6: First results of projection reconstruction using data obtained from novel type of imaging test system (50 projections within 0 – 352.8 deg using sampling step of 7.2 deg). Filtered backprojection method was used.

132

were observed in all measured projections. If one wants to distinguish wood and Fe pipe, the rotation step have to be smaller. However, as can be seen from the reconstruction even in small number of data projections (8) the reconstruction of the inner object cross-sectional image is possible. From the observed results, it can be preliminary concluded, that the iterative algorithm based on the EM method gives much better image quality.

The first testing of projection reconstruction using data obtained from an improved type of imaging test system are illustrated in Fig. 6. Data obtained from 50 projections within 0 – 352.8 deg with sampling step of 7.2 deg were used in the reconstruction by the filtered backprojection method.

3. Conclusion

Presented results show that the developed simple test system for X-CT with GaAs radiation detectors operating in single photon counting regime is applicable in dedicated CT imaging. The CT reconstruction of a simple phantoms was successfully done using standard filtered back projection algorithm and also by iterative algorithm based on the Expectation Maximization method. The later method gives better image quality. Improvement of the developed test X-CT system with SI GaAs monolithic line detectors is in progress.

Acknowledgement: Authors acknowledge M. Sekáčová, P. Boháček and J. Huran for GaAs detector technology. This work was partially supported by Agency for Promotion Research and Development under the contract No. APVV-99-P06305.

References
[1] *Advanced Imaging Techniques*, Editors: Thomas H. Newton, D. Gordon Potts, 1983.
[2] Dubecký, F., et al.: Nucl. Instr. and Meth. in Phys. Res. A 531 (2004) 314.
[3] Zaťko, B., et al.: Nucl. Instr. and Meth. in Phys. Res. A 551 (2005) 78.
[4] Pham, Van Tinh: *Výpočtová technika v hardwarovej a softwarovej realizácii počítačového toromografu pre tomografické zobrazenie vnútornej štruktúry materálu.* PhD. thesis. Technical University in Zvolen 2002.
[5] Image reconstruction from projection. HET408 Biomedical Imaging. http://marr.bsee.swin.edu.au/~dtl/het408.html
[6] Hudson, H. M., Larkin, R. S.: IEEE Trans. Med. Imaging 13 (1994) 601.

Changes of GaAs neutron detectors properties after fast neutron irradiation

Milan Ladziansky[1], Andrea Šagátová-Perďochová[1], Bohumír Zaťko[2], Vladimír Nečas[1], František Dubecký[2]

[1] Faculty of Electrical Engineering and Information Technology, Slovak University of Technology, Ilkovičova 3, SK-812 19 Bratislava, Slovakia,
[2] Institute of Electrical Engineering, Slovak Academy of Sciences, Dúbravská cesta 9, 841 04 Bratislava, Slovakia
e-mail: milan.ladziansky@stuba.sk

One of steps in the development of semi-insulating (SI) GaAs neutron detectors is study of radiation resistance to exposure of the detectors with high fluencies of neutrons (from 5 x 10^{11} n.cm^{-2} to 1 x 10^{15} n.cm^{-2}). Changes in spectra of ^{241}Am and ^{57}Co gamma radionuclide sources, measured by GaAs detectors exposed to various neutron fluencies, are observed. The energy resolution is improving with rising neutron fluency but charge collection efficiency decreased. It is probably caused by the impacting neutrons, which created new lattice defects in the material of detector. The newly created defects would partially compensate original defects, present in the material from the fabrication. However, at higher neutron fluencies a great degradation of detection performance of measured detectors occurred.

1. Introduction

Neutron digital imaging can present a complementary technique to X-ray digital radiography. X-rays are effective for imaging rather high density materials (e.g. metals) and neutrons allow evaluation of low density materials containing hydrogen - e.g. plastics [1]. Because of relatively high resistance of bulk semi-insulating (SI) GaAs material against neutrons [2] it makes this detector a suitable candidate for fabrication of a neutron imaging detector.

This contribution concerns with radiation damage created by neutrons in semi-insulating (SI) GaAs detectors. Changes in detection properties of GaAs detectors are studied with following increasing neutron dose into GaAs detectors. The detection properties of neutron bombarded detectors were evaluated from measured gamma spectra of ^{57}Co and ^{241}Am radiation sources at various bias voltages.

2. Experiment

Set of six SI GaAs monolithical samples each with 4 detectors was used in this experimental study. Each sample consists of 2 circle Schottky contacts with 1.13 mm^2 area and 2 circle Schottky contacts with 0.5 mm^2 area (Fig 1). The top Schottky contacts were formed by (AuZn) and the back, large area contact, was formed by (AuGeNi) eutectic alloys. Thickness of the SI GaAs substrate was 200 μm. Each monolithical substrate was irradiated with different neutron fluency (Tab. 1.)

Tab 1: Preview of used neutron fluencies.

Detector	Neutron fluency Φ [n. cm^{-2}]
1	5×10^{11}
2	2×10^{12}
3	5×10^{12}
4	1×10^{13}
5	5×10^{14}
6	1×10^{15}

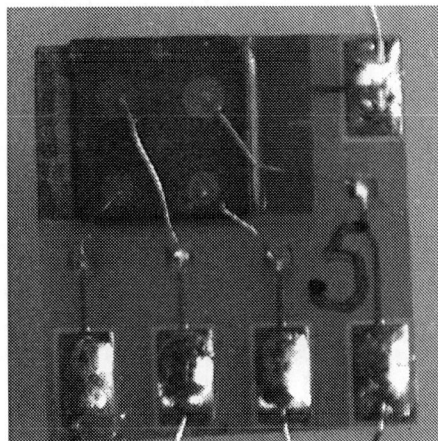

Fig 1: Semi-insulating GaAs detectors.

Detectors were irradiated in Cyclotron centre in Řež (Czech Republic). Beryllium target was bombarded with 37 MeV primary protons. In target passed Be(p,n)B reaction. The result of this reaction is continuous energetic spectrum of neutrons. The spectrum is almost constant in a range between 2 MeV and 30 MeV. Then it exponentially decreased to 37 MeV (end of spectrum).

I-V characteristics of detectors were measured before and after exposure of detectors to neutrons (Fig. 2). From the following graphs, it is evident that a great degradation in I-V characteristics in samples after their exposure to neutron fluencies greater than 10^{13} n.cm^{-2}. According to I-V characteristics we would assume that the detection properties should deteriorate at fluencies of neutrons greater than 10^{13} n.cm^{-2}.

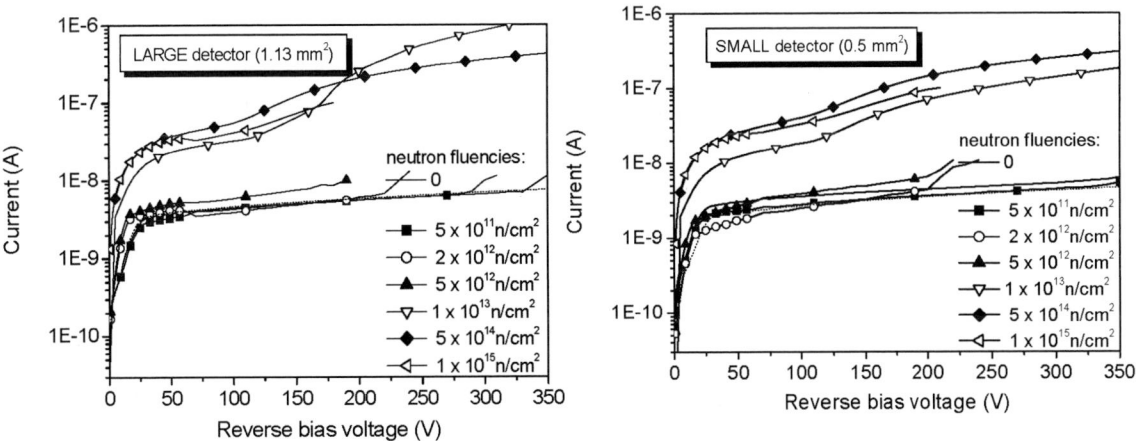

Fig 2: I-V characteristics of SI GaAs detectors irradiated by various neutron fluencies.

Changes in detection properties of examined detectors were measured by gamma radionuclide sources (^{57}Co, ^{241}Am) at reverse bias of 100, 200 and 300 V. Energy of gamma photons from ^{241}Am is 59.5 keV and from ^{57}Co is 122.1 keV. In the following graph (Fig. 3) the measured spectra of ^{241}Am of six larger detectors (contact area 1.13 mm^2) exposed to various neutron fluencies together with the spectrum measured before exposure to neutrons are shown. The spectra are observed at a reverse bias of 200 V. In the Fig. 4 spectra of ^{57}Co measured with six smaller (area 0.5 mm^2) detectors before and after exposure to neutrons are shown.

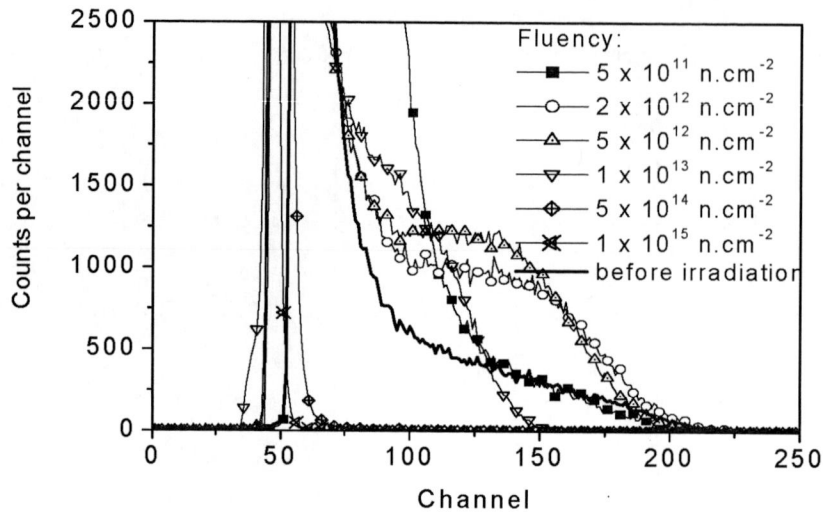

Fig. 3: Spectra of ^{241}Am measured with large detectors (area of Schottky contact 1.13 mm^2) at reverse bias of 200 V exposed to various neutron fluencies.

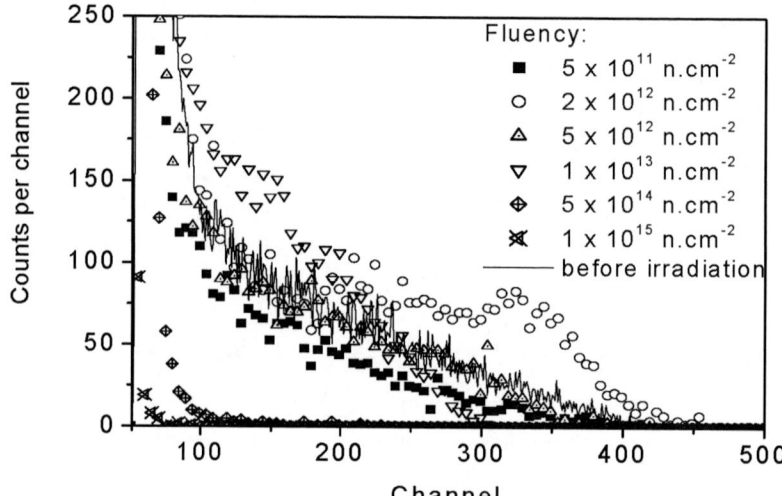

Fig. 4: Spectra of ^{57}Co measured with small detectors (area of Schottky contact 0.5 mm^2) at reverse bias of 200 V exposed to various neutron fluencies.

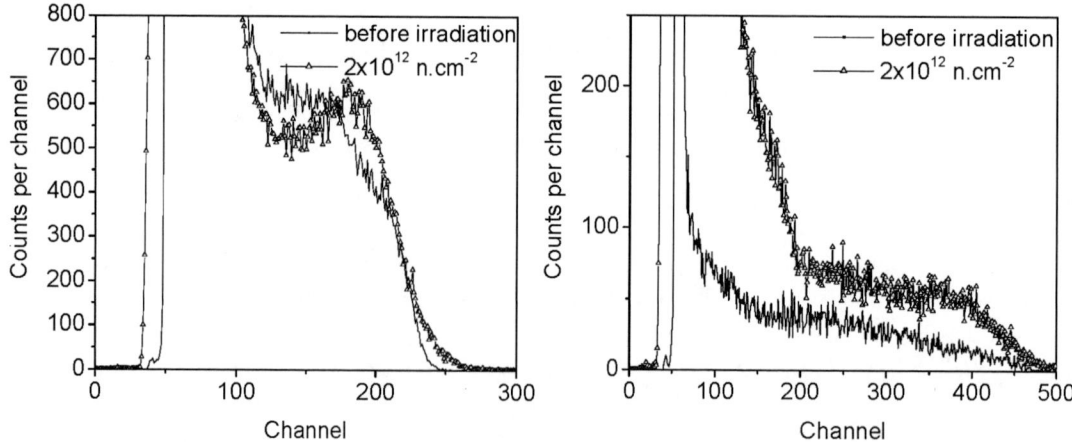

Fig. 5: Spectra of ^{241}Am (left) and ^{57}Co (right) measured with the same detector before and after its exposure to neutrons. Fluency of neutrons used was 2 x 10^{12} n.cm^{-2}, which shown improvement in detection properties. Area of Schottky contact was 1.13 mm^2. Spectra were measured at reverse bias of 300 V.

As can be deduced from observed results (shown pulse heights spectra) performance of detectors during the first phase of neutron bombardment improved as can be seen in Fig. 5. Detectors before exposure to neutrons had not very good detection performance, even no photopeak was formed in the measured pulse-height spectra (Fig. 3).

The maximal improvement in detector performance was observed at a neutron fluency of 2×10^{12} n.cm^{-2}. In such condition tendency to formation of a photopeak is evident. On other hand, position of the peak (charge collection efficiency - CCE) decreases with increasing neutron fluency. These observations are probably caused by new lattice defects in the base material or overall detector structure (electrode interfaces, surface) created by neutrons. Such new rather acceptor-like defects would partially compensate original defects present in the base material (or detectors structure) before the neutron bombardment which would result into improvement of detection performance. However, at higher neutron fluencies ($> 10^{13}$ n.cm^{-2}) a fast degradation of detection performance of the detectors occurred, as demonstrates Fig. 3,4.

3. Conclusion

Poorer detection properties of examined SI GaAs detectors have been improved by neutron exposure up to "optimal" neutron fluencies. Measurement of GaAs detectors shown that the largest improvement of detection properties was observed at a neutron fluency of 5×10^{12} n.cm^{-2}. This is probably caused by creation of acceptor-like defects in GaAs during interaction with neutrons, which could compensate defects in the original n-type material due to used growth process. This compensation caused observed energy resolution improvement. The decrease of the CCE with rising neutron fluency was not that great in comparison to the improvement of energy resolution. However, as assumed from the worsening of I-V characteristics, neutron fluency above 10^{13} n.cm^{-2} caused a great degradation of detection properties. This is probably caused by creation of a huge concentration of defects in GaAs material thus leading to degradation affecting performance of detectors.

Other experiments will be focused to GaAs detectors with better detection properties (relative energy resolution at 60 keV ~ 12.5 % and CCE ~ 97.8 %) and their behavior before and after exposure to various neutron fluencies. Experiments will also include observation of creation of deep traps in the GaAs bulk by using Photo-induced current transient spectroscopy (PICTS), which should append our knowledge in creation/decay of defects after exposure of GaAs detectors to various neutron fluencies.

Acknowledgement: This project has been supported by the Slovak Grant Agency for Science through the grant No. 2/4151/24. Authors would like to thank V. Linhart (CTU in Prague, Czech Republic) and P. Bém (NPI ASCR in Řež, Czech Republic) for assistance with neutron bombardment of investigated samples by different fluencies.

Reference
[1] Jakoubek J. et al: *CdTe hybrid pixel detector for imaging with thermal neutrons*, Nuclear Instruments and Methods in Physics Research A 563 (2006) 238–241.
[2] Morvic M. et al.: *Electrical properties of semi-insulating GaAs irradiated with neutrons*, Nuclear Instr. Methods in Physics Research, B 197 (2002) 240-246.
[3] Tu L. A. et al: *Radiation Resistance Study of Semi-Insulating GaAs-Based Radiation Detectors to Extremely High Gamma Doses*, Nuclear Physics B (Proc. Suppl.) 150 (2006) 402–406.

Control of defects and impurities at GaN and AlGaN surfaces for FET and sensor applications

Junji Kotani, Masamitsu Kaneko, Kazushi Matsuo, Tamotsu Hashizume

Research Center for Integrated Quantum Electronics (RCIQE)
Hokkaido University, Sapporo, 060-8628 Japan
*e-mail: kotani@rciqe.hokudai.ac.jp, hashi@rciqe.hokudai.ac.jp

We proposed a surface control process using ultrathin Al layer for suppressing surface donor states at AlGaN. The Ni/Au Schottky diodes fabricated on the processed AlGaN surfaces showed pronounced reduction of leakage current and clear temperature dependence of I-V curves. The AlGaN/GaN HFETs fabricated with surface control showed remarkable reduction of gate leakage currents and the improvement of stability of drain currents against high-temperature/current stress. The surface process also brought a significant decrease in dark current in the Pd/AlGaN/GaN H_2 sensor. The Pd/AlGaN/GaN diode showed the systematic shift in the C-V curve for different partial pressure of H_2. The maximum shift in the threshold voltage obtained was about 1200 mV which is much larger than 500 mV reported for the Pd-based Si MOS sensors.

1. Introduction

On the basis of excellent optical and electrical properties, significant progress has been achieved in the GaN-based devices, such as ultraviolet laser diodes, white light-emitting diodes, high electron mobility transistors (HEMTs) and high-power FETs. In addition, GaN-based materials are very attractive for various gas and chemical sensor application, because of their superb chemical stability and capability of high-temperature operation owing to widegap nature. However, these devices still suffer from surface/interface-related problems, including collapse in drain current, excess gate leakage, aging of metal contacts, etc. Among them, leakage currents through Schottky contacts not only impede device reliability but also degrade power efficiency and noise performance in such devices. To realize the excellent GaN-based devices with high-stability and high-reliability, it is essential to control the defects and impurities at GaN and AlGaN surfaces.

In this paper, the surface control process utilizing an ultrathin Al layer is proposed for suppressing the tunneling leakage currents of Schottky diodes and Schottky gates on AlGaN/GaN HFETs. It is shown that the surface process is promising for improvement of device performance in HFETs and gas sensors.

2. Surface control process using a thin Al layer

2.1 Surface-donor model for leakage current at AlGaN Schottky interface

Figure 1 shows typical I-V characteristics of Ni/Au contacts with a diameter of 600 mm on n-GaN ($n=1\times10^{17}$cm^{-3}) and AlGaN/GaN heterostructures, both grown by metal-organic vapor phase epitaxy (MOVPE). The Schottky dot was surrounded by a ring-shape Ti/Al/Ti/Au ohmic contact. In spite of the fact that these samples were prepared through the same process at the same time, much larger leakage currents appeared in the AlGaN/GaN samples, as compared with the GaN sample. In addition, the AlGaN/GaN samples showed pronouncedly poor temperature dependence of current. These results clearly indicated the enhancement of tunneling components in leakage currents through the Schottky interfaces on AlGaN/GaN heterostructures, in particular for the sample having higher Al composition.

We have reported [1] that the enhancement of tunneling transport processes by the barrier thinning due to the processing-induced surface-defect donors was the dominant mechanism associated with large leakage currents through GaN Schottky interfaces. The most possible candidate for such a surface donor was the nitrogen-vacancy (V_N) related defects [2, 3]. Based on those results, we have proposed a surface-donor model shown in **Fig. 2**. As shown in Fig. 2, the ionized high-density surface donors can reduce the width of Schottky barrier and greatly enhance the tunneling transport.

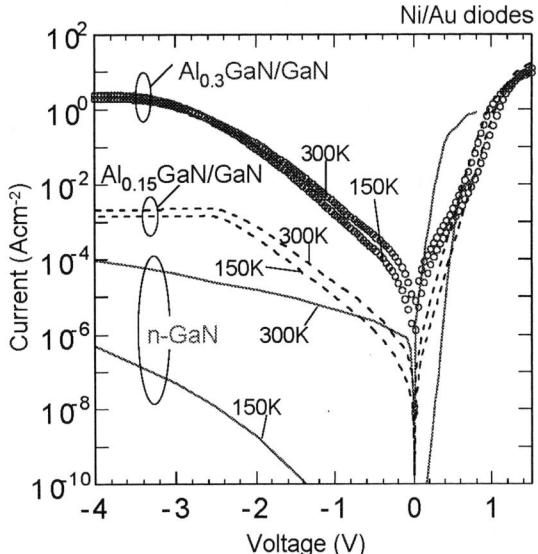

Fig.1 Typical I-V-T curves of Ni/Au Schottky contacts with a diameter of 600 μm on GaN and AlGaN/GaN heterostructures. The thickness of AlGaN barrier layer is 23 nm.

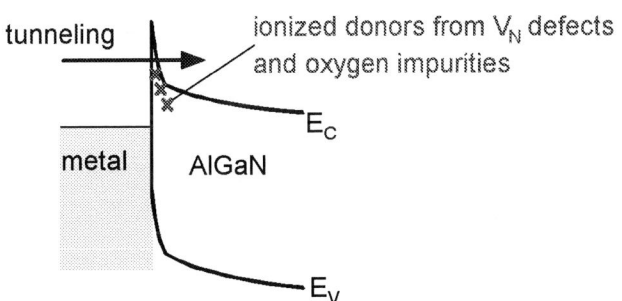

Fig.2 A surface-donor model for leakage current in GaN and AlGaN Schottky interfaces. The thinning of Schottky potential due to the ionization of V_N defects and oxygen impurities can greatly enhance the tunneling transport.

If the processing-induced V_N-related donors play a dominant role in the tunneling leakage mechanism, a similar leakage behavior could appear in the heterostructure samples. This requires the contribution of additional surface donor states to the leakage mechanism in the AlGaN Schottky interfaces.

High-density oxygen impurities were detected in AlGaN layers grown by molecular beam epitaxy [4, 5]. From a chemical analysis using a scanning photoemission spectroscopy, Jang and co-workers [6] found incorporation of high-density oxygen impurities into the AlGaN surface grown by metal-organic vapor phase epitaxy (MOVPE). Recently, a secondary ion mass spectroscopy showed existence of oxygen atoms with a density of 8 x 10^{18}cm^{-3} in MOVPE AlGaN [7]. Oxygen impurities incorporated to GaN are known to act as a shallow donor with activation energy of about 30 meV [8, 9]. Thus, we considered oxygen

139

impurity as one of the most possible candidates for additional donor states, as shown in Fig.2, which can enhance the tunneling transport at the AlGaN Schottky interfaces.

2.1 Surface control process and XPS analysis

We have developed a surface control process for suppressing tunneling leakage through AlGaN Schottky interfaces, as schematically shown in **Fig.3**. The process focused on the recovery of nitrogen vacancies (V_N) and the reduction of oxygen impurities from the AlGaN surface. After cleaning of the sample in organic solutions, the rf-excited nitrogen radicals were introduced to the AlGaN surface at 300 °C for 5 min in molecular-beam epitaxy (MBE) chamber. A base pressure of the chamber was 5×10^{-10} Torr. This process is effective in improving surface electronic properties of GaN and AlGaN [10, 11]. Then, an Al layer with a nominal thickness of 1 nm was deposited by a rate of 0.5 nm/min using a Knudsen cell at room temperature (RT), followed by in-situ annealing at 700 °C for 10 min. The deposited Al layer was then removed in a buffered HF solution. Finally, we fabricated a Ni/Au Schottky contact on the processed AlGaN surface using a standard electron-beam deposition method.

To investigate chemical properties of the AlGaN surface, we carried out in-situ X-ray photoelectron spectroscopy (XPS) analysis using Perkin Elmer PHI 1600. The XPS chamber is connected to the MBE chamber via an ultra-high vacuum (UHV) transfer chamber. **Figure 4** shows the Al2p core-level spectra from the $Al_{0.25}Ga_{0.75}N$ surface after the deposition of the 1-nm Al layer and the subsequent UHV anneal. Just after the Al deposition, a clear metallic Al peak appeared in the XPS spectrum. From the angle-resolved analysis, we

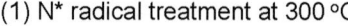

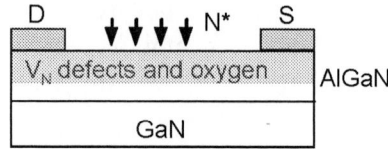

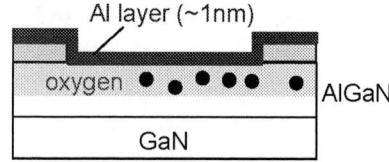

Fig.3 Flow-chart of a surface control process for suppressing tunneling leakage.

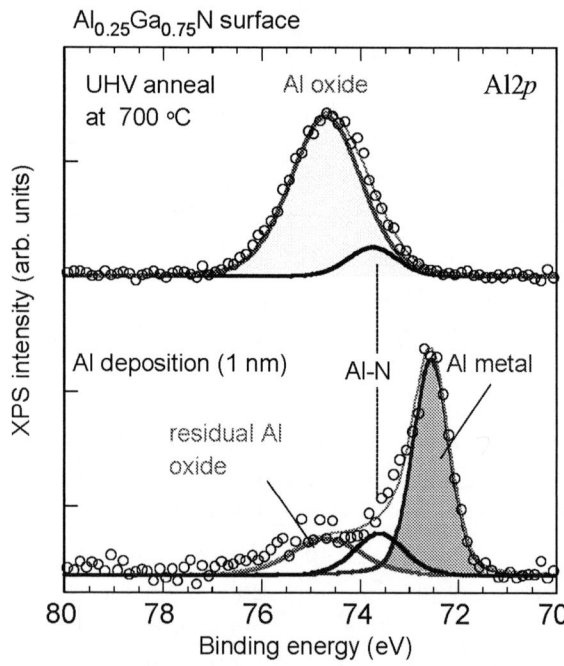

Fig.4 XPS Al2p spectra from the $Al_{0.25}Ga_{0.75}N$ surface after the deposition of an Al layer and the subsequent UHV anneal.

140

confirmed the layer formation of Al. A weak oxide peak indicated that a small amount of natural oxide remained on the AlGaN surface even after the N* radical treatment.

Surprisingly, the metallic Al peak disappeared after the UHV anneal, as shown in the upper spectrum in Fig. 4. Furthermore, we found drastic increase of the oxide peak whose energy position is very close to that of Al_2O_3. A residual oxide of AlGaN includes both Al- and Ga-oxides [11]. The Ga-oxide component seems to convert to the Al oxide during the UHV anneal, due to the high reactivity of Al and higher heat of formation of Al oxide than Ga oxide. The Ga3d spectrum (not shown here) supported this reaction. Jung, Miura and Ishida [12] reported a similar conversion of an ultrathin SiO_2 layer into an Al-oxide layer by an assist of a thin Al layer during the UHV anneal at 800 °C. In addition, it was likely that the UHV anneal induced a gettering of oxygen impurities from the AlGaN surface into the Al layer, also due to the high reactivity of Al with oxygen.

3. Electrical properties of Ni/AlGaN diodes and Schottky gates on AlGaN/GaN HEMTs

3.1 Leakage currents in AlGaN Schottky interfaces

Figure 5 compares the I-V characteristics of Ni/AlGaN Schottky diodes fabricated with and without the surface control process, respectively. The Schottky diode fabricated without the surface control process exhibited large reverse leakage currents. On the other hand, a remarkable reduction of reverse leakage current of 4~5 orders of magnitude took place by applying the proposed novel surface control process aimed at elimination of oxygen donors. Furthermore, the surface control process led to increase of temperature dependences of both forward and reverse currents, as seen in **Fig. 5**. The observed large reduction of the

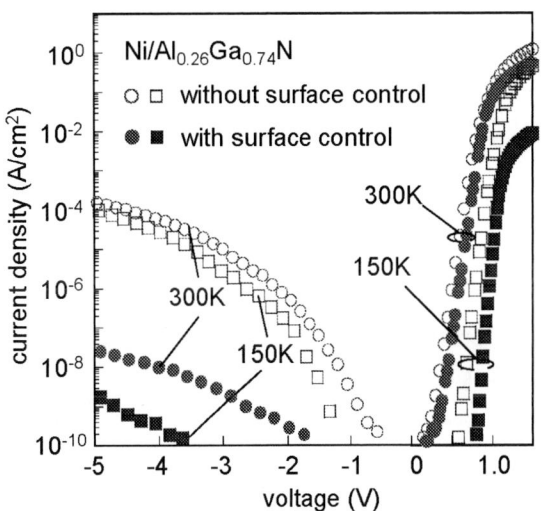

Fig.5 I-V-T characteristics of Ni/Al$_{0.26}$Ga$_{0.74}$N Schottky diodes without and with the surface control process

leakage current and increase of temperature dependences of the reverse and forward currents can be attributed to the reduction V_N-related defects and/or oxygen impurities by the proposed surface control process [13].

3.2 Gate leakage currents in AlGaN/GaN HFETs

Figure 6 shows a comparison of I-V characteristics of Ni/Au Schottky gates fabricated on the $Al_{0.25}Ga_{0.75}N$/GaN heterostructures without and with the surface control process. Again, large leakage currents and poor temperature dependence were observed for the sample without the control process. On the other hand, the Schottky gate with the surface process showed a significant reduction of leakage at RT as well as at low temperatures, as shown in the bottom curves in Fig.6. In addition, clear temperature dependence appeared in the I-V characteristics, indicating the effective suppression of the tunneling leakage component especially in the low gate bias region. It is noted that an increase in current at larger negative biases at 150 K can arise from an intrinsic tunneling effect due to rather thin AlGaN barrier thickness, which was confirmed by theoretical calculation. The result obtained reflects that the present surface process seems to reduce the V_N-related defects and/or oxygen impurities near the AlGaN surface, suppressing the turbulence of the potential distribution.

3.3 Stability of drain current against high-temperature/current stress in AlGaN HEMTs

Stability of drain current in fabricated AlGaN/GaN HFETs with and without surface control was characterized. **Figure 7** shows variations of I_D-V_{DS} characteristics of AlGaN/GaN HFETs in the high-temperature/current stress test. As for the high-temperature/current stress, the AlGaN/GaN HFETs were biased at a drain voltage of +16V and a gate voltage of +1V, with temperature set at 220 °C, for 2 hours.

After applying the high-temperature/current stress, a significant decrease in drain current of about 20 percent was detected in the AlGaN/GaN HFET without surface control, as shown in Fig. 7 (a). On the other hand, for the AlGaN/GaN HFETs with surface control, drain current was still at the same level as before stress (Fig. 7(b)). This reveals that the surface control process is effective in stabilizing drain current against high-temperature/current stress, besides reducing gate leakage current.

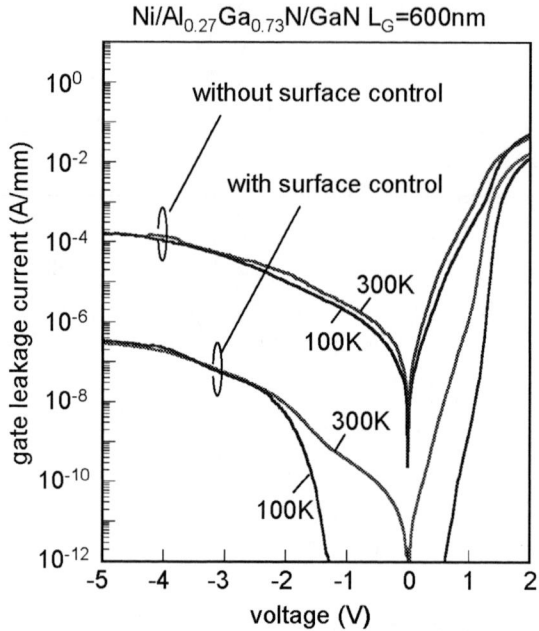

Fig.6 I-V characteristics of Ni/Au gates on $Al_{0.27}Ga_{0.73}N$/GaN heterostructures without and with the surface control process

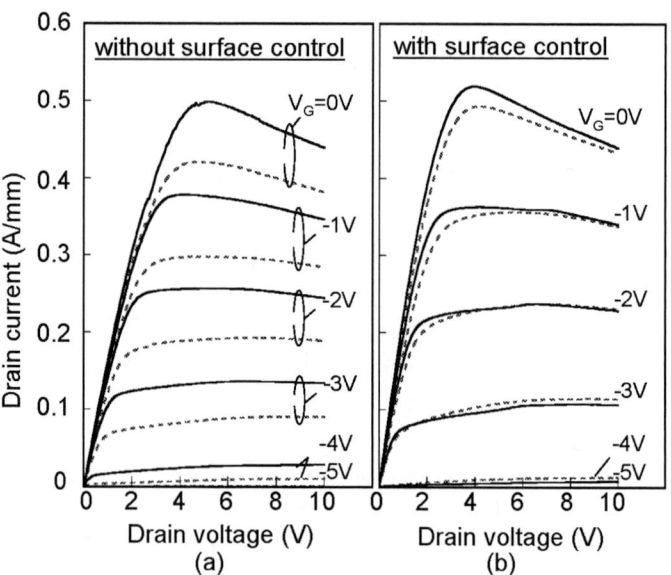

Fig.7 Variation of I_D-V_{DS} characteristics of $Al_{0.25}Ga_{0.75}N$/GaN HFETs (a) without and (b) with the surface control process in high-temperature/current stress test. The solid lines and broken lines represent I_D-V_{DS} characteristics measured before and after applying stress, respectively.

4. Hydrogen gas sensor using Pd Schottky contact on AlGaN/GaN

We have applied the surface control process to the H_2 sensor based on the Pd-AlGaN/GaN structure. As shown in **Fig. 8**, circular Pd Schottky diodes with Ti/Al/Ti/Au ohmic ring electrodes were fabricated on an AlGaN/GaN heterostructure. Pd Schottky contacts with a thickness of 75 nm and a diameter of 600 mm were formed by electron-beam deposition after surface treatment in HF solution. Gas sensing characteristics were measured in a chamber with a base pressure of 0.05 Torr by rotary pump system. The diode was fixed on a ceramic heater and high-purity H_2 gas was introduced into the chamber. Current-voltage (I-V) and Capacitance-voltage (C-V) characteristics were measured by Agilent 4156A semiconductor parameter analyzer and Agilent 4192A impedance analyzer.

The change in the I-V characteristics caused by exposure to H_2 are shown in **Fig. 9** for the Pd/AlGaN/GaN diode with the surface control process. The current increase of nearly five orders magnitude took place by an introduction of H_2 for 1 Torr. A significant reduction of dark current after the surface control process gives rise to the high sensitivity obtained.

The C-V characteristics of the diodes also showed pronounced change upon exposure to hydrogen in vacuum. As shown in **Fig. 10**, the Pd/AlGaN/GaN diode showed the systematic shift in the C-V curve for different partial pressure of H_2, keeping the shape of the curve approximately the same. The maximum shift in the threshold voltage obtained was about 1200 mV which is much larger than 500 mV reported for the Pd-based Si MOS sensors. In the AlGaN/GaN heterostructures, the threshold voltage is determined by the Schottky barrier height (SBH), the potential off-set at the AlGaN/GaN interface (DEC) and the potential for the depletion of 2-dimensional electron gas (2DEG). When H_2 is introduced to the Pd surface, hydrogen molecules

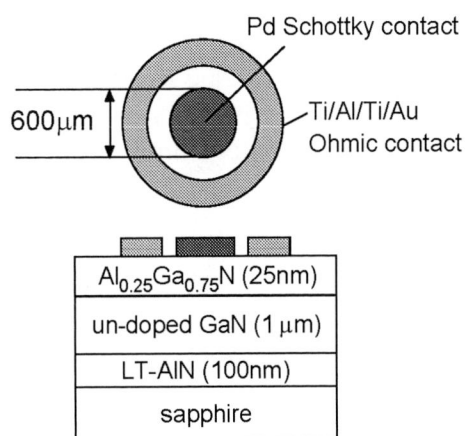

Fig.8 H_2 sensor structure using the Pd-AlGaN/GaN contact.

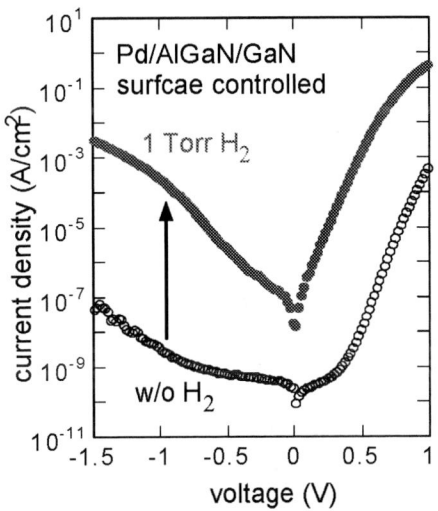

Fig.9 Change in I-V characteristics of the Pd/AlGaN/GaN diode without and with H_2.

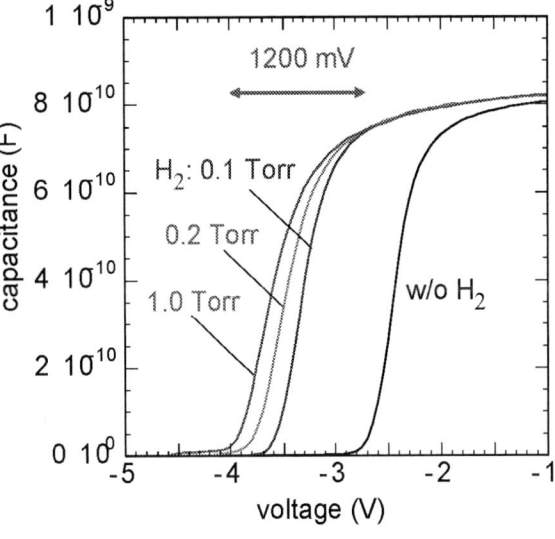

Fig.10 Change in C-V characteristics of the Pd/AlGaN/GaN diode without and with H_2.

dissociate owing to catalytic action of Pd. Then, hydrogen atoms diffuse through the Pd bulk layer, and the adsorbed hydrogen atoms at the AlGaN surface can form dipoles, resulting in the reduction of SBH. This large shift of threshold voltage is very promising for the voltage-detecting sensor with a high sensitivity.

5.   Summary

(1) We proposed a surface control process for suppressing tunneling leakage currents of AlGaN Schottky interfaces. The process consisted of the nitrogen radical treatment, the deposition of an ultrathin Al layer, the UHV annealing and finally the removal of the Al layer. The Ni/Au Schottky diodes fabricated on the processed AlGaN surfaces showed pronounced reduction of leakage current and clear temperature dependence of I-V curves, indicating the effective suppression of the tunneling leakage component in the current transport process. Thus, the present surface process can improve the AlGaN Schottky interface properties, probably due to the reduction of the V_N-related defects and oxygen impurities.

(2) We have fabricated the Schottky gate on the AlGaN/GaN HFETs using the proposed surface process. A remarkable reduction of gate leakage currents was achieved and clear temperature dependence of leakage currents appeared. Furthermore, the improvement of stability of drain currents against high-temperature/current stress was achieved. These results show that the surface control process is effective in improving the AlGaN Schottky interface properties.

(3) The H_2 gas-sensing characteristics of Pd Schottky diodes formed on AlGaN/GaN heterostructure were investigated. By introducing the surface control process, the pronounced reduction of dark current was achieved, resulting in high sensitivity to H_2. The Pd/AlGaN/GaN diode also showed the systematic shift in the C-V curve for different partial pressure of H_2, keeping the shape of the curve approximately the same. The maximum shift in the threshold voltage obtained was about 1200 mV which is much larger than 500 mV reported for the Pd-based Si MOS sensors.

Acknowledgement

One of the authors (J. K) thanks a support by Grant-in-Aid for Japan Society for the Promotion of Science (JSPS) Fellows.

References

[1] T. Hashizume, J. Kotani and H. Hasegawa, Appl. Phys. Lett. **84**(2004)4884.

[2] T. Hashizume and R. Nakasaki, Appl. Phys. Lett. **80**(2002)4564.

[3] T. Hashizume and H. Hasegawa, Appl. Sur. Sci. **234**(2004)387.

[4] C.R. Elsass, T. Tates, B. Heying, C. Poblenz, P. Fini, P.M. Petroff, S.P. DenBaars and J.S. Speck, Appl. Phys. Lett. 77(2000)3167.

[5] S.T. Bradley, S.H. Goss, L.J. Brillson, J. Hwang and W.J. Schaff, J. Vac. Sci. Technol. B **21**(2003)2558.

[6] H.O. Jang, J.M. Baik, M.K. Lee, H.J. Shin and J.L. Lee, J. Electrochem. Soc. **151**(2004)G536.

[7] M. L. Nakarmi, N. Nepal, J. Y. Lin, and H. X. Jiang, Appl. Phys. Lett. **86**(2005)261902.

[8] K. H. Ploog and O. Brandt, J. Vac. Sci. Technol. A **16**(1998)1609.

[9] A. J. Ptak, L. J. Holbert, L. Ting, C. H. Swartz, M. Moldovan, N. C. Giles, T. H. Myers, P. Van Lierde, C. Tian, R. A. Hockett, and S. Mitha, A. E. Wickenden, D. D. Koleske and R. L. Henry, Appl. Phys. Lett. **79**(2001) 2740.

[10] T. Hashizume, S. Ootomo, S. Oyama, M. Konishi and H. Hasegawa, J. Vac. Sci. Technol. B **19**(2001)1675.

[11] T. Hashizume, S. Ootomo, T. Inagaki and H. Hasegawa, J. Vac. Sci. Technol B **21**(2003)1828.

[12] Y.-C. Jung, H. Miura and M. Ishida, Jpn. J. Appl. Phys. **38** (1999)2333.

[13] J. Kotani, M. Kaneko, H. Hasegawa and T. Hashizume, J. Vac. Sci. Technol. B **24** (2006)2148.

Comparison of AlGaN/GaN MSM varactor diodes based on HFET and MOSHFET layer structures

M. Marso, A. Fox, G. Heidelberger, P. Kordoš*, and H. Lüth

Institute of Bio- and Nanosystems (IBN-1) and cni - Center of Nanoelectronic Systems for
Information Technology, Research Centre Jülich, D-52425 Jülich, Germany
* Institute of Electrical Engineering, Slovak Academy of Sciences, Bratislava, SK-84104
Bratislava, Slovakia and Department of Microelectronics, Slovak Technical University,
SK-81219 Bratislava, Slovakia
e-mail: m.marso@fz-juelich.de

In this comparative study we investigate the performance of AlGaN/GaN based MSM varactor diodes based on HFET and MOSHFET layer systems. Device fabrication uses standard HFET fabrication technology, allowing easy integration in MMICs. Devices with different electrode geometries are characterized by DC and by S-parameter measurements up to 50 GHz. The HFET based varactors show capacitance ratios up to 14. The MOSHFET based devices, on the other hand, exhibit lower capacitance ratios and poorer stability because of the insulation layer between electrodes and semiconductor.

1. Introduction

The metal-semiconductor-metal diode above a two dimensional electron gas (MSM-2DEG) has shown its potential as varactor with large voltage-dependent capacitance swing [1]. Due to similar fabrication technology, the MSM-2DEG can be processed simultaneously with HFET devices (Fig. 1) – a huge advantage for integrated circuit design. Furthermore, the capacitance swing of this varactor diode based on an HFET layer structure is much larger than for conventional varactor diodes, and it can be tuned by the electrode geometry in contrast to conventional pn, Schottky or heterostructure diodes where the capacitance ratio is only defined by the layer structure [5, 6]. In this work we compare the varactor properties of MSM-2DEG devices based on the HFET layer system (HFET MSM) with similar devices based on a MOSHFET system (MOSHFET MSM) (Fig. 2).

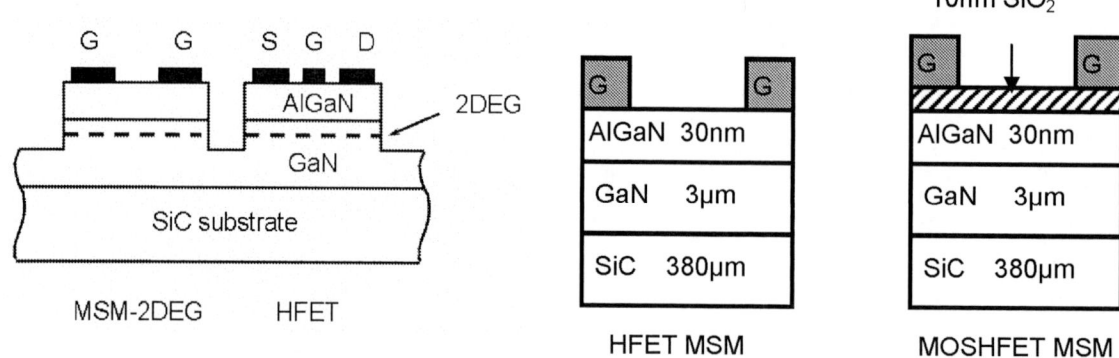

Fig. 1: Layer structure and schematic
diagram of the MSM-2DEG
integrated with a HFET.

Fig. 2: Layer system and schematic diagram of
the investigated devices.
G: gate electrode.

The MSM-2DEG diode consists of two back-to-back connected Schottky diodes above a 2DEG layer structure, as depicted in Fig. 1. The capacitance of the unbiased device is determined by the electrode area and by the distance to the 2DEG channel that acts as equipotential plane. An applied bias voltage depletes the channel below the reverse biased electrode. For large bias voltages the depletion region penetrates deeply into the device and the capacitance drops to a very low value within a small voltage range. The transition voltage V_{TRANS} from high to low capacitance depends on the 2DEG carrier density and is strongly related to the threshold voltage of the corresponding transistor. In fact, the transistor threshold voltage corresponds to the gate-to-source voltage that is needed to completely deplete the channel. Therefore, the transition voltage V_{TRANS} of the MSM-2DEG is the sum of the threshold voltage of the reverse-biased contact and the voltage drop at the forward-biased contact.

2. Device fabrication

For device processing an $Al_{0.28}Ga_{0.72}N/GaN$-based HFET layer stack grown by MOVPE on semi-insulating SiC substrate is used. The devices are fabricated with standard HFET fabrication technology using the gate metallization for the electrodes (Fig. 2). A SiO_2 layer with nominal thickness of 10 nm is deposited by PECVD at 300°C prior to gate metallization for the MOSHFET based devices. The two-finger gate electrodes are defined by electron beam lithography. Unless otherwise stated, the electrode dimensions are 50 µm, 2 µm and 1 µm for finger width, length and separation, respectively. Gate diodes consisting of a large gate electrode with 25x25 µm² area and an ohmic contact are fabricated as reference. These reference diodes are prepared without contact pads that are needed to characterize the much smaller MSM-2DEG devices.

3. Results and discussions

Figure 3 shows the DC behavior of MSM devices and of reference gate diodes based on the HFET and the MOSHFET layer structures. The HFET gate diode shows a typical Schottky diode characteristic with a reverse current in the 10 to 50 nA region and a forward current that increases from 10 nA to 1 mA in a bias range from 1 to 1.5 V. Because the MSM-2DEG consists of two gate diodes connected back-to-back, the forward current of the forward-biased electrode of the MSM-2DEG is identical to the reverse current of the reverse-biased contact. Therefore, the applied bias voltage drops mainly across the reverse-biased contact, minus the well-defined voltage drop of 1 to 1.2 V across the forward-biased electrode.

Figure 3:
DC characteristics of the
MSM-2DEG devices
(electrode length = 2 µm,
spacing = 1 µm)
and of the reference gate diodes
(area = 25x25 µm²)

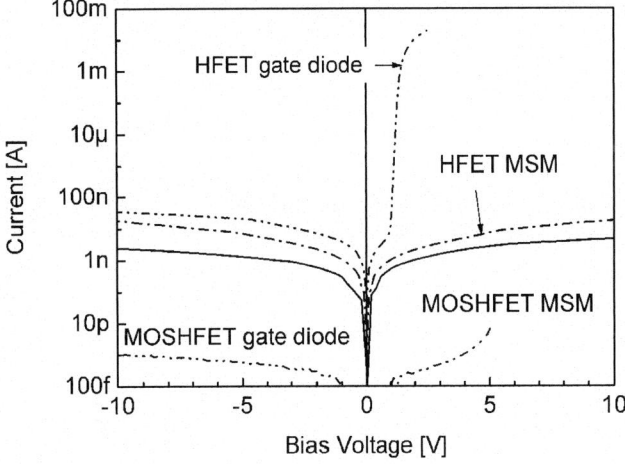

147

The MOSHFET reference gate diode shows essentially the leakage current in the pA range through the SiO_2 insulation layer. The current of the MOSHFET MSM device, on the other side, is governed by the current through the GaN buffer layer between the contact pads that connect the gate electrodes on the mesa with the buffer layer. The (although low) buffer conductivity creates a parasitic resistance in the Giga-Ohm range parallel to each MOSHFET gate diode. The current through this parallel resistance is several decades larger than the intrinsic MOSHFET gate current. This behavior has a large impact on the device performance because the voltage distribution between forward and reverse biased diode is governed by the parasitic parallel resistances rather than by the intrinsic diodes. It results in a more or less symmetric voltage distribution, affected by the different geometries of the contact pads, by the pad-to-semiconductor resistances, by the contribution of parasitic surface currents, by the trap influence in the buffer, etc. Obviously the relation between the bias voltage and the voltage drop of the reversely biased contact of the MOSHFET MSM is not as well-defined as for the HFET MSM diode.

The capacitance-voltage characteristics are evaluated by two-port S-parameter measurements with an HP Network Analyzer 8510C in the frequency range from 500 MHz to 50 GHz. The measurements are fitted to a common varactor equivalent circuit model of the device with varactor capacitance C, parallel conductance G, series resistance R and inductance L (inset of Fig. 5). Figure 5 shows the capacitance-voltage relationship of the MSM-2DEG devices. The capacitance of the HFET MSM is nearly constant up to the transition voltage of 5 V that corresponds to the threshold voltage of -4.1 V for HFET transistors in this layer system [8], added by the voltage drop of 1V of the forward-biased diode (Fig. 3). The capacitance of the unbiased device corresponds well with half the value of a plate capacitance with the area of one electrode and with the 30 nm thick AlGaN layer as dielectric [2]. Above the transition voltage where the reverse-biased electrode has completely depleted the 2DEG channel, the capacitance drops below 10 fF within a small voltage range.

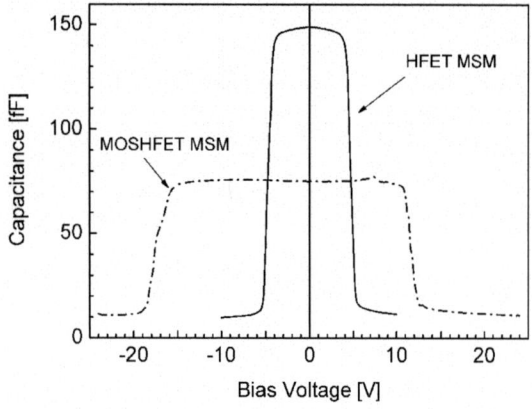

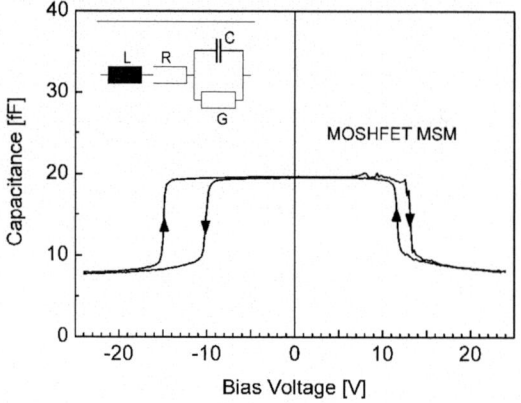

Figure 4:
Capacitance-voltage characteristics of the investigated devices. Electrode width and length are 50 μm and 2 μm, respectively.

Figure 5:
Hysteresis of the capacitance-voltage characteristics of the MOSHFET MSM. Electrode width and length: 50 μm / 0.5 μm. Inset: varactor equivalent circuit.

The capacitance characteristics of the MSM-2DEG based on the MOSHFET structure show large differences compared to the HFET devices: The low-voltage capacitance is lower because of the SiO_2 dielectric layer between Schottky electrode and semiconductor. The dielectric layer also increases the threshold voltage of the MOSHFET transistor from 4.1 V to 9.8 V [8]. Furthermore, the characteristic is asymmetric, i.e. the transition voltage V_{TRANS} depends on the bias polarity. It even depends on the measurement direction, as depicted in Fig. 2 for a device with 0.5 μm electrode length. The origin of both asymmetry and hysteresis lies in the undetermined and unstable distribution of the bias voltage between forward and reverse-biased contact that is governed by the parasitic currents through the buffer.

4.   Conclusions

We have fabricated MSM diodes above a two dimensional electron gas based on HFET and MOSHFET AlGaN/GaN layer systems. Fabrication of these devices is fully compatible with the transistor fabrication process. HFET based devices show the suitability for use as varactor diodes with large capacitance ratio and transition voltages defined by the layer structure and the 2DEG properties. The MOSHFET based devices, on the other hand, show a smaller capacitance swing, a hysteresis in the capacitance-voltage characteristics and a transition voltage that is strongly influenced by the parasitic buffer currents. Therefore, we conclude that in order to maintain the improvements of the transistor performance by the dielectric layer for the MSM-2DEG, the semiconductor layer stack as well as the dielectric material must be further optimized.

Acknowledgement

The authors would like to thank Dr. Juraj Bernát for sample preparation.

References

[1]:   M. Horstmann, K. Schimpf, M. Marso, A. Fox, and P. Kordoš, Electron. Lett., **32**, 763, 1996.

[2]:   M. Marso, M. Horstmann, H. Hardtdegen, P. Kordoš, and H. Lüth, Solid-State Electron., **41**, 25, 1997.

[3]:   M. Marso, M. Wolter, P. Javorka, A. Fox, and P. Kordoš, Electron. Lett., **37**, 14761478, 2001.

[4]:   Y. C. Pao, J. Franklin and C. Yuen, J. Cryst. Growth, **127**, 892, 1993.

[5]:   G. Simin, A. Koudymov, Z.-J. Yang, V. Adivarahan, and J. Yang, M.A. Khan, IEEE Electron Device Lett., **26**, 56, 2005.

[6]:   M. Marso, J. Bernát, M. Wolter, P. Javorka, A. Fox, and P. Kordoš, Advanced Semicon. Dev. & Microsystems, ISBN: 0-7803-7276-X, pp. 295298, 2002.

[7]:   V. Duez, X. Mélique, O. Vanbésien, P. Mounaix, F. Mollot, and D. Lippens, Electron. Lett., **34**, 1860, 1998.

[8]:   P. Kordoš, G. Heidelberger, J. Bernát, A. Fox, M. Marso, and H. Lüth, Appl. Phys. Lett., **87**, 143501 1-3, 2005.

Dark current of AlGaAs/GaAs n-QWIP
prepared on patterned (001) GaAs substrate by MOVPE

*Štrichovanec P., Kúdela R., Vávra I., and, [2]R. Srnánek, Novák J.

[1] Institute of Electrical Engineering, Slovak Academy of Sciences, Bratislava, Slovakia
[2] Faculty of Electrotechnical Engineering and Computer Science, Bratislava, Slovakia
* Corresponding author: eleks3ch@savba.sk, Phone: +421-2-5477 5806, Fax: +421-2-54775816

A 400 μm x 400 μm sized square quantum-well infrared photodetector with 30 periods of the GaAs/Al$_{0.3}$Ga$_{0.7}$As multiple quantum well has been prepared on patterned (001) GaAs substrate. The absorption spectrum at room temperature with normal incidence geometry is peaked at 9 μm with a full width at half maximum (FWHM) of 3,7 μm. Dark current with influence of 300K background was measured at room temperature and at 77K.

1. Introduction

A multi-quantum well (MQW) heterostructure designed to detect infrared (IR) light is the core of a quantum well infrared photodetector (QWIP). It is well known that n-QWIPs do not absorb radiation incident normal to the surface unless the infrared radiation has an electric field component normal to the layers of the MQW heterostructure (growth direction) [1]. Thus, various light coupling techniques, such as those based on 45° edge coupling, random reflectors, corrugated surfaces, two-dimensional grating structures, etc., have been used to couple the normal incidence radiation into the QWIPs. Epitaxial growth on patterned substrates is employed for the development of advanced semiconductor devices [2]. Main parts of experiments was done for quantum wire laser structures, have been fabricated by metalorganic chemical vapor deposition (MOCVD) in a single-step growth run on patterned GaAs substrates [3–5]. To develop such structures, it is of great relevance to understand how various growth aspects depend on growth temperature for different growth species on various crystal planes, which includes the migration of growth species between the planes [6].

2. Experiment

We prepared a patterned (001) GaAs substrate by wet etching via [1-10]-oriented stripes. Triangular-profile ridges were formed via controlled excessive underetching.
The etching solution H_3PO_4:H_2O_2:H_2O led to anisotropic etching with smooth surface morphology and for [1-10]-oriented stripes V-shape grooves with (111)A quasi-facet were revealed. The duration of the etching process was set to achieve a total undercutting of the etching mask. This excluded the (001)-related surfaces from the pattern.
We also applied another etching process based on a sacrificial AlAs/GaAs heterostructure [7]. The method makes it possible to prepare ridges or grooves with controllable tilt of the side facets. Using this process we prepared ridges with 30°-tilted facets via [1-10]-oriented mask pattern and simple V-shape grooves confined with to 45°-tilted facets via [110]-oriented mask. The process based on the sacrificial layer was terminated when the layer was etched away (Fig.1).
After the patterning, the substrates were MOVPE-overgrown with the MQW heterostructures, witch resulted in corrugated heterostructures. The growth conditions, such as temperature T=700°C and ratio V/III=167 for the MQW (191 for contact layers) were applied to achieve sharp heterostructure interfaces and high doping density in the well layer. The MQW structure

1-4244-0396-0/06/$25.00 ©2006 IEEE

consisted of 30 periods of 5-nm-GaAs quantum well layers (Si doped $n = 1 \times 10^{18}$ cm^{-3}) and of 40-nm-thick undoped barrier Al$_{0.3}$Ga$_{0.7}$As layers. This photosensitive MQW structure was sandwiched between 0.5-μm-thick GaAs top and bottom contact layers doped with Si to $n = 1 \times 10^{18}$ cm^{-3} to obtain an n–i–n photoconductive QWIP device. Single devices with area 400μm x 400μm were defined by wet chemical etching through a resist mask. Ohmic contacts were processed using photolithography and the evaporation of a Au/Sn metal layer on the top of the mesas and on the bottom areas between them.

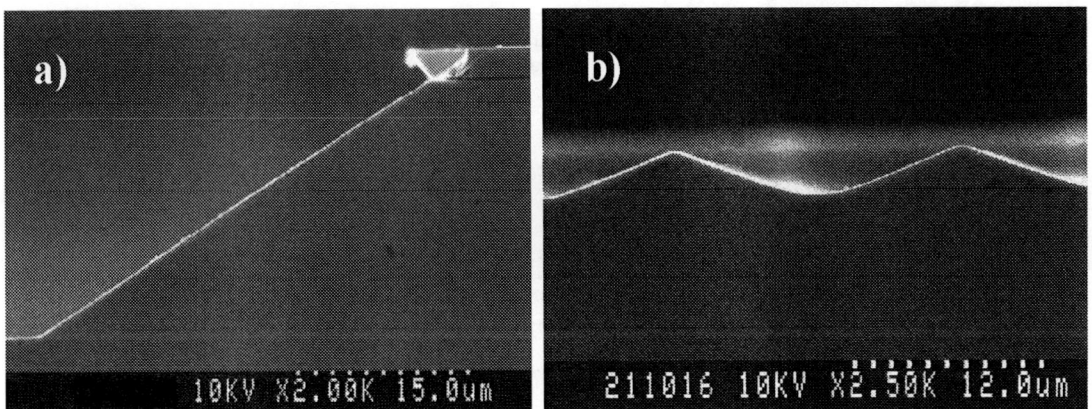

Fig. 1 Cross-section SEM of 20μm deep mesa-ridges revealed using the sacrificial heterostructure with stripe mask oriented in [1-10] direction **a)** and ridges with 30°-tilted facets via 15-μm-wide [1-10]-oriented mask stripes **b)**.

The layer thickness and morphology were analyzed using SEM, TEM. The MQW heterostructure on the ridges reavealed through the sacrificial heterostructure showed a minimal deviation of thickness on the side facets. The ratio between the thickness at top part of the heterostructure and at the side facets on the „30°-ridges" amounted to 0,72 for GaAs and 0,75 for AlGaAs. The surface roughness of the 30°-side facets because the side facets were far from the nearest well-defined crystallographic plane that may be revealed during this type of chemical etching process (the angle between the (211) and (001) planes is 35°). For the „45°-ridges", the ratio was changed to ~8 for the GaAs well and to 8.75 for the AlGaAs barier (Fig.2).

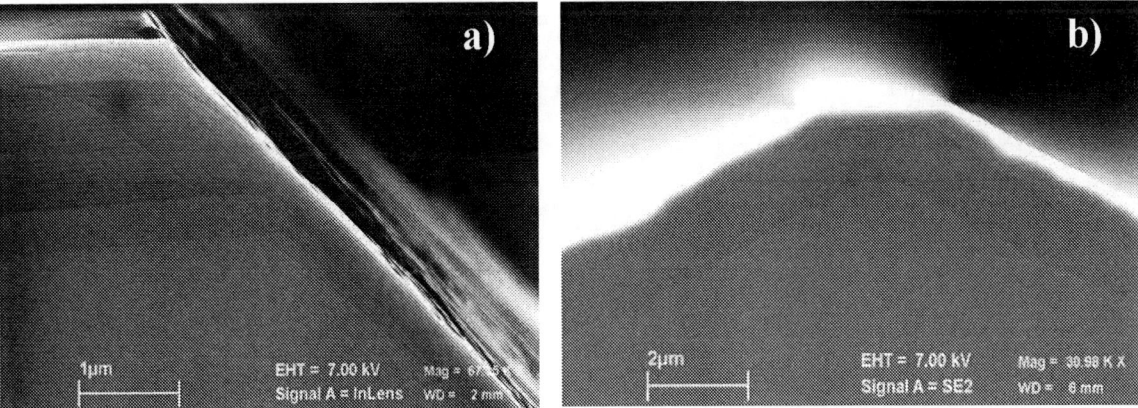

Fig. 2 Cross-section SEM picture of MQW heterostructure on GaAs mesa-ridges confined to facets tilted at 45° **a)** and MQW heterostructures on ridges tilted at 30° sidewall **b)**.

Normal incidence absorption measurments (Fig. 3b) showed a response peaked to 8.7 μm, with a shoulder around 7 μm and with a tail to longer wavelength for nonplanar sample. A relatively high absorption was observed at about 11 and 12 μm. For planar sample we measured only weak normal absorption, comparable to those obtained for GaAs substrate, resulted from no additional coupling mechanism (45° facets, diffraction grattings etc.)

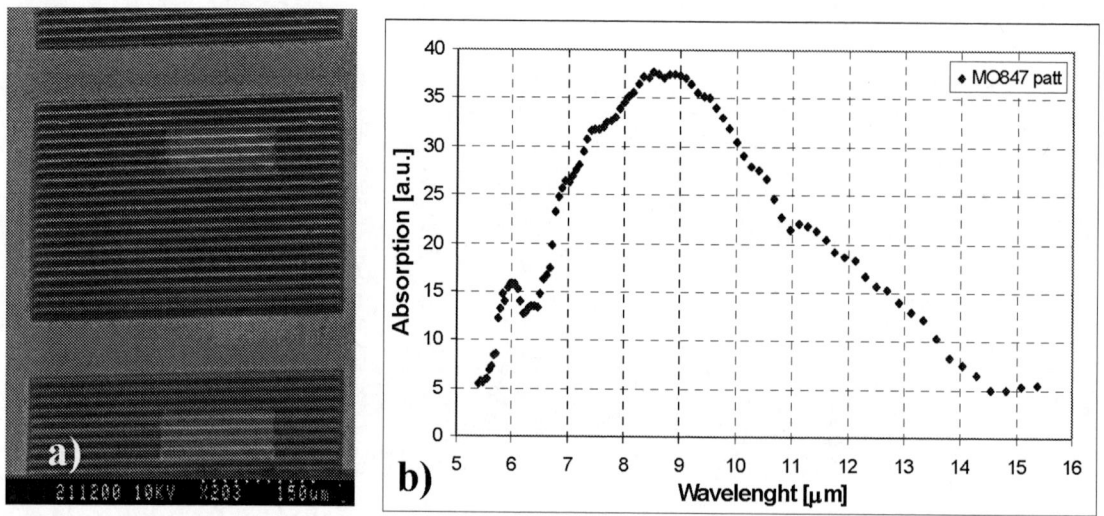

Fig. 3 The QWIP devices on patterned GaAs substrate with square area 400x400μm **a)** and absorption of normal incident IR light at room temperature (measured with Specord 200 setup) **b)**.

We measured the dark current of the device at low temperature to determine carrier transport properties of the non-planar MQW heterostructure. The dark was correlated with current-voltage characteristics of a planar MQW based QWIP (Fig.4). The asymmetrical I-V characteristic indicates the non-uniform doping in the quantum wells and the impurity out-diffusion during growth [8,9].

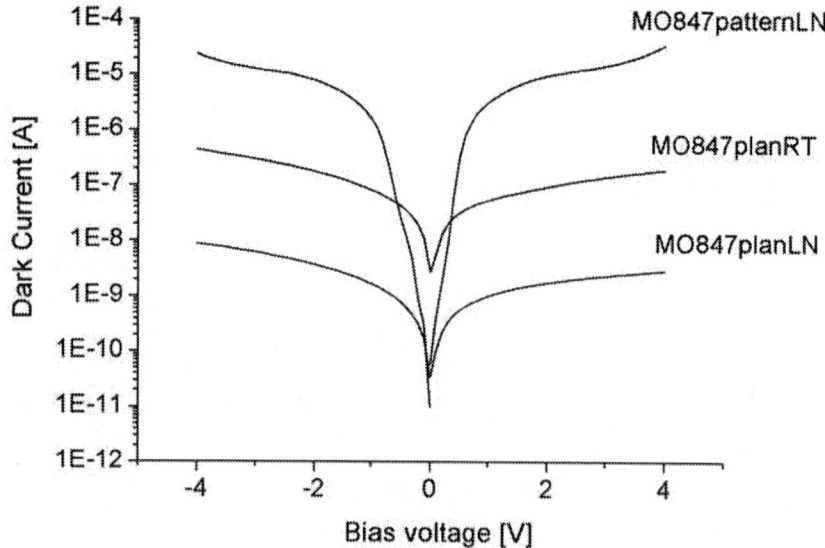

Fig. 4 Dark current of QWIP device on planar and patterned GaAs substrate at 77K (LN) and at room temperature (RT).

152

3. Conclusion

We prepared a QWIP heterostructure on patterned GaAs substrate with the aim to fabricate a normal incident device without additional gratings or corrugations. The sacrificial layer prepared on top of the (001) substrate allowed for the revelation of mesa-ridges confined to side facets with variable tilt. Influence of pattern surface properties on growth conditions plays important role of MQW heterostructure quality. We demonstrated a Si doped n-type QWIP device prepared on the [1-10]-oriented ridges confined to facets tilted at 30°. The normal incidence absorbtion spectra is peaked to range 8-9μm. The higher dark current of device with nonplanar MQW layer resulted from different barier thickness and quality of heterostructrure interfaces. We showed an alternative approach to making a normal incidence QWIP device by demonstrating the feasibility to form non-planar MQW heterostructures as the active region of such a device.

Acknowledgement

The authors gratefully acknowledge the support provided by I. Kostič and A. Šatka concerning the SEM measurements. This research was jointly sponsored under projects VEGA 2/6096/26, APVV-51-050602 and CENG programme.

References

[1] L. C. West and S. J. Eglash, Appl. Phys. Lett. **46**, (1985) 1156
[2] Maria P.P. de Castro, N.C. Frateschi, J. Bettini, M.M. de Carvalho, Journal of Crystal Growth **193** (1998) 510–515
[3] M. Okayasu, A. Kozen, J. Temmyo, S. Uehara, Optoelectron-Dev. Technol. **3** (1988) 73.
[4] A. Shima, H. Kizuki, A. Takemoto, S. Karakida, M.Miyashita, Y. Nagai, T. Kamizato, K. Shigihara, A.Adachi, E. Omura, M. Otsubo, IEEE J. Quantum Electron.1 (1995) 102.
[5] Z.L. Liau, S.C. Palmateer, S.H. Groves, J.N. Walpole, L.J.Missaggia, Appl. Phys. Lett. **60** (1992) 6.
[6] C.J. Chang-Hasnain, Y.A. Wu, G.S. Li, G. Hasnain, K.D.Choquete, C. Caneau, L.T. Florez, Appl. Phys. Lett. **63** (1993) 1307.
[7] V. Cambel, D. Gregušová, R. Kúdela, J. Appl. Phys 94/7, (2003) 4653
[8] E. Luna, A. Guzman, J.L. Sanchez-Rojas, E. Calleja, E. Munoz, Infrared. Phys. Technol. **44** (2003) 383
[9] E. Luna, A. Guzman, E. Munoz, Infrared. Phys. Techol. **47** (2005) 22

Study of electrically active defects in GaAs/InAs/GaAs QDs structures by DLTS and TEM

M. Prezioso, E. Gombia, R. Mosca, L. Nasi, A. Motta, P. Frigeri, G. Trevisi, L. Seravalli, S. Franchi

IMEM Institute-C.N.R., Parco Area delle Scienze 37a, 43010 Fontanini-Parma, Italy,

The apparent C-V profiles and the deep levels in GaAs/InAs/GaAs quantum dot nanostructure, have been investigated by space charge spectroscopy techniques (C-V and DLTS). Accumulation peaks and/or depletion of free carriers at the QDs plane are observed under the considered growth parameters. It is shown that, both in the cap and at the QD-layer/cap interface, the deposition of InAs QDs induces deep levels which exhibit a logarithmic dependence of the DLTS signal amplitude from the pulse width. Transmission Electron Microscopy (TEM) analysis shows extended defects in the cap and near the QD/cap interface, which have been correlated to electrical measurement results.

1. Introduction

In recent years quantum dots (QD) have attracted a great deal of attention for technological applications in novel optoelectronic devices such as lasers and optical memory structures. The characteristics of these devices strongly depend on both the quantum levels of the dots and the deep levels due to point or extended defects induced by QDs. Several papers have been devoted to the study of deep-level defects in InAs/GaAs QDs [1,2] However, since the deep levels features in QDs structures are very sensitive to the growth parameters and growth techniques, further investigations are required to better understand their nature and dependence on the growth procedures. In this work the apparent free carrier profiles and the deep levels in GaAs/InAs/GaAs quantum dot nanostructures, grown under different conditions, have been investigated by space charge spectroscopy techniques (C-V and DLTS). The results of the electrical characterization have been compared with Transmission Electron Microscopy (TEM) analysis in order to identify the origin of the detected traps.

2. Experimental

The investigated structures were grown in a Varian Gen II Modular MBE system on n$^+$GaAs substrates. They consist of a 1 μm thick n-GaAs buffer and a 10 nm thick undoped GaAs lower spacer, grown by Molecular Beam Epitaxy (MBE) at 600 °C, an InAs layer deposited at 460 °C capped by a GaAs upper spacer both grown by Atomic Layer MBE (ALMBE) and finally a 0.9 μm thick MBE-GaAs cap with the same growth temperature and doping used for the buffer.

1-4244-0396-0/06/$25.00 ©2006 IEEE

When the coverage of the InAs layer is above the critical value of 1.6 monolayers (ML), the growth of a continuous two dimensional InAs layer turns to the formation of self-assembled three dimensional QDs according to the Stransky-Krastanov growth mode. Different samples have been prepared by changing: i) the InAs coverage (2.4 and 3.0 ML) ii) the growth temperature of the undoped GaAs upper spacer from 360 to 460 °C and iii) the doping level N of the GaAs layers from 2×10^{16} to 1×10^{17} cm^{-3}.

After the formation of the AuGeNi ohmic contact on the backside of n$^+$ GaAs substrate, circular Au Schottky contacts, 400 μm in diameter, were prepared by photolithography and metal deposition on the cap layer. In samples with higher doping level, the GaAs cap has been partially etched before the deposition of the Au Schottky contact in order to investigate by C-V and DLTS the buffer or the region near the QDs. C-V measurements have been carried out by a 4192A HP impedance analyzer while DLTS were performed by a high sensitivity lock-in type spectrometer.

TEM analysis were carried out in Jeol 2000FX operated at 200 kV. Planar and cross sections were prepared by conventional mechano-chemical procedures followed by room temperature Ar ion milling.

1. Results and discussion

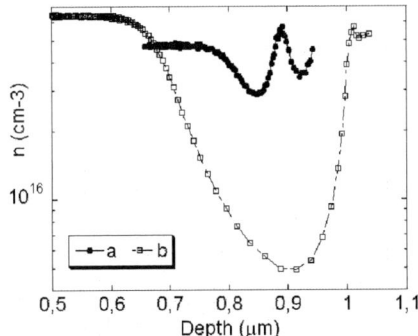

Fig.1. CV profiles showing a) strong depletion and b) accumulation peak.

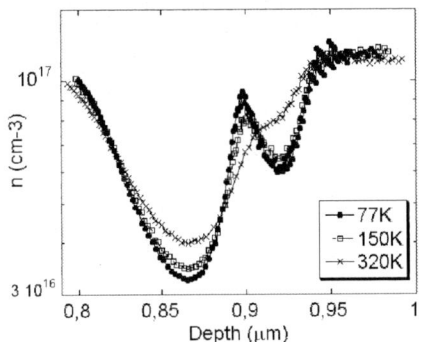

Fig.2. CV profiles, taken at different temperatures, of a quantum dot structure with highly

Fig.1 shows the C-V profiles of two QD structures with an InAs coverage of 3 ML and undoped upper spacer grown at 360 °C. Curve a) exhibits a typical accumulation peak, centred at about the depth of the InAs QDs while curve b) is characterized by a strong depletion.

It has been suggested that [3] the carrier reduction around the InAs plane is due to a high density of acceptor like levels and that the carrier accumulation is related to electrons in the QD and/or the wetting layer. The possible origin of different C-V profiles could be the different acceptor concentration and then the relative distance between the Fermi level E_F and the QDs levels, which is mainly determined by the local free carrier concentration. Indeed, if E_F is located under the QD ground state level, accumulation peaks cannot be observed since the QDs are always empty. This hypothesis seems to be confirmed by C-V measurements, carried out on samples with higher doping level ($N=1 \times 10^{17}$ cm^{-3}) (Fig.2), which show a depletion at RT and an accumulation peak at low temperatures, that has been attributed to the shift of E_F across the QD level

155

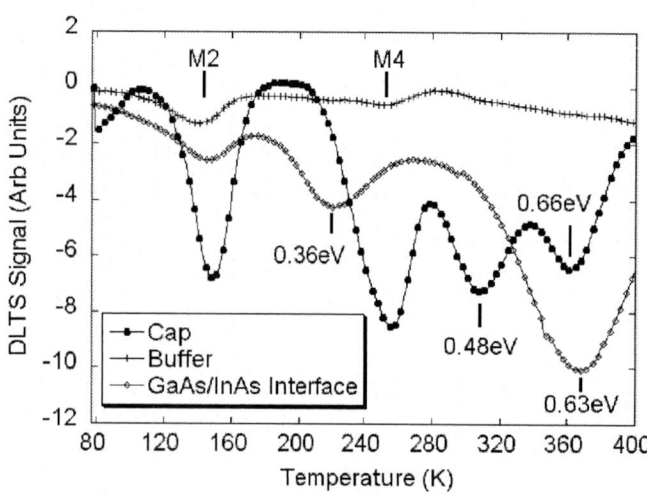

Fig.3. DLTS spectra related to buffer, cap, and interface regions. Rate windows is 400 s-1.

distribution due to the temperature decrease.

Depletions at the QD/GaAs interface have been also observed in samples with lower coverage (2.4 ML) and with upper spacer layers grown at higher temperature (up to 460 °C).

In order to identify the electrically active defects responsible of the carrier depletion, an accurate DLTS investigation of the buffer, cap and InAs/GaAs interface showed that (see Fig.3): i) the buffer layer exhibits only low concentrations $(1-3 \times 10^{13}$ cm$^{-3})$ of the so called M2 and M4 levels, which are typical of MBE-GaAs, ii) the cap layer, far from the interface, reveals higher concentrations of M2-M4 levels and two additional DLTS peaks corresponding to activation energies of 0.48 and 0.66 eV from the bottom of the conduction band; iii) the interface region is characterized, in addition to M2, by DLTS peaks, that in different samples, show activation energies in the 0.30-0.36 and 0.63-0.67 eV ranges. In particular the DLTS peak at 0.48 eV in the cap and the peaks at the interface show a logarithmic dependence of the DLTS signal amplitude from the pulse width suggesting that they could be associated to extended defects [4].

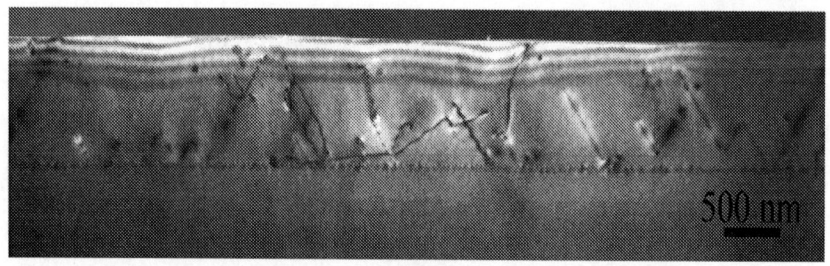

Fig.4. Cross section TEM micrograph showing threading dislocations and SFs starting from the QDs and reaching the free surface.

TEM analysis, carried out on the same samples, shows stacking faults (SFs) and threading dislocations starting from the QDs and reaching the free surface irrespective of the presence of accumulation or depletion in the C-V profiles (Fig.4).

Moreover, additional crystal defects strictly confined in the QDs region and identified as stacking fault tetrahedra, are observed only in samples with strong depletion in the C-V profiles. Unlike the other SFs, the stacking fault tetrahedra are closed defects which do not extend more than about 100 nm from the interface where they originate, as clearly seen in Fig.5.

Fig.5. Cross section TEM micrograph showing the stacking fault tethraedra confined at the QDs/cap interface.

These results suggest that: i) acceptor states induced by SF tetrahedra could originate the strong reduction of free carriers near the QDs/cap interface, ii) the level at 0.48 eV in the cap, far from the interface, could be associated to SF and/or threading dislocations extended over the whole depth of the layer and iii) the DLTS peaks at the interface could be attributed to SF tetrahedra located near the QDs plane.

!. Conclusions

MBE-grown InAs/GaAs quantum dots structures have been investigated by space charge spectroscopy techniques and TEM analysis. Strong depletions of free carriers near the InAs layer have been correlated with the presence of stacking faults which extend up to 100 nm above the QDs/cap interface. Moreover, the presence of SF tetrahedra near the interface and SF and threading dislocations extending up to the surface have been associated to DLTS peaks, showing logarithmic dependences of the signal amplitude from the pulse width.

Acknowledgements
The work was partially supported by the "SANDiE" Network of Excellence of EU, contract no NMP4-CT-2004-500101

References

[1] M. Kaniewska, O. Engström, A. Barcz, M. Pacholak-Cybulska, Material Science and Engineering C 26, 871 (2006)
[2] C. Walther, J. Bollmann, H. Kissel, H. Kirmse, W. Newmann, W.T. Masselink, Appl. Phys. Lett.76, 2916 (2000)
[3] E. Gombia, R. Mosca, S. Franchi, P. Frigeri, C. Ghezzi, Mat. Sci. and Eng. C 26, 867 (2006)
[4]T. Wosinski, J. Appl. Phys. 65, 1566 (1989)

Technology related issues regarding fabrication of AlGaN/GaN-based MOSHFETs with GdScO₃ as dielectric

G. Heidelberger[1], M. Roeckerath[1], R. Steins[1], M. Stefaniak[1], A. Fox[1], J. Schubert[1],
N. Kaluza[1], M. Marso[1], H. Lüth[1], and P. Kordoš[2,3]

[1] Institute for Bio- and Nanosystems and cni - Center of Nanoelectronic Systems for
Information Technology, Research Centre Jülich, D-52425 Jülich, Germany,
e-mail: g.heidelberger@fz-juelich.de

[2] Institute of Electrical Engineering, Slovak Academy of Sciences, SK-84104 Bratislava,
Slovakia

[3] Department of Microelectronics, Slovak Technical University, SK-81219 Bratislava,
Slovakia

Starting out from our well established process for AlGaN/GaN HFETs we discuss ways to enrich the process in order to fabricate Metal-Oxide-Semiconductor HFETs (MOSHFETs) with a Gadolinium Scandate (GdScO₃) insulation layer. In particular, adequate processing orders, various etching procedures and possible drawbacks of the GdScO₃ deposition process on ohmic contacts are discussed.

Making use of the gained knowledge we fabricated GdScO₃-MOSHFETs for the first time. Compared to a conventional HFET the new device shows a higher saturation drain current and a lower gate leakage current. Nevertheless, the potential insulating properties of GdScO₃ are not fully exploited yet and further optimization of the deposition process is needed.

1. Introduction

AlGaN/GaN-based Metal-Oxide-Semiconductor Heterojunction FETs (MOSHFET) using dielectrics such as SiO_2 or Al_2O_3 have been demonstrated to be superior to conventional HFETs, showing lower gate leakage currents, higher saturation drain current and higher rf-power performance [1, 2].

One disadvantage of today's MOSHFETs arises from the fact, that the capacitance of the dielectric layer is in series with the capacitance of the AlGaN-depletion-region, thus the total capacitance will always be less than the smaller capacitance – which is the latter. As a consequence, the channel properties can not be controlled as effectively for MOSHFETs as for HFETs. For today's dielectric materials, the decrease of total capacitance is significant, e.g. for a MOSHFET with 10 nm SiO_2 dielectric layer and 30 nm AlGaN-layer the total capacitance falls down to 50 % of the value of a conventional HFET [3].

To overcome this problem, we investigate dielectrics with higher dielectric constant such as HfO_2 and GdScO₃ (GSO), which have dielectric constants beyond 20. In this work, we will give a survey of our efforts to modify our standard HFET process to fabricate MOSHFETs with a GSO-dielectric and present results of the first GSO-MOSHFET realized.

2. Technological Issues

For our HFETs as well as for our MOSHFETs we use AlGaN/GaN-layers epitaxially grown by MOVPE on sapphire substrates. The standard fabrication process consists of etching mesa-tables, fabrication of ohmic contacts (optical lithography, deposition of Ti/Al/Ni/Au and subsequent annealing at 900 °C for 30 sec), gate contacts (e-beam

1-4244-0396-0/06/$25.00 ©2006 IEEE

lithography, Ni/Au deposition) and contact-pads (optical lithography and deposition of Ti/Au).

To incorporate the dielectric layer deposition into the standard HFET process one can chose between two reasonable options: either the dielectric is deposited before mesa-etching (advantage: ideal deposition conditions; disadvantage: difficult task to etch dielectric to open window for ohmic contacts) or the dielectric is applied just ahead of gate-electrode processing (advantage: hardly any distortions of the dielectric structure, easy task of etching the dielectric to enable pad contact; disadvantage: high temperature deposition process may affect ohmic contact behaviour).

Our favoured technology to deposit GSO is electron beam evaporation. This method utilizes a conventional deposition chamber which provides high vacuum conditions ($1-5 \times 10^{-6}$ mbar). For deposition of stoichiometric GSO thin films a stoichiometric evaporation source is used, which is made from a mixture of the metal oxides. The process temperature is 600 °C and the duration of deposition is 5 min. For further details concerning the deposition process see [4].

During the adaptation of our process it showed that the most challenging task is to etch the dielectric, since hardly any experiences regarding this issue have been reported so far. Therefore we will mainly focus on the first technological option, in particular we address the

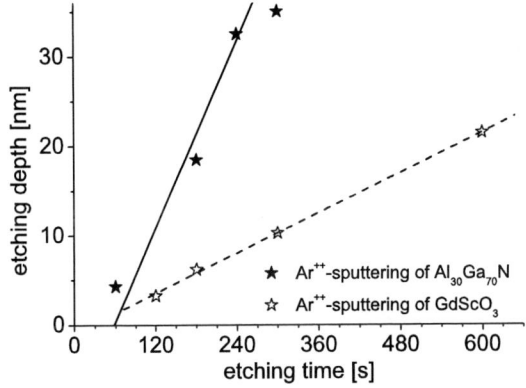

Fig. 1: Etching rates for GdScO₃ and AlGaN using Ar⁺⁺ IBE.

Fig. 2: Comparison of contact resistances of conventional Ω-contacts and Ω-contacts based on etched AlGaN material

situation where GSO has already been deposited on AlGaN, the mesa-tables have been etched and the photo resist mask for ohmic contacts has been structured. In order to fabricate ohmic contacts, a window needs to be etched into the GSO layer. This requires that the defined distance between source and drain contacts remains unchanged, i.e. underetching is to be minimized, and the behaviour of the ohmic contacts must be comparable to those of conventional HFETs, i.e. in the ideal case selectivity and/or controllability of the etching process is such that only GSO is removed.

In a first approach, wet etching by means of various HCl-solutions was performed, but severe underetching of up to 1 μm was observed. Considering the fact that the distance between source and drain electrodes varies from 2 μm to 5 μm, the underetching is beyond any acceptable tolerance range. Trying to use a buffered HCl-solution with different pH-values turned out to be ineffective as well. As a conclusion, wet etching was ruled out as a potential option for our MOSHFET technology.

In a second attempt, dry etching by means of Ar^{++}-sputtering was investigated. As expected, critical underetching was not observed. But although the Ar^{++} process shows a highly linear, thus controllable etching rate of 2.2 nm/min (Fig. 1), the lack of selectivity inherent to dry etching methods became an issue: To be sure to remove GSO entirely one has to take the risk of etching into the AlGaN layer. And since AlGaN is etched relatively fast with approximately 8 nm/min it is likely that a significant part of the AlGaN layer is removed as well.

Under these circumstances, we had to further investigate if the subsequently processed ohmic contacts were still comparable to those fabricated in the standard process. Therefore we prepared samples where 75 % and 100 % of the AlGaN layer was etched away and processed ohmic contacts according the standard process. TLM measurements showed that the contact resistances were in the range of conventional ohmic contacts. To illustrate the deviations even on a single probe two distinct reference contacts are shown in Fig. 2. DC-measurements of the fabricated devices confirmed the fact that the contact resistances are in the same range as for conventional HFETs although the AlGaN-layer has been etched as well. Thus, we conclude that using the first technological option, i.e. deposition of GSO before mesa-etching, yields comparable MOSHFETs if Ar^{++}-sputtering is used as etching technology.

With regard to the second technological option, i.e. deposition of GSO before gate-processing, the major question is whether ohmic contacts degrade if they are exposed to high temperature during the GSO-deposition process. We compared the properties of conventionally fabricated ohmic contacts with those being exposed to conditions usually encountered during GSO-deposition, i.e. temperature of 600 °C for 5 min. It showed that the behaviour of the contacts does not change significantly.

3. Fabricated Devices: Results and Discussion

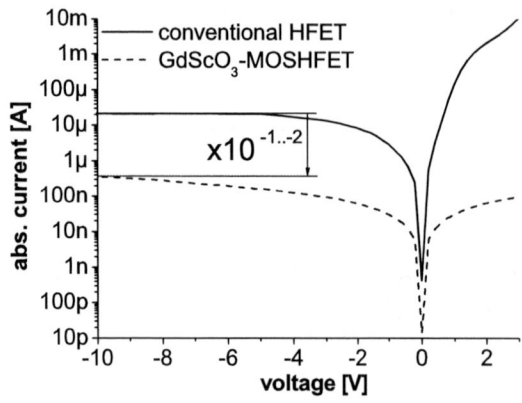

Fig. 3: Gate leakage current of conventional HFET and GSO-MOSHFET.

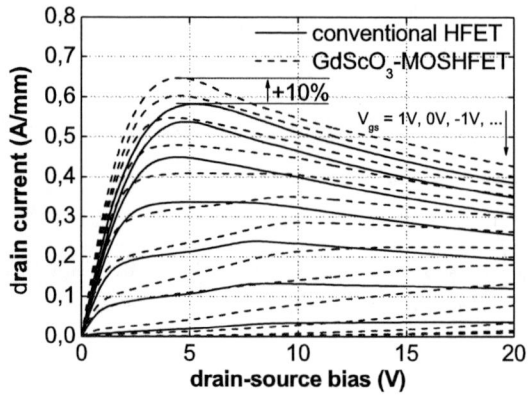

Fig. 4: Output characteristic of conventional HFET and GSO-MOSHFET, both with gate length of 700 nm and gate width of 300 μm.

Encouraged by these results, we fabricated MOSHFETs with GSO as dielectric layer for the first time. By separating the gate from the AlGaN the gate leakage currents can be reduced by one to two orders of magnitude (see Fig. 3). This result can be considered a good first attempt. But since SiO_2-MOSHFETs are capable of reducing leakage currents by up to four orders of magnitude [1, 3] this result is still to be improved by optimizing the GSO layer.

DC-measurements showed comparable contact resistances, as can be seen in Fig. 4 by comparing the slopes of the drain current within the linear regime. Moreover, the peak drain current was increased by 10 % at a gate-to-source bias of 1 V. For the MOSHFET, an overshot of drain current at the first drain-to-source voltage sweep is observed (not depicted here) which might be due to oxide related interface traps. The gate-source capacitance (C_{gs}) is 11 % smaller in the MOSHFET case and justifies the use of a high-k material as dielectric: if SiO_2 would have been used C_{gs} would have been 50 % smaller. The peak extrinsic

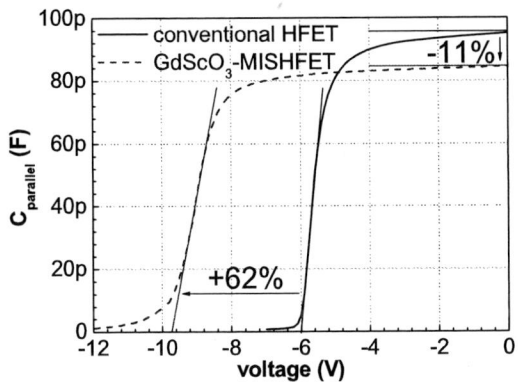 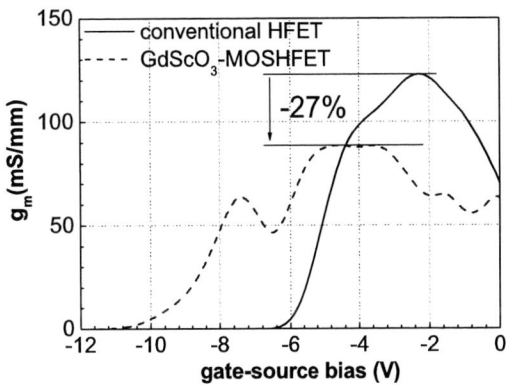

Fig. 5: Gate-source capacitance of conventional HFET and GSO-MOSHFET.

Fig. 6: Extrinsic transconductance of conventional HFET and GSO-MOSHFET.

transconductance (g_m) of the MOSHFET is about 27 % less than that of the conventional HFET, this gives a hint to the rf-performance one can expect: the cut-off frequency (f_T) is proportional to the ratio of g_m and C_{gs}, thus for GSO-MOSHFETs f_T will be lower than for the conventional HFET.

4. Conclusion and Outlook

Either of the described modifications of the standard HFET technology and the development of a suitable GSO etching process enables us to fabricate AlGaN/GaN-based GSO-MOSHFETs. Now, further investigation can focus on the optimization of the deposition process of GSO itself, since the results concerning electrical insulation of the dielectric layer is still to be optimized.

References

[1] J. Bernát, D. Gregušová, G. Heidelberger, A. Fox, M. Marso, H. Lüth, and P. Kordoš, *Electronics Letters*, **41**, 11, 2005, pp. 667 – 668.

[2] T. Hashizume, S. Ootomo, and H. Hasegawa, *Applied Physics Letters*, **83**, 14, 2003, pp. 2952 – 4.

[3] P. Kordoš, G. Heidelberger, J. Bernát, A. Fox, M. Marso, and H. Lüth, *Applied Physics Letters*, **87**, 14, 2005, pp. 143501-1 – 3.

[4] M. Wagner, T. Heeg, J. Schubert, S. Lenk, S. Mantl, C. Zhao, M. Caymax, and S. De Gendt, *Applied Physics Letters*, 88, 2006, 172901-1 – 3.

Investigation of GaN/ ZnO heterostructures properties

J.Kováč, J. Škriniarová, P. Kúdela, I. Novotný, J. Bruncko, D,Donoval,
J. Jakabovič, M. Michalka, Ľ. Jánoš, A.Vincze, D. Haško

Microelectronics Dept., Slovak University of Technology
& International Laser Centre, Ilkovičova 3, 812 19, Bratislava, Slovakia
e-mail:jaroslav kovac@stuba. sk

We report on the fabrication and measurement of electrical and optical properties of GaN/ZnO heterostructures. The I-V characteristics of the fabricated diodes revealed the ohmic and rectifying behavior under different conditions of ZnO films deposition. The sputtered ZnO films on n type GaN shows ohmic character and are promising for the transparent contact formation. Formation of p-type ZnO film was find out from measured I-V characteristics of n GaN/ZnO heterostructures prepared by pulsed laser deposition and could be a promising for potential application in optoelectronic devices.

1. Introduction

Recently, GaN and ZnO both wide-band-gap semiconductors, have attracted considerable interest because of their potential application in blue and ultraviolet optoelectronic devices because of its direct wide- band gap of 3.4 eV and large exciton binding energy ~ 60 meV [1,2]. The similarity of GaN and ZnO in terms of energy gap, crystallographic structure and high thermal stability make them attractive candidates for forming heterojunctions. The growth processes naturally produce ZnO/GaN interfaces and the crystalline nature of grown layers depends on the electronic properties of ZnO/GaN interfaces. This is because due to the difference of valence electron number between ZnO and GaN, i.e. the valency mismatch, the Ga-O donor and Zn-N acceptor bonds appear at the interface, which often produces charged defects such as vacancy. The ZnO/GaN interface shows the type-II band alignment with an average valence-band offset of 1.6 eV, where both the valence-band top and the conduction-band bottom of ZnO are located below those of GaN [3]. The undoped ZnO typically exhibits n- type conductivity due to the presence of native defects, which makes acceptor doping of ZnO difficult. However, ZnO n-type conductivity is relatively easy to realize via excess Zn or with Al, Ga or In doping. To make ZnO - based semiconductor optoelectronic devices, p-type ZnO is needed, while candidates of p-type dopant introduce deep acceptor levels. The most promising dopants for p-type material are the groups V elements. The lack of p-type behavior in p-doped ZnO has been addressed theoretically within the context of N-N complex formation [4]. P-type ZnO:N has been reported for films grown by PLD [5,6] and sputter deposition [7]. Measurements of the electrical properties of high-resistance zinc oxide (ZnO) are strongly influenced by the sample annealing.

2. Experimental

ZnO films were deposited by RF sputtering and pulsed laser deposition (PLD) on different doped GaN layers epitaxially grown on sapphire substrates with (UD n-type $1x10^{16}$ cm^{-3},n –type $2x10^{18}$ cm^{-3} and p-type $1x10^{17}$ cm^{-3}). At first a thin unintentionally doped polycrystalline ZnO films were prepared in a planar RF sputtering diode system

1-4244-0396-0/06/$25.00 ©2006 IEEE   162

Perkin Elmer 2400/8L from hot-pressed ceramic ZnO target (99.99 % purity, in 20 cm diameter) in Ar atmosphere. The sputtering power was 650 W and the sputtering time varied from 12 to 30 minutes and accordingly thicknesses of ZnO films were in range of 500 nm. The film thickness was measured by Talystep instrument. The typical AFM image of the deposited ZnO film is shown in Fig. 1. In the second set ZnO films were prepared by PLD. During PLD process the vacuum chamber was pre-evacuated before deposition (approx. 10^{-3} Pa) and pure oxygen atmosphere at the level of 35 Pa and substrate temperature of 400° C was maintained during the deposition. The target (ZnO sintered pellet of high purity) was ablated by laser pulses at 355 nm (third harmonic generation of Nd:YAG laser) with energy fluency density of 2,5 $J.cm^{-2}$. The focusing lens rotated with subtle eccentricity for purpose to get more homogenous ablation of the target and growth of the films. After deposition the sample were cooled in pure oxygen at atmospheric level pressure without annealing. The thickness of ZnO films were in the range of 300 nm.

After ZnO films growth, Al contacts 200 nm thick were deposited using lift-off technique on the top of the ZnO surface, followed with mesa etching of ZnO layer (ϕ= 100 μm) for realizing the devices (Fig.2). For processing the etching solution was used, consisted of acetic acid: phosphoric acid: de-ionized water in ratio of 2:1:75. After etching Al bottom contacts were deposited on GaN layer. The electrical and optical properties of devices were measured before (without, w/o) and after rapid thermal annealing in temperature range of 500-600 °C using N_2 gas atmosphere. The I-V characteristics were measured using Keithley 2400 and 238 source meter in dark and under illumination by halogen lamp. For photocurrent measurement broadband halogen lamp, SPM2 monochromator and lock-in preamplifier was used.

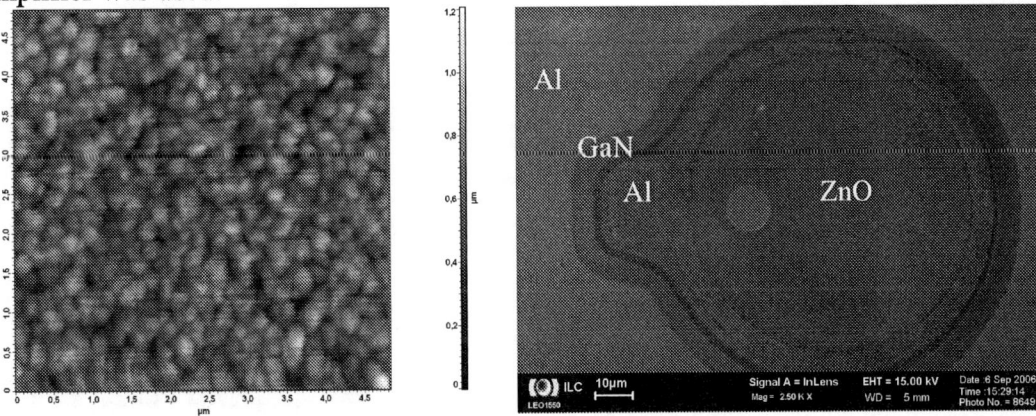

Fig.1 AFM image of sputtered ZnO structure   Fig. 2 SEM image of mesa etched ZnO device

3.   Results and discussion

As a first the properties of sputtered n-GaN/ZnO heterostructures were investigated. The doped n GaN sample shows high ohmic conductivity without annealing. The I-V characteristic of the low doped n GaN sample without annealing shows ohmic contact character in dark as shown in Fig.3. The annealing process by using ambient gases N_2 changes the I-V characteristics to nonlinear character. This is caused by the modification of ZnO layer conductivity and Al/AlGaN layer formation at GaN interface as was confirmed by SIMS measurements. The measured results confirmed n type conductivity of ZnO layer without annealing, which can be modified by annealing in different atmosphere creating interface defects and nonlinearities in I-V characteristics. This property was confirmed on p GaN/ZnO structures where p-n junction properties were achievable after annealing. The I-V characteristic of sample without annealing shows very high resistance as shown in Fig.4, which changed to diode characteristic after annealing in N_2 atmosphere at 600 °C. Because of the interface ZnO layer shows high resistance, high driving voltage is needed to reach higher current. The GaN/ZnO structures prepared by PLD deposition at substrate temperature 400 °C in oxygen atmosphere shows formation of p-type ZnO film. This was confirmed

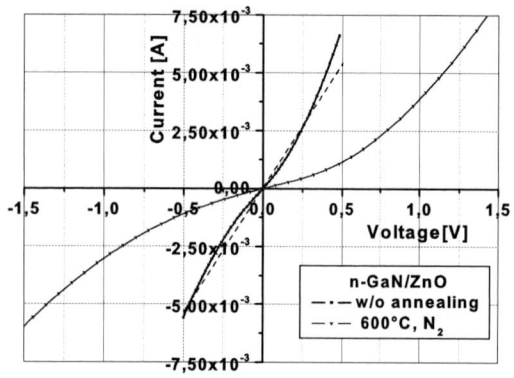

Fig. 3 I-V characteristics of sputtered
n ZnO/n GaN w/o and with annealing

Fig.4 I-V characteristics of sputtered
n ZnO/p GaN without and with annealing

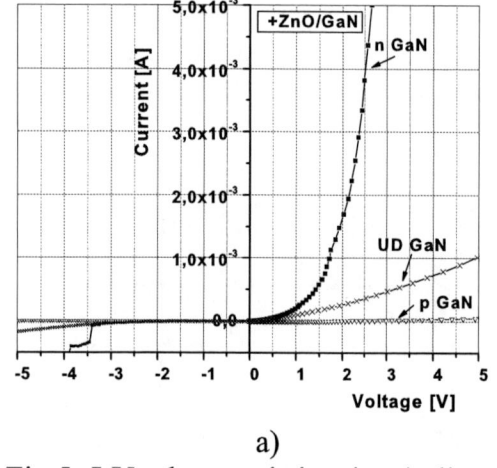

a)

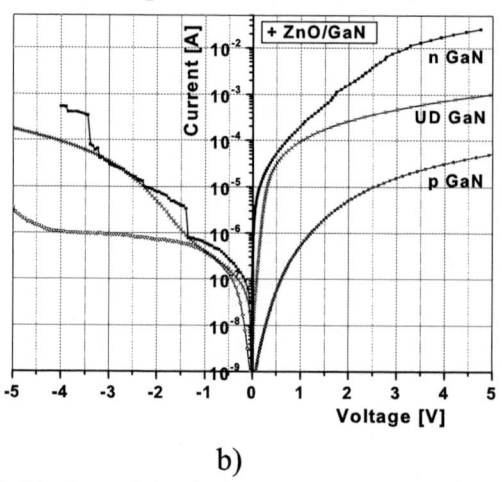

b)

Fig.5 I-V characteristics in a) linear and b) logarithmic scale of PLD p ZnO/GaN
without annealing on different doped GaN

by measurement of I-V characteristics on different type GaN layers as shown in Fig. 5a in linear and Fig. 5b logarithmic scale, respectively. The I-V characteristic on n type doped GaN shows typical diode characteristic. Similar characteristics show undoped GaN layer with higher resistance in forward bias. The p doped GaN sample shows very high resistivity due to Schottky contact formation at Al/p GaN interface, as was confirmed by the test measurements. This behaviour of p type ZnO formation correspond to previous Hall measurements
of the PLD deposited ZnO films, where under certain growing conditions the p type layer formation was measured. Because of photosensitivity of the structures the photocurrent spectroscopy was employed for the structures investigation. In the photocurrent spectra of ZnO/GaN samples dominate the optical transition around 3,4 eV (Fig.6), which is characteristic for band to band absorption in both GaN and ZnO layers. The photons with higher energy are absorbed near GaN/ZnO interface and a sharp spike at the spectral characteristic indicate the formation of excitons at the interface. The shift of the absorption

edge for sputtered and PLD ZnO films indicate the different structure properties. The SIMS depth profile of ZnO, Ga and Al of PLD of prepared ZnO/GaN structure without and after annealing at 550 °C is shown in Fig.7. The GaN/ZnO interface shows relatively sharp interface with slight interpenetration after annealing. The top Al contact penetrate into the ZnO layer already after the deposition and diffuse deeply into ZnO film after annealing

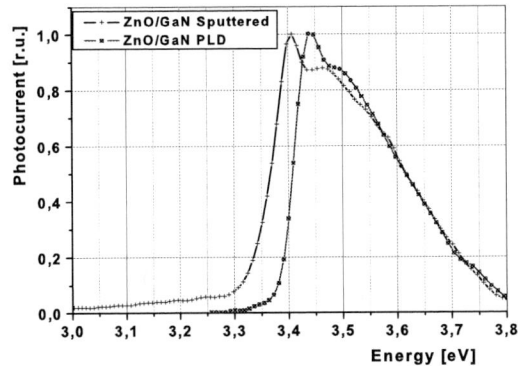

Fig.6 Photocurrent spectra of different
ZnO/GaN devices

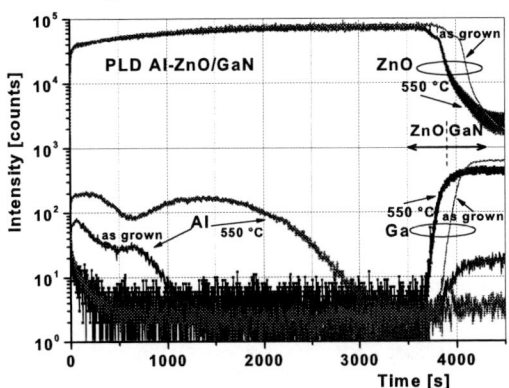

Fig.7 SIMS profile analysis of Al/ZnO/GaN
structure prepared by PLD

creating n type doping. This was revealed also by measurement of I-V characteristics of the annealed samples that shows high leakage current in comparing with not annealed sample.

4. Conclusions

The properties of different doped GaN/ ZnO heterostructure prepared by RF sputtering and PLD technique have been investigated. The sputtered ZnO films on n type GaN shows ohmic character and are promising for the transparent contact formation, applicable in optoelectronic devices while ZnO/pGaN structures shows diode characteristics. Formation of p-type ZnO film was find out from measured I-V characteristics of GaN/ZnO heterostructures prepared by PLD. These indicate that under certain PLD growing conditions p-type ZnO film can be prepared and is promising for potential application in optoelectronic devices.

Acknowledgement

This work was supported by Slovak Grant Agency projects 1/3108, 3076, 3098/06 and 1/2041/05.

References

[1] E. Kaminska, A. Piotrovska, K. Golaszewska, R. Kruszka, A.Kuchuk, J.Shade, A.Winiarski, J.Jasinski, *Journal of Alloys and Compounds,* **371,** 129, 2004

[2] Ya. I. Alivov, Ű. Özgűr, S. Doğan, C. Liu, Y. Moon, X. Gu, V. Avrutin, Y. Fu, H. Morkoc, *Solid-State Electronics,* **49,** 1693, 2005

[3] T.Nakayama , M. Murayama, *Journal of rystal Growth ,* **214/215,** 299, 2000

[4] D. P. Norton, Y. W. Heo, M. P. Ivill, K. Ip, S. J. Pearton, M. F. Chrisholm, T. Steiner, *Nanotoday,* **June,** 34, 2004

[5] K. Vanheudsen, C. H. Seager, W. I. Warren, D.R.Tallant, *Appl. Phys. Lett.* **68,** 403,1996

[6] Xin-Li Guo, Hitoshi Tabata, Tomoji Kawai, *Journ. Crystal Growth,* **223,** 135, 2001

[7] K. K. Kim, H. S. Km, D. K. K. Hwang, J. H. Lim, *Appl. Phys Lett.,* **83,** 63, 2003

Preparation and properties of AlGaN/GaN MOSHFETs with MOCVD Al_2O_3 as gate oxide

R. Stoklas[1], K. Čičo[1], D. Gregušová[1], J. Novák[1], and P. Kordoš[1,2]

[1] Institute of Electrical Engineering, Slovak Academy of Sciences, SK-84104 Bratislava, Slovakia

[2] Department of Microelectronics, Slovak Technical University, SK-81219 Bratislava, Slovakia

The I-V and C-V measurements were used for characterization of AlGaN/GaN HFETs and MOSHFETs. Deposited Al_2O_3 gate oxide yielded an increase of the sheet carrier density from 6.9×10^{12} to 7.25×10^{12} cm^{-2} and subsequent increase of the drain current density up to 40% and increase of the extrinsic transconductance up to 37%, respectively. The gate leakage current of 10^{-5} A/mm at -10 V is about 3 orders of magnitude lower than that of HFETs. This results indicate the suitability of a thick Al_2O_3 gate oxide for improvement of electronic properties of GaN-based MOSHFETs.

1. Introduction

Wide band-gap semiconductor GaN is a suitable material for high-frequency, high-power and high-temperature applications. The GaN-based HFETs can operate at temperatures around 600 °C and maintain high current densities (>1A/mm). However, surface effects are supposed to be responsible for the degradation of their high-frequency performance (current collapse, DC/RF dispersion). One way to suppress the deficiency is realized by gate insulation. Encouraging results have been presented recently on GaN-based MOSHFETs, e.g. an output power of 7.5 W/mm at 2 GHz [1] and 6.7 W/mm at 7 GHz [2] on a non-field-plated devices and 20 W/mm at 2 GHz [1] on field-plated device. However, the question of a suitable gate dielectric is still open. A significantly lower device performance was observed if a highly insulating layer was used instead of a semi-conducting one [1].

In this paper, AlGaN/GaN MOSHFETs with "high-κ" Al_2O_3 gate oxide are under study. Performance of MOSHFETs is compared with HFETs prepared on similar material structure and advantages of MOSHFETs are demonstrated.

2. Experiment

The layer structure consisted of a 1-μm-thick GaN and a 28-nm-thick AlGaN (23% Al) layers, both nominally undoped, grown by MOCVD on sapphire substrate. The device processing consisted of conventional HFET fabrication steps. As an Ohmic contact of devices, a Ti/Al/Ni/Au multilayer was deposited on the AlGaN surface by rapid thermal annealing at 850°C for 30s in N_2 ambient. The Schottky gate metalization consisted of a Ni/Au double layer patterned by optical lithography. HFETs with a gate width of 70μm and gate lengths of 2 and 4μm were prepared. FatHFETs with variable gate dimensions 5, 10 and 20x250 $μm^2$, and 40x500 $μm^2$ and were processed simultaneously with the HFET devices. An MOCVD technique was used to deposit 9- and 14-nm-thick Al_2O_3 gate oxide to create MOSHFETs. The dielectric Al_2O_3 layers were deposited at 600°C. The thickness of the Al_2O_3 layers was evaluated by x-ray reflectivity measurements. The devices were characterized by the DC (output and transfer), pulsed $I-V$, and $C-V$ measurements. $I-V$

1-4244-0396-0/06/$25.00 ©2006 IEEE

measurements were performed using HP 4140B and Keithley 2400 devices. HP 4140B was used as the gate source, and Source/Meter Keithley 2400 was used as the drain source. The C-V measurements were performed using Agilent 4284A LCR meter. The fatHFETs were used to characterize the carrier transport properties of the layer structure used.

3. Results and Discussion

The $C-V$ measurements were used to characterize the layer structure. Fig. 1 shows $C-V$ characteristics of the fat-HFETs and MOS fat-HFETs with variable gate-oxide thickness (f = 1MHz). The measurements were realized on fat-HFETs and MOS fat-HFEts with an active area of 40x500 μm^2. The capacitance of HFET $C_0^{HFET} = 295$ nF/cm^2 was assigned from the $C-V$ dependence. The AlGaN thickness of 27 nm was evaluated from the capacitance $C_0^{HFET} = 295$ nF/cm^2 on the HFETs using the equation

$$d_{AlGaN} = \frac{C_0^{HFET} A}{\varepsilon_{AlGaN}} \tag{1}$$

where d_{AlGaN} is the AlGaN thickness, A is the gate area, and ε_{AlGaN} is the AlGaN dielectric constant. The capacitances of the MOSHFETs $C_0^{MOSHFET} = 215$ and 188 nF/cm^2 for 9- and 14-nm-thick of the Al$_2$O$_3$ layer were assigned from the $C-V$ dependences, respectively. Values of dielectric constant of Al$_2$O$_3$ $\varepsilon_{Al_2O_3} = 8.1$ and 8.2 were evaluated from the MOSHFET capacitances using the following equation

$$\varepsilon_{Al_2O_3} = \frac{C_0^{MOSHFET} A}{\left(d_{AlGaN} + d_{Al_2O_3} \right)} \tag{2}$$

This is in very good agreement with published data for MOCVD grown Al$_2$O$_3$. The sharp capacitance transition from depletion to 2DEG accumulation indicates a high quality of gate dielectric. The threshold voltages of the MOSHFETs $V_{th} = -4.27$ V and -4.5 V were obtained for 9- and 14-nm-thick Al$_2$O$_3$ respectively, and $V_{th} = -3.85$ V was obtained for the HFETs. The sheet carrier density and drift mobility in the channel were evaluated from channel conductivity and $C-V$ measurements on the fat-HFETs using

$$n_s = \frac{1}{q} \int C_n dV \tag{3}$$

$$\mu_d = \frac{L_G}{q.w_G} \frac{g_{CH}}{n_s} \tag{4}$$

where C_n is the normalized capacity, L_G is the gate length, g_{CH} is the output conductivity, and w_G is the gate width. Values of sheet carrier densities $n_s = 6.9 \times 10^{12}$ and 7.25×10^{12} cm^{-2} were obtained for the HFET and MOSHFET, respectively. Higher n_s on the MOS-structure indicates the presence of a passivation effect [3]. Slightly higher μ_d for given V_G on the MOSHFETs compared with the HFETs and the same μ_d-n_s dependence with 825 cm^2/Vs peak drift mobility were found.

The existence of high leakage current is a well-known problem of AlGaN/GaN HFETs. It can be partially suppressed by gate insulation. Another alternative is an enhancement of the Schottky barrier height by an undoped GaN cap on the top of heterostructure. It is already known that MOS-based devices exhibit much lower gate leakage current than simple HFETs [2]. The gate leakage current of the MOSHFETs was found to be 3 orders of magnitude lower (~10^{-5} A/mm at −10 V) than that of the HFET, as shown in Fig.

2. On the other hand, this is a partially higher value than that reported for similar devices with highly insulating gate oxide.

The measurement of the HFET output characteristics and the transistor transconductance g_m are basic tools for the HFET DC performance evaluation. The maximum drain current density of 0.36 A/mm at $V_G = 1$ V and peak extrinsic transconductance of 84 mS/mm were obtained on 2 μm gate length of MOSHFETs with 9-nm-thick semiconductive Al_2O_3 gate oxide. An increase of drain current I_{DS} of the GaN-based MOSHFETs (up to 40% higher saturated drain current at $V_G = 1$ V (Fig. 3)) was probably caused by an increase of the sheet carrier density n_s and/or carrier velocity v. The self-heating and current collaps can be responsible for a decrease of drain current of the HFETs and MOSHFETs at higher drain voltage (>15 V). An increase of the extrinsic transconductance g_m (up to 37% of peak values) was found, as demonstrated in Fig. 4. The results are in contradiction to mostly reported lower values because of a larger gate-channel separation. However, the effect observed can be explained if the gate oxide acts as a semi-conductive electrode [4].

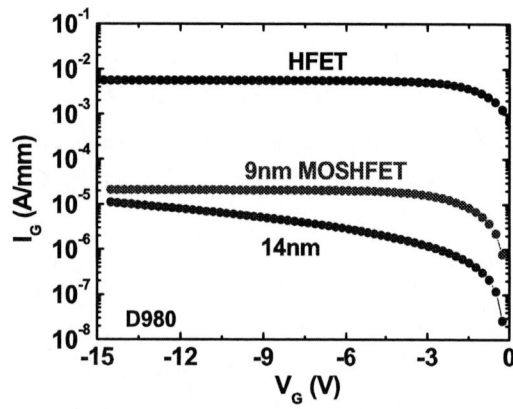

Fig. 1: Typical $C-V$ characteristics of AlGaN/GaN fatHFETs without and with 9-nm and 14-nm-thick Al_2O_3 gate oxide

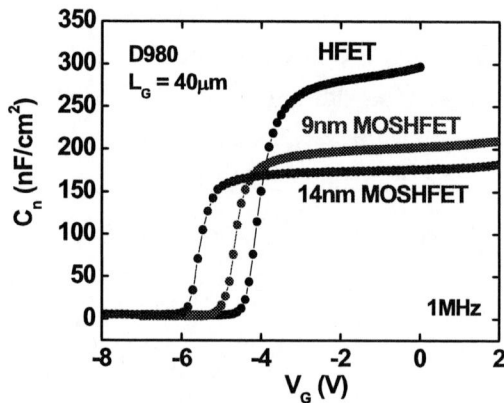

Fig. 2: Gate leakage current on AlGaN/GaN HFET and MOSHFETs ($L_G = 2$ μm)

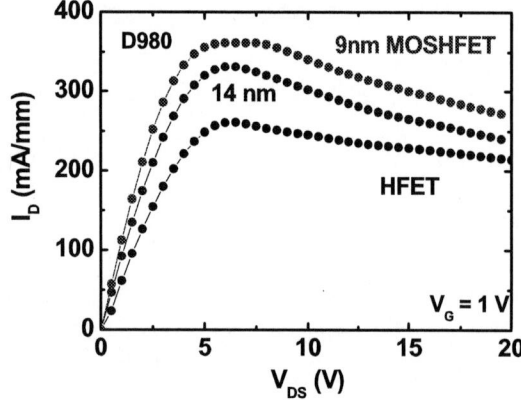

Fig. 3: Comparison of the output characteristics at $V_G = 1$V for the AlGaN/GaN HFETs and MOSHFETs

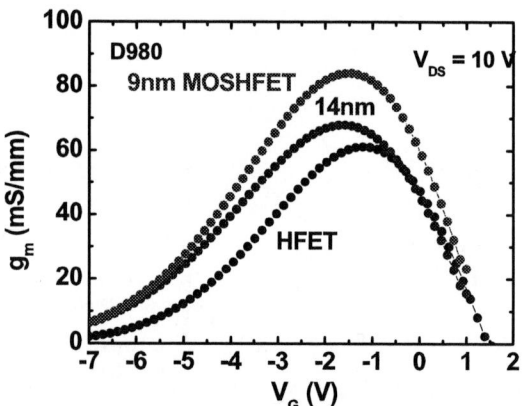

Fig. 4: Extrinsic transconductance of the HFET and MOSHFETs with with 9-nm and 14-nm-thick Al_2O_3 gate oxide.

4. Conclusions

In summary, the *I-V* and *C-V* measurement was used for the characterization of AlGaN/GaN HFETs and MOSHFETs. The sheet carrier density increased from $n_s = 6.9 \times 10^{12}$ cm^{-2} for HFETs to 7.25×10^{12} cm^{-2} for MOSHFETs with 14-nm-thick Al_2O_3 gate oxide. Higher n_s on the MOS-structure indicates a passivation effect [3]. The present MOSHFETs exhibited favorable characteristics when compared with the HFETs. The gate leakage current of 10^{-5} A/mm at -10 V is about 3 orders of magnitude lower than that of the HFETs. The drain current and extrinsic transconductance of the MOSHFETs were up to 40% and 37% higher than those of the HFETs, respectively. The AlGaN/GaN MOSHFETs with semi-conductive Al_2O_3 gate oxide showed an improved performance compared that of the HFETs and they are promising candidates for reliable high-frequency and high-power devices.

Acknowledgement

This work was supported by VEGA Agency under contracts No. 2/6099/26 and 1/2041/05, and by APVT Agency under contract No. 20-026104.

References

[1] A. Koudymov, V. Adivarahan, J. Yang, G. Simin, and M.A. Khan, IEEE Electron Device Lett. **26**, 704 (2005).

[2] P. Kordoš, G. Heidelberger, J. Bernát, A. Fox, M. Marso, and H. Lüth, Appl. Phys. Lett. **87**, 143501 (2005).

[3] D. Gregušová, J. Bernát, M. Držík, M. Marso, J. Novák, F. Uherek, and P. Kordoš, phys. stat. solidi (c) **2**, 2619 (2005).

[4] D. Kikuta, J.P. Ao, and Y. Ohno, Solid-St. Electron. **50**, 316 (2006).

Post-metallization H$_2$ annealing of electrically active defects in Ta$_2$O$_5$/nitrided Si stacks

A.Paskaleva, E.Atanassova

Institute of Solid State Physics, Bulgarian Academy of Sciences, 72 Tzarigradsko Chaussee, 1784 Sofia, Bulgaria, e-mail: paskaleva@issp.bas.bg; elenada@issp.bas.bg

The effect of H$_2$ post-metallization annealing (PMA) on the electrical behavior of sputtered Ta$_2$O$_5$ layers on nitrogen ion implanted Si is investigated. The high densities of oxide charge, interface and slow states typical of as-deposited stacks are strongly reduced by one to two orders of magnitude after annealing. H$_2$ treatment affects both bulk Ta$_2$O$_5$ and interfacial layers but is more efficient in annealing electrically active defects in nitrided layer. The effect of defect annealing on the leakage currents and conduction mechanisms is also discussed.

1.   Introduction

According to the last edition of the International Technology Roadmap for Semiconductors (ITRS) Ta$_2$O$_5$ is one of the key high-*k* dielectric players for the next generations of dynamic random access memories (DRAMs) [1]. Our previous studies have shown that Ta$_2$O$_5$ with excellent parameters - permittivity as high as 37; leakage current < 10^{-8} A/cm^2 at 1 MV/cm; oxide charge ~10^{10} cm^{-2}; and breakdown fields ~ 4.5 MV/cm could be obtained by rf sputtering of Ta in Ar+O$_2$ mixture [2,3]. As many other transition metal oxides, however, Ta$_2$O$_5$ is thermodynamically unstable in contact with Si and forms lower-*k* interfacial layer. Nitridation of Si surface [4] before deposition of Ta$_2$O$_5$ may be an efficient way to prevent or at least to minimize the growth of SiO$_2$-like interfacial layer; to increase the dielectric constant of the stack and to improve its electrical performance. In this paper, the influence of post-metallization annealing in H$_2$ on Ta$_2$O$_5$/nitrided Si stacks is studied. As is known H$_2$ is commonly used annealing ambient in Si technology. At present, however, the question how does hydrogen behave in high-*k* dielectrics is open. Systematic results for the impact of hydrogen on the system Ta$_2$O$_5$/Si are also missing. Here, the electrical characteristics of Ta$_2$O$_5$ − based capacitors before and after hydrogen heating are compared and the origin of the observed strong annealing of electrically active defects is discussed.

2.   Experimental Procedure

n-type Si wafers with resistivity of 4 Ωcm were implanted with nitrogen ions with an energy of 1 keV and a dose of 3x10^{16} cm^{-2} using plasma immersion ion implantation. After the implantation tantalum pentoxide film (d=30 nm) was deposited by the reactive sputtering of Ta target in a mixture of Ar+10 % O$_2$ [2,3]. For electrical characterisation top Al electrodes were evaporated and structured by photolithography to form MOS capacitors. A post-metallization annealing (PMA) in H$_2$ at 450 °C for 30 min was performed. Standard capacitance-voltage (C-V) measurements were performed to extract the effective dielectric constant ε_{eff}, interface and oxide charges. The leakage currents and conduction mechanisms in the structures were evaluated by measuring I-V curves at different temperatures (25-110^0 C).

1-4244-0396-0/06/$25.00 ©2006 IEEE

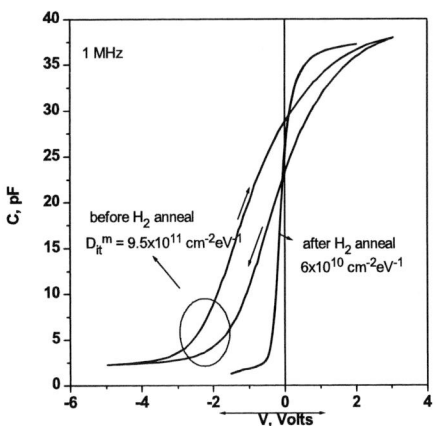

Tab. 1. Parameters extracted from C-V measurements before and after H_2 annealing

	Q_f (cm^{-2})	ΔV_{fb} (V)	ε_{eff}
before PMA	1.3×10^{12}	-0.82	14.5
after PMA	2.1×10^{11}	<0.01	14.3

Fig.1 Comparison of C-V curves before and after H_2 annealing.

3. Results and Discussion

The as-deposited samples have high oxide charge density, Q_f ($>10^{12}$ cm^{-2}) which is strongly frequency dependent; large flat band voltage hysteresis, ΔV_{fb}, revealing the existence of a high density of slow states ($>10^{12}$ cm^{-2}). The curves are strongly stretched-out along the voltage axis indicative for a high density of interface states (Fig. 1, Tab. 1). A dramatic improvement of C-V curves is observed after H_2 annealing – very well behaved and steep curves are obtained (Fig. 1). All kind of charged defects – oxide charge, slow and interface states are strongly reduced by H_2 annealing. The reduction of oxide charge is about one order of magnitude; the interface states – more than one order of magnitude and the slow states – two orders of magnitude (Fig. 1, Tab. 1). So, the post-metallization H_2 treatment affects more effectively the interface and near interfacial region (i.e. charges in the nitrided layer) resulting in a stronger decrease of slow and interface states. The possible origin of defects in nitrided layer are the unsaturated and/or broken N-bonds which act as electron traps and the efficient trapping of hydrogen at these bonds results in a strong decrease of electron trapping in the nitrided layer. As discussed in [5] network N atom ($Si_3 \equiv N$) can trap hydrogen more efficiently than can network O atom ($Si_2 = O$). The results give no evidence for occurrence of structural changes which are permittivity detectable, (permittivity remains unaltered) after H_2 annealing (Tab.1). The AFM topographic images show that the Ta_2O_5 films deposited on nitrided surface are more rough than those deposited on bare Si (Fig.2) – the surface roughness values (peak-to-valley, Z_{range}, and root mean square, RMS) are higher by a factor of about 4-5 for the former. It is suggested that H_2 annealing may smooth the surface and this can be additional reason for the observed reduction of electrically active centers in the stacks.

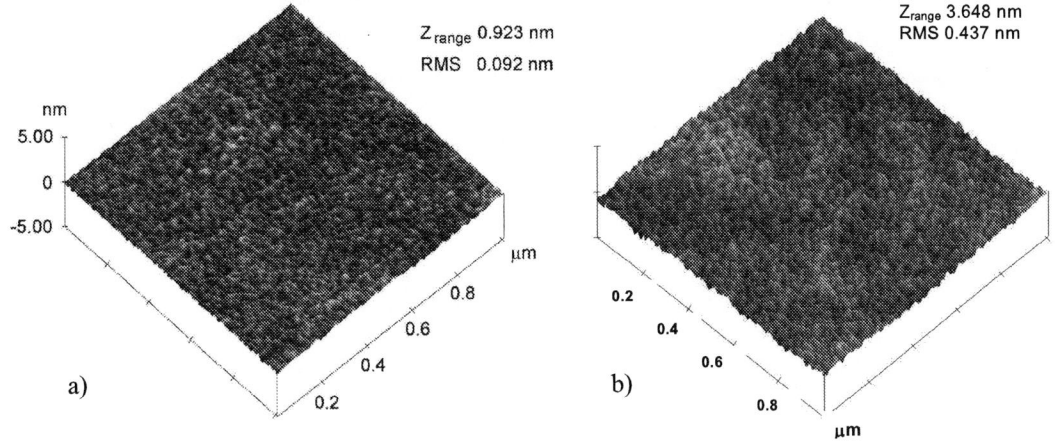

Fig.2 AFM topographic images of Ta_2O_5 deposited on: a) bare Si; b) nitrided Si.

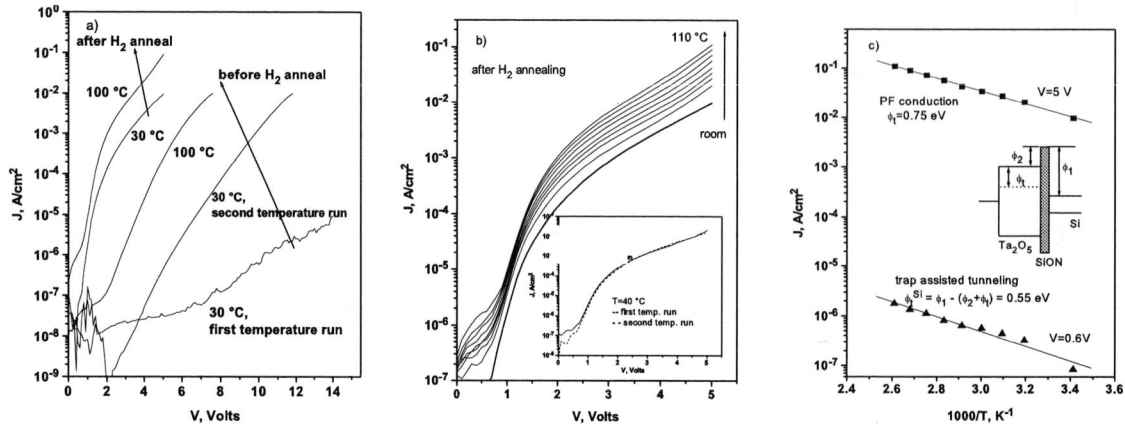

Fig.3 a) I-V curves measured at 30 and 100 °C before and after H_2 heating; **b)** Temperature dependent I-V curves of samples after H_2 annealing. The inset shows the curves measured at 40 °C during the first and second temperature run; **c)** Arrhenius plot of current at different biases. The inset shows the band diagram and energy location of traps.

The annealing of electrically active defects, however, comes at the expense of the degraded leakage current (Fig.3a) - the current increases strongly after PMA. There is a significant difference also in the shape and temperature dependence of I-V curves. The non-annealed structures reveal very strong temperature dependence – the current increases more than five orders of magnitude as the temperature changes from room to 110 °C. The curve measured at 30 °C after the sample is cooled down from 110°C is significantly shifted to lower voltages (Fig.3a). The lower leakage current in as-deposited stacks is assigned to the existence of a high density of negatively charged traps located in the nitrided layer. This negative charge modifies the cathode field and hinders electron injection from the accumulation layer in this way causing the conduction to start at higher fields. A temperature stimulated detrapping of electrons is observed at T>80 °C. The process is not reversible, i.e. once emptied the traps are not filled again that results in a substantial shift of I-V curves after cooling down from 110 °C (Fig.3a). No shift of the I-V curves is observed after cooling down annealed samples (see the inset of Fig.3b), hence no irreversible trapping/detrapping processes take place. Therefore, despite of the increased level of leakage current annealed stacks reveal improved stability with respect to trapping processes.

The temperature dependence of I-V curves after H_2 annealing is generally significantly weaker and the shape of the curves is substantially different implying a change of the main conduction mechanism induced by H_2 treatment. Three different regions of temperature dependence are clearly visible (Fig.3b). The conduction in the first two regions (V<1.5 V) is consistent with the trap-assisted tunneling (TAT) process, i.e. the electrons tunnel through the nitrided layer to traps in Ta_2O_5 and then from these traps to the conduction band of Ta_2O_5. The current governed by TAT process is strongly temperature dependent at low voltages, and much weakly temperature dependent at higher voltages [6]. From Ahrrenius plot of current at 0.6 V (Fig.3c) we obtained for $\phi_t^{(Si)} = \phi_1 - (\phi_2+\phi_t) = 0.55$ eV, where ϕ_1 – barrier height between Si and SiO_xN_y, ϕ_2 – energy difference between Si/SiO_xN_y and Si/Ta_2O_5 barrier heights, ϕ_t – energy position of traps with respect to conduction band (CB) edge of Ta_2O_5. $\phi_t^{(Si)}$ is the energy position of traps above Si CB edge (see the inset of Fig.3c). For V> 1.5 V stronger temperature dependence is again observed. The assessment of the curves reveals that the dominant conduction mechanism in this field range is PF emission from traps and from the temperature dependence of the current at V = 5 V (Fig.3c) the trap level is estimated to be at $\phi_t = 0.75$ eV. A question arises whether the traps responsible for TAT and PF emission are one and the same. In this case the sum of ϕ_t and $\phi_t^{(Si)}$ gives the Si/Ta$_2$O$_5$ barrier height, i.e. 1.3 eV. This value is significantly higher than theoretically estimated conduction band offset of Ta$_2$O$_5$ with respect to Si of 0.4 eV [7]. However, the effective offset between Si and Ta$_2$O$_5$ with SiO$_2$ interfacial layer can increase to as much as 1 eV [8]. A value of 1.5 eV for this offset is assumed in [6], i.e. there is not a unique value of the band offset and it can be modulated by interface chemistry. The value obtained in our case is in-between that gives us a reason to suggest that the traps promoting TAT and PF emission most likely are the same. Therefore we can conclude that at voltages up to 1.5 V TAT through traps located at about 0.75 eV below CB edge of Ta$_2$O$_5$ is the dominant conduction mechanism. At higher voltages the PF emission from the same traps governs the current. Another point to be stressed is the change of the energy position of the traps promoting PF conduction after H$_2$ annealing. In our earlier work [9] it was found that the traps responsible for PF emission in the samples before annealing are situated at about 1 eV below CB edge. It seems that H$_2$ annealing changes the trap energy level from 1 eV to 0.75 eV. This change is another possible explanation for the higher leakage current in PMA structures. Therefore, we conclude that H$_2$ annealing affects also electrically active defects in Ta$_2$O$_5$. The possible reaction is the trapping of hydrogen at non-bridging Ta-O sites and formation of neutral Ta-OH centers. This leads to a change of the energy position of the traps, i.e. the hydrogen heating makes the traps more shallow. This reaction could also account for the decrease of positive oxide charge.

4.   Conclusion

In summery, H$_2$ PMA is very effective in annealing N-related electrically active defects in Ta$_2$O$_5$/nitrided Si stacks. It affects also bulk traps in Ta$_2$O$_5$ by changing their energy location, making them more shallow. The observed weaker temperature dependence and improved stability of I-V curves after annealing are consistent with the strong reduction of these electrically active defects. However, this comes at the expense of the degraded leakage current. The results give no evidence for occurrence of structural modifications in the stack after H$_2$ annealing, affecting ε (e.g. an increase of interfacial layer thickness, reduction of Ta$_2$O$_5$, crystallization effects, etc.).

Acknowledgement

This work was partly supported by the Bulgarian Ministry of Education and Sciences in the frame of project F1508.

References

[1] Intern. Technol. Roadmap for Semicond. http://public.itrs.net.
[2] E. Atanassova, and **A.** Paskaleva, *Microel. Reliab.* **42**, 157, 2002.
[3] E. Atanassova, N. Novkovski, A. Paskaleva, and M. Pecovska - Gjorgjevich, *Solid St. Electr.* **46**, 1887, 2002.
[4] E.P. Gusev, H-C. Lu, E. Garfunkel, T. Gustafsson, and M.L. Green, *IBM J. Res. Develop.* **43**, 265, 1999.
[5] S.S. Tan, T.P. Chen, C.H. Ang, and L. Chan, *Appl. Phys. Lett.* **82**, 269, 2003.
[6] M. Houssa, M. Tuominen, M. Naili, V. Afanas'ev, A. Stesmans, S. Haukka, and M. Heyns, *J.Appl.Phys.* **87**, 8615, 2000.
[7] J. Robertson, and C.W. Chen, *Appl. Phys. Lett.* **74**, 1168, 1999.
[8] G. Lucovsky, and J. Phillips, *Appl. Surf. Sci.* **166**, 497, 2000.
[9] A. Paskaleva, E. Atanassova, and M. Georgieva, *J. Phys. D: Appl. Phys.* **38**, 4210, 2005.

2D electron transport through potential barrier prepared by LAO on shallow GaAs/Al$_x$Ga$_{1-x}$As/InGaP heterostructure

J. Martaus*, V. Cambel, R. Kúdela, D. Gregušová, J. Šoltýs

Institute of Electrical Engineering, Slovak Academy of Sciences, Bratislava, Slovak Republic
*e-mail: elekmajo@savba.sk

We have studied transport of a two-dimensional electron gas through a potential barrier prepared on shallow GaAs/Al$_x$Ga$_{1-x}$As/InGaP heterostructure by local anodic oxidation (LAO) with an AFM tip. The potential barrier height after LAO was 55 meV, and it increased to 270 meV after oxide line removal at 300 mV, at room temperature. Barriers with this height can be used for room temperature nanometre-sized structures and devices fabrication.

1. Introduction

Local anodic oxidation (LAO) realized by the tip of an atomic force microscope (AFM) allows for the formation of a small oxidized dots and narrow oxidized lines on a sample surface. The dots and lines are elementary building blocks for nanometre-sized structures and devices. LAO can be applied to metallic layers, bulk semiconductors, and semiconductor heterostructures to prepare quantum wires, quantum point contacts and other mesoscopic objects. The oxidation of a sample is caused by a negative voltage applied to the AFM tip with respect to the sample. If the local electric field reaches 10^9 V/m [1], molecules from the water bridge located between the tip and the sample [2], are disintegrated to H$^+$ and OH$^-$ ions. Subsequently, the OH$^-$ ions are transported and accelerated by the local electric field towards the sample, where they react with the surface and form the oxides.

An optimal layer configuration for LAO is represented by a shallow δ–doped GaAs/Al$_x$Ga$_{1-x}$As heterostructure with a two-dimensional electron gas (2DEG) prepared by molecular beam epitaxy (MBE). We use a shallow δ–doped GaAs/Al$_x$Ga$_{1-x}$As/InGaP heterostructure with a 2DEG prepared by organo-metallic vapor phase epitaxy (OMVPE).

In this paper we study a 2D electron transport through the potential barrier in a 2DEG prepared by the AFM LAO. We also study the transport after the oxide line removal by wet etching and discuss these results.

2. Experiment, results and discussion

A shallow δ–doped GaAs/Al$_x$Ga$_{1-x}$As/InGaP heterostructure with a 2DEG positioned 31 nm below the surface was grown by OMVPE in an AIX200 horizontal low-pressure IR-heated reactor. The heterostructure grown on semi-insulating GaAs (100) substrate consisted of a 500 nm GaAs buffer layer, a 17 nm AlGaAs spacer, a Si δ-doping $8,6 \cdot 10^{12}$ cm^{-3}, a 12 nm AlGaAs cover layer and a 2 nm strained InGaP cap layer.

In the second step, Hall bars with 5-μm-thick arms were defined on the heterostructure by optical lithography. Milling with neutralized argon ions was used for the mesa definition, and standard Ni-AuGe Ohmic contacts were alloyed to access the 2DEG. Sample with the Hall bars was mounted onto chip holder and wired.

The final step was the LAO. We formed an oxide line across one arm of a Hall bar by AFM Topometrix Explorer using a Si tip in non-contact mode. The set point of the oscillating tip was set on 15 %. The tip speed during oxidation was 100 nm·s^{-1}. To realize the LAO

1-4244-0396-0/06/$25.00 ©2006 IEEE   174

experiment, AC voltage was used at a frequency of 1 kHz and amplitude of 25 V. The ambient humidity was 58 %.

The mean values of height and width of oxide line were 9 nm and 185 nm respectively. By removing the oxide line, we experimentally verified that the depth of the oxidation was approximately 1,5 times that of the oxide height above the sample surface [3]. The mean value of the depth was 15 nm. It means that we oxidized through the InGaP cap layer, AlGaAs cover layer, and also affected the Si δ-doping.

After the LAO, the device was divided into two symmetric 2DEG regions by a potential barrier $q\Phi$. In the following, the potential barrier is referred to as 'barrier'. Figure 1 shows temperature dependent IV characteristics after oxidation and after the oxide line removal (points).

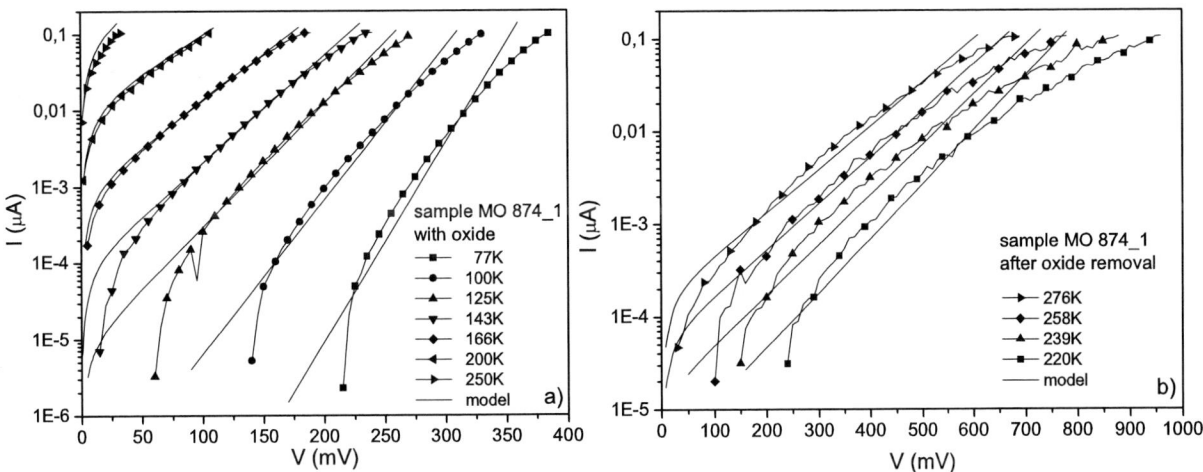

Fig.1: *IV characteristics (points) compared with theoretical model (lines):* **a)** *after LAO,* **b)** *after oxide line removal.*

If the voltage is applied across the barrier, the current flows from the source to the drain, and the voltage drop is almost entirely located in the barrier region. In general, a barrier height is a function of voltage applied, i.e. $\Phi = \Phi(V)$. If the transport through the barrier is dominated by thermally activated carriers, one [4] can express the current density as

$$J = qN_S\sqrt{(k_B T/2\pi m)} \cdot \exp(-[q\Phi(V)/k_B T]) \cdot (1 - \exp[-(qV/k_B T)]), \qquad (1)$$

where q is the elementary charge, N_S is the sheet electron density, k is Boltzman constant, m is the effective electron mass, and T is the temperature. The term exponentially dependent on voltage is the backward current from the drain to the source.

Figure 2 a, b shows the current through the barrier measured as a function of temperature for variable voltage applied after LAO and after the oxide line removal. The current depends exponentially on the inverse temperature for any applied voltage. This means that the transport through the barrier is thermally activated, i.e. the barrier is thick enough to suppress a contribution from the tunneling current. The slope of the curves is different at different voltages, which means that the barrier height Φ depends on the voltage applied. We extracted [4] a $\Phi(V)$ dependence from the relation (1):

$$\Phi(V) = \left(\frac{k_B}{q}\right) \cdot \left(\frac{T_1}{T_2}\right) \left(\ln\left(\frac{I_1}{I_2}\right) - \frac{1}{2}\ln\left(\frac{T_1}{T_2}\right) - \ln\left[\left(1 - \exp\left[-\frac{qV}{k_B T_1}\right]\right) \middle/ \left(1 - \exp\left[-\frac{qV}{k_B T_2}\right]\right) \right] \right), \qquad (2)$$

where T_1 and T_2 define the temperature range used for computing, and I_1 and I_2 are currents which pertain to a given voltage. The results are shown in Figure 2 d.

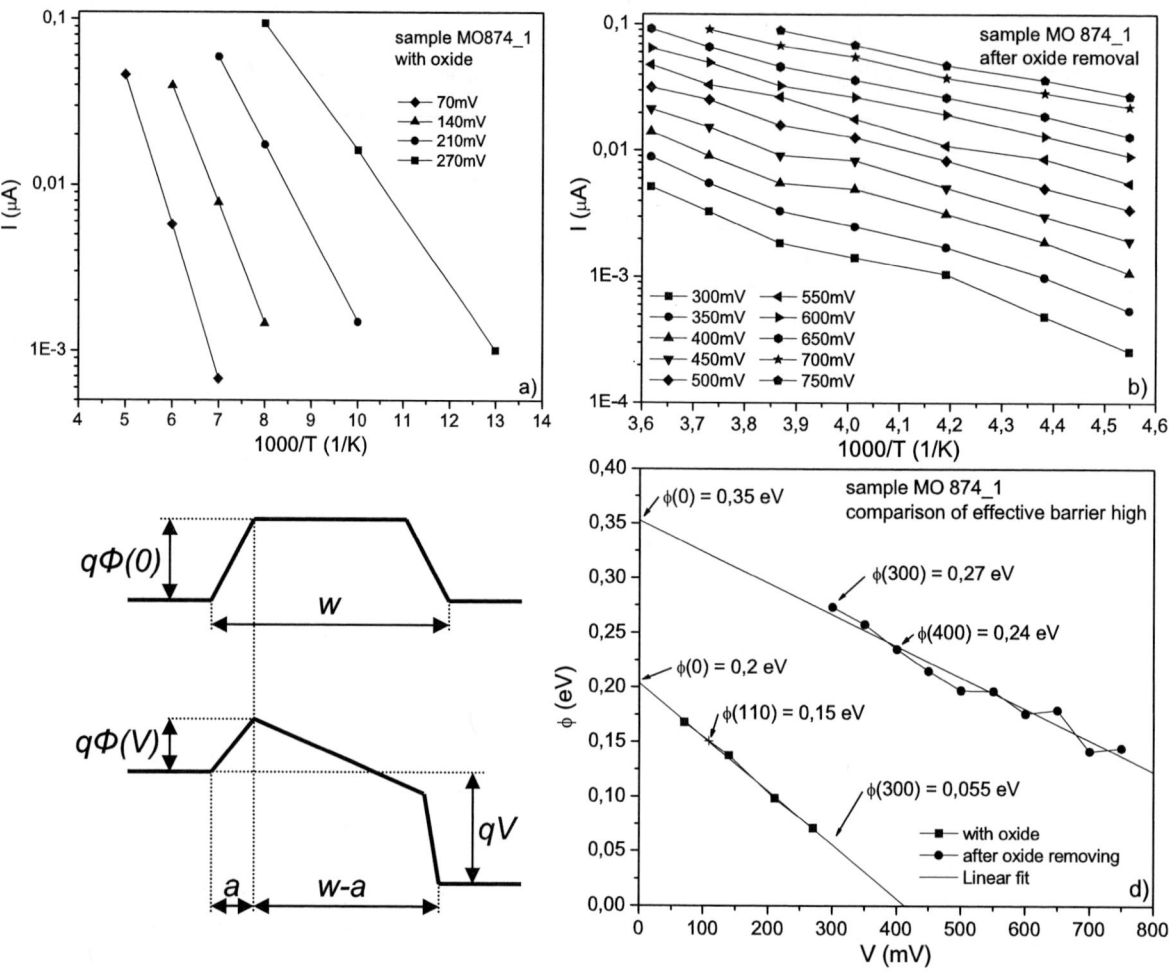

Fig.2: *Current versus inverse temperature dependence **a)** after LAO, **b)** after the oxide line removal. The slope of the curves defines the effective barrier height. **c)** Deformation of taper shaped barrier by applied voltage. **d)** Comparison of the effective barrier height after LAO and after oxide line removal.*

An effective barrier height after LAO was 55 meV, and it increased to 270 meV after oxide line removal at 300 mV, at room temperature (see Fig. 2d). Difference between the effective barrier heights after the LAO and after oxide line removal could be caused also by a change of the sheet electron density N_S [13]. We suppose that the InGaP cap layer was affected during the oxide line removal by wet etching. It led to a higher density of the surface states and a decrease in the barrier thickness between the surface and the 2DEG, subsequently to a decrease of the N_S, and finally to an increase of the effective barrier height. We partly verified this conjecture by *IV* measurements on an unaffected arm of the Hall bar before and after etch process step.

Assuming, that the barrier had a tapered shape (see Fig. 2c), we modified the relation (1) by taking into account the influence of the forward and backward currents. We estimated the lateral dimension of inclined edge to $a \approx 25$ nm, which is the distance between Si δ-doping and GaAs/AlGaAs interface plus the distance between the interface and the 2DEG.

To model the experimental curves, we had to chose N_S and the ratio a/w, both for the situation after the LAO and after the oxide line removal. The barrier height after the LAO decreased from $\Phi(0) = 200$ meV by $\Delta\Phi = 50$ meV at 110mV, and from $\Phi(0) = 350$ meV by $\Delta\Phi = 110$ meV at 400 mV after the oxide line removal (see Fig. 2d). Since the voltage drop was almost entirely located in the barrier region, part of the voltage drop over inclined edge denoted as 'a' satisfies the relation $aV/w \approx \Delta\Phi$, which yields $w \approx 50$ nm after LAO ($N_S = 1{,}9 \cdot 10^{16}$ m^{-2}), and $w \approx 78$ nm after oxide line removal ($N_S = 6{,}5 \cdot 10^{15}$ m^{-2}). The barrier for the forward current was reduced by $q\Phi(V) = q\Phi(0) - qV(a/w)$, while the barrier for the backward current was increased by $q\Phi(V) + qV = q\Phi(0) + qV(1 - a/w)$, so the backward current was suppressed. After the modification, one can express the current density as [14]:

$$J = qN_S \sqrt{\frac{k_B T}{2\pi m}} \cdot \left[\exp\left(-\frac{q\Phi(0) - qV(a/w)}{k_B T} \right) - \exp\left(-\frac{q\Phi(0) + qV(1 - a/w)}{k_B T} \right) \right]. \quad (3)$$

Figure 1 a, b compares the data measured and the theoretical model for transport through the potential barrier.

Although a simple model was considered, it gives a reasonable explanation of the experimental data. However, it is based on a simplified assumption, and it does not give a lucid insight into the carrier transport through the potential barrier in the 2DEG.

3. Conclusion

We have studied transport of the 2DEG through a potential barrier prepared by AFM LAO on shallow GaAs/Al$_x$Ga$_{1-x}$As/InGaP heterostructure. The barrier height was increased from 55 meV after the LAO to 270 meV after the oxide line removal at 300 mV, at room temperature. We have compared the data measured and the simple theoretical model for the 2D transport through the potential barrier. We have shown the possibility of the potential barrier preparation with the effective height suitable for the room temperature nanometre-sized structures and devices. The oxide lines will be used for the definition of quantum wires and single electron transistor devices in the future.

4. Acknowledgment

This work was supported by the project APVV-51-045705, and Centre of Excellence CENG at the IEE SAS, Bratislava.

References

[1] R. Garcia, M. Calleja, and H. Rohrer *J. Appl. Phys.* **86** 1898 (1999)
[2] S. Gómez-Monivaz, J. J. Sáenz, M. Calleja and R. García, *Phys. Rev. Lett.* **91** (2003)
[3] R. Held, T. Vancura, T. Heinzel, and K. Ensslin, M. Holland, W. Wegscheider *Appl. Phys. Lett.* **73** (2) (1998)
[4] V. Cambel, J. Šoltýs, J. Martaus, and M. Moško *J. de Physique IV* **132** (2006) 171-175
[5] J. Šoltýs, *dissertation*, Institute of Electrical Engineering SAS, Bratislava, July 2005

Recent Developments in Microsystem- and Nanodevice-TCAD

Gerhard Wachutka

Institute for Physics of Electrotechnology, Munich University of Technology
Arcisstrasse 21, 80290 Munich, Germany.
e-mail: wachutka@tep.ei.tum.de

The rapid progress in microsystems and nanodevice technology is increasingly supported by problem-specific modeling methodologies and dedicated simulation tools. These do not only enable the detailed visualization of fabrication processes and operational principles, but they also assist the designer in making decisions with a view to finding optimized micro- and nanostructured devices, components, and systems under technological, economical, and application-specific constraints, including customer specifications. Currently strong efforts are being made towards comprehensive simulation platforms for the predictive simulation of complete microsystems. In this way, "virtual fabrication" and "virtual experimentation, characterization, and test" will replace the real-world development processes of today by "virtual prototyping" on the computer [1].

In this context, we discuss the most important aspects and practical methodologies for setting up physically-based and consistent nanostructure, microdevice, and full system models for the effort-economizing and yet accurate numerical simulation of mechatronical microsensors and microactuators and complete microsystems built up of them. In this modeling framework [2], we demonstrate the consistent treatment of coupled fields and coupled energy and signal domains (briefly, even though not very precisely termed "multi-physics modeling"), which is required for deriving micromechatronical macromodels on the system level from the continuous field level (so-called "order reduction process"), and we also address some important issues to be focussed on for the reliable validation and accurate calibration of the models.

One important aspect is the fact that the dynamical behavior of many microsystem components is strongly affected by inherently continuous, distributed physical interactions such as viscous damping effects. They have to be carefully taken into account during the design and optimization process to obtain a realistic and reliable description of the device and system operation. In order to include these effects already at a very early stage of the design process efficiently and accurately in simulation models, system-oriented modeling methods are indispensable [3].

Following this general methodology, we in particular propose a "full system mixed-level simulation" scheme for continuously distributed interaction effects in microstructures, which enables their incorporation in system-level macromodels of entire microsystems in a natural, physically-based and flexible way, while it also allows for structural units with complex geometry and the consistent coupling to other energy and signal domains. A multitude of computational results obtained for elementary and complex microstructures such as highly perforated plates [4] are in excellent agreement with accurate 3D-Navier-Stokes FEM calculations and, thus, corroborate the practicality and quality of this approach to predictive simulation.

Another aspect refers to the control circuitry of microsystems. It is a popular approach to employ control algorithms with feedback loops for increasing the sensitivity and robustness of microsystems [5,6]. Nowadays, such control strategies are supposed to take also into account that the dynamical behavior of the mechanical transducer elements is affected by variations or fluctuations of external parameters (e.g. air pressure and humidity). In order to compensate such disturbing effects, it seems very promising to use adaptive algorithms for the control of the sensor and actuator elements, so that the impact of fluctuating ambient conditions on the system operation can be largely compensated by a proper corrective adjustment of the respective operating conditions. Again, a fast and yet accurate full system macromodel is needed for the design, test, and optimization of such control architectures

The adequate formal representation of the full system description is provided in terms of a Kirchhoffian network description in combination with an appropriate analog hardware description language such as VHDL-AMS, SpectreHDL, Verilog-A a.o. This makes it possible to code the models of all the individual system components in a generic and uniform way and to assemble the full system model by linking the constituent parts on the same descriptional level.

The availability of a easy-to-use predictive "CAD tool box" for microstructures and systems, which provides the functionality necessary for automated parameter identification by closed-loop simulation and, thus, the capability of easy model calibration, is largely desired, already attempted, and in part realized. They must be based on the above-mentioned modeling methodologies for coupled continuous-field "multi-physics" models and multi-energy domain compact models as the essential parts of a comprehensive strategy for bottom-up and top-down microsystem development. The practicability and efficiency of such a hierarchical approach has already been demonstrated in the MEMS community by numerous examples.

Now efforts must be made to transform these results in robust, easy-to-use software packages which are ready for use in existing professional CAD frameworks. This implies that, on the device level, software tools must be developed which allow for the efficient interfacing of different single-effect simulators in such a way that new advanced coupling schemes can be realized by a flexible control of the solution process.

On the system level, libraries of compact models have to be established, preferably in a simulator-independent generic hardware description language. In addition, fast and reliable order reduction and parameter extraction techniques for compact models, using simulation results from the continuous field level as well as measured data, are required. The corresponding software tools are the indispensable prerequisite for statistical modeling, which in turn addresses such important issues as fabrication yield and reliability.

References

[1] G. Wachutka, G. Schrag, R. Sattler: "Predictive Simulation of Microdevices and Microsystems: The Basis of Virtual Prototyping", *Proc. of 24th International Conference on Microelectronics* (MIEL 2004), Niš, Serbia, May 16-19, 2004, pp. 71-78.

[2] G. Wachutka, "Coupled Field Modeling of Microdevices and Microsystems", in *Simulation of Semiconductor Processes and Devices 2002* (Proc. of SISPAD 2002), Kobe, Japan, Sept. 4-6, 2002, pp. 9-14.

[3] P. Voigt, G. Schrag, and G. Wachutka, "Methods for model generation and parameter extraction for MEMS", in *Simulation of Semiconductor Processes and Devices*, K. De Meyer and S. Biesemans, Eds., Wien, New York: Springer, 1998, pp. 149-152.

[4] G. Schrag, G. Wachutka: "Accurate System-Level Damping Model for Highly Perforated Micromechanical Devices", *Sensors and Actuators* **A111**, 2004, pp. 222-228.

[5] M. Handtmann, R. Aigner, R. Nadal, G. Wachutka, "Methodology of Macromodeling Demonstrated on Force Feedback Σ/Δ-Architectures", *Proc. of 3rd Int. Conf. on Modeling and Simulation of Microsystems* (MSM-2000), San Diego, CA, U.S.A., March 27-29, 2000, pp. 138-141.

[6] R. Khalilyulin, T. Hauck, G. Wachutka: "Adaptive Control for Reducing the Effect of Damping on the Output Signal of Microgyroscopes", *Proc. of NSTI-Nanotech 2006 Conference*, Boston, MA, U.S.A., May 8-11, 2006, pp. 562-565.

Analysis of device geometry on the ruggedness of power DMOS transistor supported by 3-D modeling and simulation

Andrej Vrbicky[1], Daniel Donoval[1], Juraj Marek[1], Ales Chvala[1] and Peter Beno[2]

[1] Department of Microelectronics, Slovak University of Technology in Bratislava
Bratislava, Slovakia
[2] ON Semiconductor Piestany, Slovakia

The influence of geometrical dimensions on the properties of power DMOSFET's has been studied by 3-D numerical modeling and simulation. The results of 3-D simulation provide a very effective way for the identification of failure mechanism and location of device hot-spots. The analysis of the influence of the geometry of one device cell including the position of ohmic contact to a p-type well on the turn-on of the parasitic bipolar transistor and corresponding device ruggedness is straightforward.

1. Introduction

The device ruggedness and trade off between the series on-resistance R_{don} and breakdown voltage V_{br} are the most important parameters of power DMOSFET devices used in switching applications [1]. By reducing the device size the energy handling capability is becoming very important issue which should be addressed together with the trade off between the series on-resistance and breakdown voltage. Optimization of device geometry and fabrications steps to suppress the opening of the parasitic bipolar transistor (BJT) and subsequent catastrophic device failure is the primary objective of any laboratory. There exist many approaches how to improve transistor properties by special design of the device structure (CoolMOS, Trench, VDMOS) [2-4] and/or process technology [5].

3-D numerical simulation has been used to model the loading conditions relating to unclamped inductive switching (UIS) test. To study the internal structure behavior and parasitic effects dependent on considerably self-heating due to conductive losses the non-isothermal equations using the thermodynamic model must be incorporated into the device simulation. The device structure doping, geometry and boundary conditions were finely tuned to model the electrical and thermal properties of actual device.

2. Device structure

The schematic view of the half of one device cell studied in this work in 3-D representation is shown in Fig.1. While the lateral geometrical dimensions under investigation (w_s, w_p and z_l) are in μm range the vertical dimension is set to 250 μm to model properly the thermal resistivity and corresponding heat transfer in the real structure. The ISE DEVISE tool [6] was used to design the individual layers with corresponding doping profile in n^+- type source, p- type well and n– type drain. The ohmic contact to the p-type well 0.8 μm width is located at the boundary of the each cell. To suppress the numerical instabilities in 3-D simulations the very dense mesh was generated particularly in regions with very steep doping profiles and high ionization rates, e.g. at the reverse biased p- well to n- drain junction and n^+- source to p- type well particularly in vicinity of the MOSFET channel. The latter is very important during the opening of the parasitic BJT when large current density formerly distributed homogeneously through the bottom of a cell is sink towards the middle of the full single cell located in maximum distance from ohmic contact to p- well.

1-4244-0396-0/06/$25.00 ©2006 IEEE

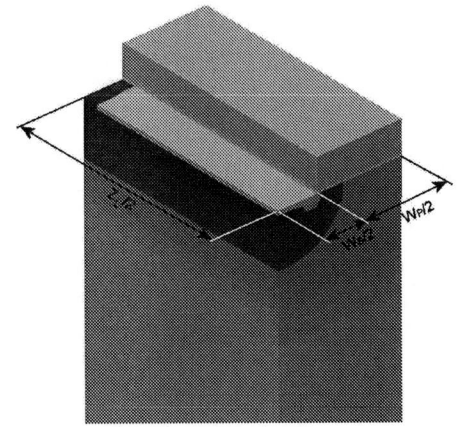

Fig. 1. Schematic view of the analyzed structure of the half of single cell

20   30   40   50   60   70   80   90
Time (μs)

Fig. 2. Time dependence of V_{ddbr} and T_{max} for different cell half-width

3. Experimental results and discussion

The simulated dependences of breakdown voltage V_{ddbr} and T_{max} for different cell half-width using device simulator DESSIS [7] are presented in Fig. 2. The test conditions comprise current source (I = 55 A) connected to drain contact, gate and source contacts connected to zero potential. The maximum device sustainable energy (ruggedness) can be expressed by

$$E_{as} = V_{ddbr} I_{ds} t_{bjt}$$

where V_{ddbr} is drain-to-source breakdown voltage during avalanche, I_{ds} is drain-source current and t_{bjt} is time when the parasitic BJT is opened. The values of V_{ddbr} = 29.7 V and threshold voltage V_{th} = 1.98 V are almost constant and independent on the studied device geometry. From Fig. 2. we can see that with increasing half width of one cell the parasitic BJT is opened earlier at smaller current (energy) due to increasing value of p- well series resistance on which the voltage drop is generated which results in decreasing device ruggedness (Fig. 3). As the frequency of ohmic contacts to p- well decreases with increasing cell width the effective channel width is longer for the device with the same area and the R_{don} decreases with z_l also. The increase of device ruggedness with increase of w_s due to decrease of p-well series resistance is paid by considerable increase of R_{don} due to JFET like effect under shorter poly-Si and underlying region (channel) for vertical current flow towards drain contact (Fig. 4.).

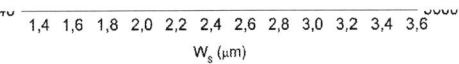

1,4 1,6 1,8 2,0 2,2 2,4 2,6 2,8 3,0 3,2 3,4 3,6
W_s (μm)

Fig. 3. Max E and R_{don} vs cell width

Fig. 4. Max E and R_{don} vs w_s: w_p ratio

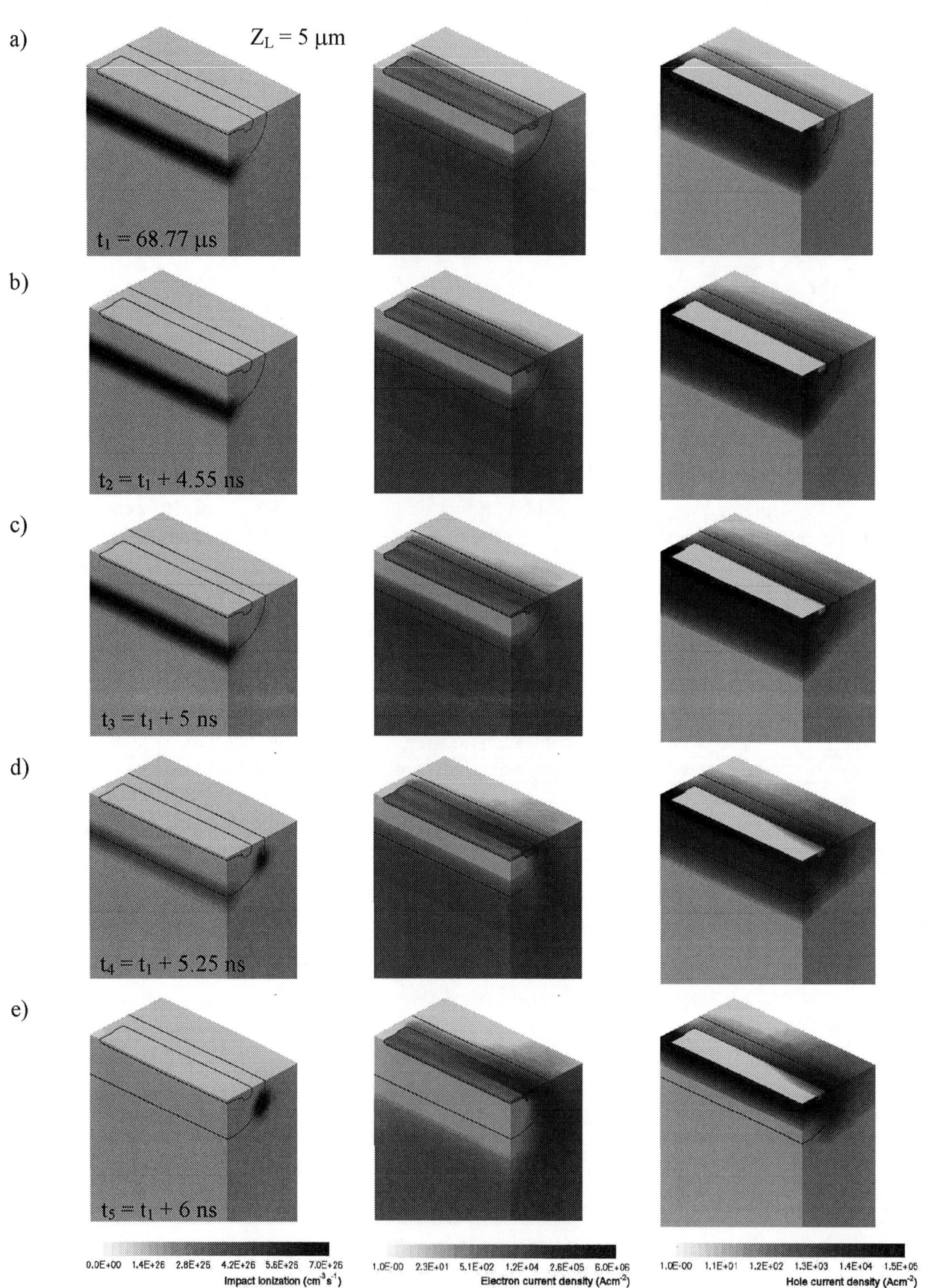

Fig. 5. Impact ionization and electron and hole current density at time slots indicated in Fig. 2.

The internal device behavior represented by the impact ionization, electron and hole current densities for different time slots indicated in Fig. 2. is shown in Fig. 5. It can be nicely seen that with rising time the values of V_{ddbr} and T_{max} increase in transistor off state due to a significant self heating effect in avalanche regime with the highest impact ionization homogeneously distributed at the reverse biased p-n junction at the bottom of the p- well. Subsequently the very fast transition (in few ns) of the location of avalanche breakdown to the central part of one cell (boundary at right cross section in Fig. 1.) located farthest from the ohmic contact to p- well (boundary at left cross section in Fig. 1.) due to the opening of the parasitic BJT is confirmed by the highest value of the impact ionization and dominant current flow to n^+- source region [8]. It corresponds to a drop of breakdown voltage (Fig. 2.) due to the changed location and mechanism of diode breakdown to bipolar transistor breakdown. The small kink in temperature can be attributed to this effect also, the current is flowing in smaller volume where higher current density and Joule heat is generated.

4. Conclusions

The influence of device geometry on the electro-thermal properties of the vertical power DMOS transistor was analyzed by 3-D modeling and simulation. The simulated characteristics and extracted values of the device ruggedness and series on-resistance R_{don} dependent on the ratio of cell length w_s to w_p and cell width z_l contribute to the optimization of the device design. The visualization of the internal properties of analyzed structure allows better identification and understanding of different mechanisms of breakdown and location of critical regions responsible for device failure. Although very high attention was given to the proper structure design and fine mesh generation to obtain the resulting 3-D simulations in a reasonable time the extracted values of device ruggedness suffers by the non negligible numerical instabilities particularly for different w_s to w_p ratio.

Acknowledgements

This work was supported by the Slovak Research and Development Agency under the contract APVV-20-055405 and the grant VEGA 1/2041/05 of the Slovak Ministry of Education.

References

[1] Fischer K., Shenai K.: Dynamics of Power MOSFET Switching Under Unclamped Inductive Loading Conditions, IEEE Trans. on Elec. Dev., Vol. 43, 1996, 1007 – 1015

[2] Coe: High voltage semiconductor device, US-Patent 4,754,310, (1988)

[3] Chang, et al.: Vertical double diffused metal oxide semiconductor VDMOS device with increased safe operating area and method, US-Patent 4,801,986, (1989)

[4] Shutten et al.: Lateral bidirectional notch FET with extended gate insulator, US-Patent 4,546,367, (1985)

[5] Chien F., Lai M., Su S., Tu K., Cheng Ch.: High Ruggedness Power MOSFET Design by Self-Align p+ Process, IEICE Trans. Electron, Vol. E88-C, No.4, 2005, 694 – 698

[6] DEVISE - ISE, User manual, ver. 10.0, ISE Zurich, 2004

[7] DESSIS - ISE, User manual, ver. 10.0, ISE Zurich, 2004

[8] Vrbicky,A., Donoval, D., Marek, J., Chvala, A., and Beno, P., Analysis of Electro - Thermal Behaviour of DMOS Transistor Structure during UIS Test, In: Proceedings of 8th International Seminar on Power Semiconductors ISPS'06, Prague 2006, pp. 111 - 116

New 600V Lateral Superjunction Power MOSFETs Based on Embedded Non-Uniform Column Structure

K. Permthammasin*, G. Wachutka*, M. Schmitt**, and H. Kapels**

* Institute for Physics of Electrotechnology, Munich University of Technology,
Arcisstrasse 21, 80290 Munich, Germany
e-mail: komet@tep.ei.tum.de, wachutka@tep.ei.tum.de
** Infineon Technologies AG, Am Campeon 1-12, 85579 Neubiberg, Germany

New lateral power MOSFETs employing two different, non-uniform, column-shaped superjunction (SJ) structures are proposed for use in smart power ICs that require voltage ratings up to 600V. Using three-dimensional device simulations, the basic electrical characteristics of the new SJ MOSFETs have been evaluated, together with the effect of charge imbalance on the device performance. The simulation results show that the proposed devices exhibit excellent robustness against doping fluctuations, improve the specific on-resistance by as much as 47% compared to conventional RESURF LDMOS structures with similar voltage ratings, and can compete with existing other charge compensation devices.

1. Introduction

The rapid development in power semiconductor device technology observed in recent years has led to an innovative device concept termed superjunction (SJ). The SJ concept is based on the principle of charge compensation in the voltage-sustaining drift zone of power devices. Recent studies have shown that, by implementing a SJ structure in vertical power MOSFETs, the physical device limitations known as silicon limit can be overcome [1]. Regarding the application of the SJ principle to lateral power MOSFETs in bulk silicon technology, a primary concern stems from the occurrence of a progressively widening of the substrate depletion region from source onwards to drain under reverse bias, which affects the charge balance in the SJ structure [2]. In an attempt to suppress this substrate effect, we have developed a new lateral SJ double diffused MOS (SJ-MOSFET) transistor. The device is designed for smart power applications, where the voltage ratings are limited to about 600V. Employing three-dimensional device simulation, we studied the operation and performance of the device and investigated the sensitivity of basic device parameters to the charge imbalance in the SJ structure.

2. Device Concept

As sketched in Fig. 1, the proposed SJ device structure is identical to a RESURF LDMOSFET built on a p-substrate, except that a number of p-type column-shaped semiconductor "islands" are embedded into the n-drift zone at an equal distance to form the SJ structure. The column diameters are proportionally enlarged with increasing distance from the drain; the nearer the column lies to the drain, the smaller is its diameter. Because of the non-uniformity of the columns, the concentration of positively charged donor impurities in the drift zone will, under reverse bias condition, decrease along a line from the source to the drain, thereby counterbalancing the concentration gradient of negative charges in the p-substrate, which arises from the substrate depletion effect. Two design variants of the pattern of the column arrays building the SJ structure have been investigated: an orthogonal and a hexagonal pattern. Each of the SJ structures has been optimally designed to have a breakdown voltage as high as 600V by adjusting the device dimensions as well as the doping concentration in the columns, N_A, for a given doping concentration of the n-drift layer N_D.

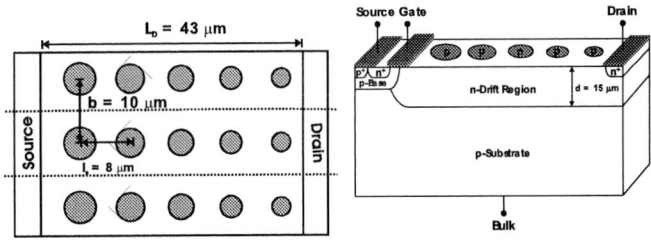

(a) Orthogonal SJ

3. Performance Characteristics

In smart power applications, the SJ-MOSFETs operate as switch between on- and off-state. The steady-state operation of the devices is primarily characterised by the specific on-resistance $R_{DS(on)} \cdot A$ during the on-state and the breakdown voltage $V_{(BR)DSS}$ during the off-state. $R_{DS(on)} \cdot A$ is determined by the product of the active chip area, A,

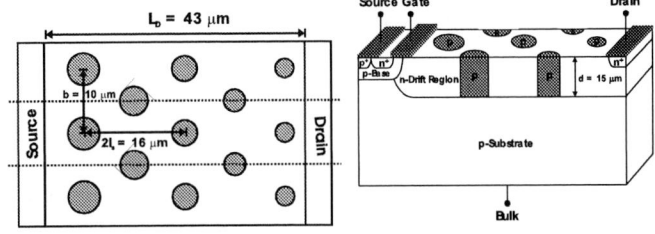

(b) Hexagonal SJ

Fig. 1: *Top view and perspective view of a unit cell structure of the proposed SJ-MOSFETs.*

and the inverse of the slope of the current-voltage characteristics at the gate-source voltage $V_{gs} = 10V$ in the ohmic region shown in Fig. 2(a), yielding 10.16 $\Omega \cdot mm^2$ for the orthogonal SJ and 10.94 $\Omega \cdot mm^2$ for the hexagonal SJ. Evidently, the orthogonal SJ handles more current with less power dissipation than the hexagonal SJ, in spite that N_D is kept equal for the two design layouts. This is because the hexagonal column pattern leads to an apparently smaller effective cross-sectional area for electron flow through the n-drift zone, which is reflected in an increase of the on-resistance $R_{DS(on)}$ of the drift zone. Fig. 3 displays the current flow through the drift region near the surface for each of the two design variants.

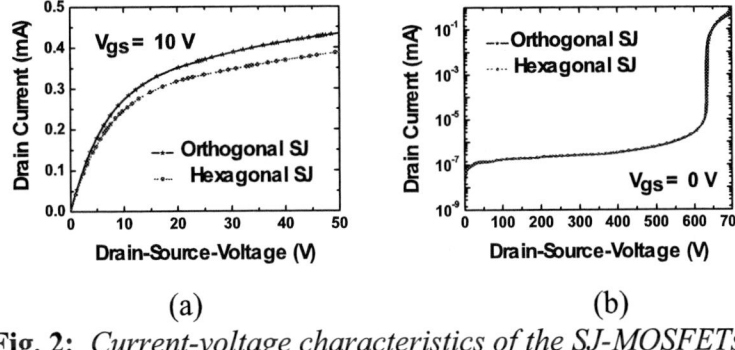

(a)

(b)

Fig. 2: *Current-voltage characteristics of the SJ-MOSFETs: (a) for gate-source voltage $V_{gs} = 10$ V; (b) for $V_{gs} = 0$ V.*

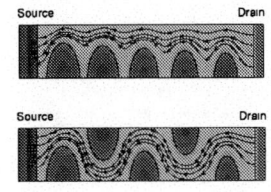

Fig. 3: *Current flowlines through the drift zone near the surface: Top: orthogonal SJ; Bottom: hexagonal SJ.*

From the current-voltage characteristics in Fig. 2(b), where the gate and source are shorted together ($V_{gs} = 0V$), the drain voltage at which significant drain current begins to flow constitutes an estimate of $V_{(BR)DSS}$. We find a $V_{(BR)DSS}$ of around 632V for the orthogonal SJ and of around 634V for the hexagonal SJ. Fig. 4 shows contour plots of the electrostatic potential for each of the SJ structures close to breakdown. The contour lines in both SJ structures are crowded together at the interface between drift-region and substrate near the drain end, indicating the position at which avalanche breakdown occurs.

For evaluating the switching performance of the SJ-MOSFETs, the gate charge Q_g is an appropriate parameter because it expresses the total amount of charge required to turn on the devices. Using a clamped inductive switching circuit for testing, we obtain the gate charge characteristics of the two device structures as illustrated in Fig. 5. The dependence on the drain-source voltage V_{ds} is also included in the charge curves, as this is the relation relevant for the transient behaviour. At the full gate voltage $V_{gs} = 15V$, both column patterns yield a value of Q_g of around 110 nC. Obviously, Q_g and, hence, the switching speed are hardly affected by the column configuration. Because high Q_g is associated with low $R_{DS(on)}$, the product

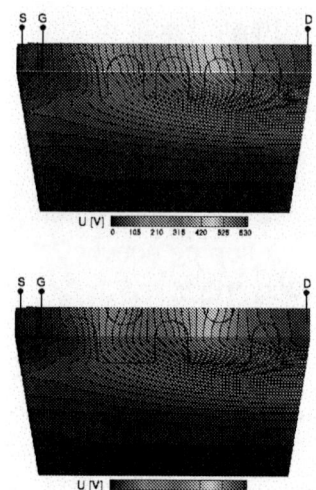

Fig. 4: *Potential contours at breakdown: (Top) orthogonal SJ; (Bottom) hexagonal SJ.*

$R_{DS(on)}\cdot Q_g$ provides a useful figure of merit. According to this, the orthogonal SJ features an $R_{DS(on)}\cdot Q_g$ that is by 7% better than that of the hexagonal SJ.

In Fig. 6 we compare the $R_{DS(on)}\cdot A$ vs. $V_{(BR)DSS}$ trade-off relation of the orthogonal SJ with results reported in other work [4, 5] as well as with the theoretical silicon limit of RESURF devices [6]. The diagram reveals that the orthogonal SJ achieves a $R_{DS(on)}\cdot A$ which is by 47% lower than that of the RESURF devices with comparable voltage ratings, so that its $R_{DS(on)}\cdot A$ comes closer to the silicon limit of conventional RESURF structures. Compared to the other compensation devices considered, the $R_{DS(on)}\cdot A$ of the orthogonal SJ is at least by 4% superior in a wide range of $V_{(BR)DSS}$.

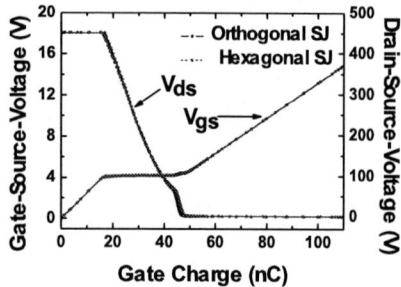

Fig. 5: *Gate charge characteristics under clamped inductive switching.*

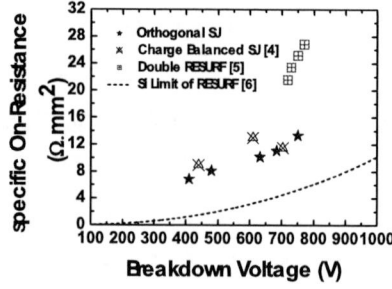

Fig. 6: *Trade-off relation between specific on-resistance and breakdown voltage.*

4. Charge Imbalance Effects

Both quantities, $V_{(BR)DSS}$ and $R_{DS(on)}\cdot A$, depend on how completely the charge in the p-columns is compensated by that in the n-drift region. Particularly, for a given value of N_D, there exists an optimum value of N_A, for which $V_{(BR)DSS}$, attains its maximum, and vice versa. Doping variations of about ±10% from the optimum level have commonly to be accepted in view of the tolerances achievable in today's manufacturing processes [3]. The relative doping variation is referred to as "charge imbalance". For each SJ structure considered, the variation of $V_{(BR)DSS}$ and $R_{DS(on)}\cdot A$ with the charge imbalance in the p-columns, ΔN_A, and the charge imbalance in the n-drift zone, ΔN_D, is plotted in Fig. 7. Evidently, the changes in $V_{(BR)DSS}$ are

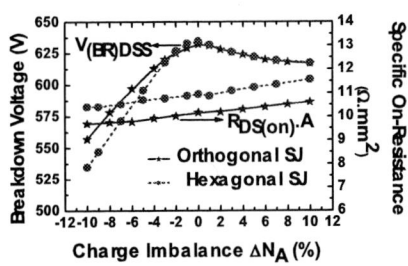

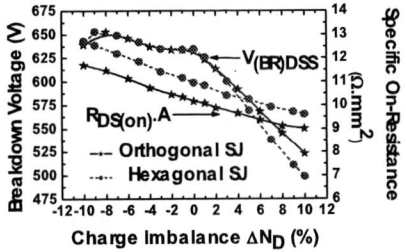

Fig. 7: *Sensitivity of the breakdown voltage and specific on-resistance to the charge imbalance in the p-columns (top) and in the n-drift layer (bottom).*

not symmetric under a change of sign of the charge imbalance. For $\Delta N_A < 0$ or $\Delta N_D > 0$, respectively, $V_{(BR)DSS}$ falls off much more rapidly than in the opposite case, where $\Delta N_A > 0$ or $\Delta N_D < 0$ holds. As we maximize $V_{(BR)DSS}$ of the two SJ designs by varying the values of N_A with respect to a fixed value of N_D, a lower ΔN_D results, to some extent, in an increase of the peak value of $V_{(BR)DSS}$, obeying the relation $V_{(BR)DSS} \propto 1/N_D$. The asymmetric dependence of $V_{(BR)DSS}$ can be attributed to a displacement of the location of avalanche breakdown from the optimum point near the substrate interface. Considering that all deviations of $V_{(BR)DSS}$ from its peak value, as caused by a charge imbalance in the range of $\pm 10\%$, stay below 17% for the orthogonal SJ and below 21% for the hexagonal SJ, respectively, the sensitivity of the blocking voltage of the two design patterns is comparable and fairly insignificant. This means that, in terms of blocking capability, both SJ structures are relatively robust against doping fluctuations. With regard to the dependence of $R_{DS(on)}{\cdot}A$ on charge imbalance, there is a slight increase of $R_{DS(on)}{\cdot}A$,

when ΔN_A is varied from -10% to +10%. In contrast to that, when ΔN_D is varied from -10% to +10%, $R_{DS(on)}{\cdot}A$ will sharply decrease. The reason is that increasing ΔN_D is equivalent to lowering $R_{DS(on)}$. On the other hand, increasing ΔN_A has the effect that the column diameters get apparently enlarged, because more p-type impurities thermally diffuse in the radial direction during the fabrication process. This, in turn, reduces the effective cross-sectional area of the drift zone, leading to a higher $R_{DS(on)}$.

5. Conclusions

Two different SJ structures, the orthogonal and the hexagonal SJ, were implemented in 600V-rated lateral SJ-MOSFETs on bulk silicon to tackle the substrate depletion effect. While both design variants are comparable in terms of switching operation, voltage blocking capability, and sensitivity of the breakdown voltage to charge imbalance, the orthogonal SJ outperforms the hexagonal one with respect to the current handling capability. In comparison to RESURF devices with similar voltage ratings, the orthogonal SJ exhibits a 47% reduction of the specific on-resistance; thus, it can be considered as a competitive candidate for state-of-the-art lateral compensation devices in smart power ICs.

References

[1] G. Deboy et al., in *Proc. IEDM*, 1998, pp. 683-686.
[2] S. G. Nassif-Khalil and C. A. T. Salama, in *Proc. ISPSD*, 2002, pp. 81-84.
[3] P. M. Shenoy, A. Bhalla, and G. M. Dolny, in *Proc. ISPSD*, 1999, pp. 99-102.
[4] M. H. Kim et al., in *Proc. ESSDERC*, 2002, pp. 367-370.
[5] M. H. Kim et al., in *Proc. ISPSD*, 2001, pp. 347-350.
[6] R. P. Zingg, in *Proc. ISPSD*, 2001, pp. 343-346.

FEM Simulation and Characterization of Microcantilevers Resonators

Margarita S. Narducci, Eduard Figueras, Isabel García, Luís Fonseca and Carles Cané

Centro Nacional de Microelectrónica de Barcelona, CNM-IMB(CSIC),
Campus UAB, Bellaterra, 08193 Barcelona, Spain
e-mail: margarita.narducci@cnm.es

This work has been focused on the design, simulation, fabrication and characterisation of cantilevers structures that include the excitation element and a bending sensor. Structures have been fabricated with different dimensions and geometries, its operation verified and their resonance frequency and quality factor (Q) measured. Device performance was compared with simulation predictions obtained using Finite Element Analysis (FEA) with ANSYS. For example, structures with dimensions in the range of $200 \times 150 \mu m^2$ to $400 \times 300 \mu m^2$ shown the first mode of resonance frequency laid between 440 KHz and 90 KHz and the quality factor between 770 and 1100.

1. Introduction

Microcantilevers have been widely studied and have received a lot of attention as an important category of MEMS resonators. This extensive research and development is trying to enable new and versatile applications in sensor technology and telecommunications. All this is possible because of their outstanding features, including: high sensibility, precision and stability. Other advantages about this kind of resonator are their simple structure, reduce size and power consumption besides repeatability and potential for direct integration with electronics [1].

In this work cantilever structures have been designed, fabricated and the frequency response analyzed. The device tests have been performed by measuring the frequency of resonance and quality factor and finally these results have been compared with previous works.

2. Design

The resonator was designed as a structure formed by 3 cantilevers with a length L and a width W each, that are hold together by means of an extra, square mass (see Figure 1). The total extra mass for the whole structure is:

$$m_{ex} = \rho_{Si} \times h_{Si} \times L_m \times 3W_m .$$ (1)

The extra mass performs a platform necessary for the deposition of a sensitive layer. With the appropriated layer de structure will work as a sensor.

Cantilevers were driven at their mechanical resonance in flexural mode by two heaters (electro-thermal principle) placed on each external beam and the resonance frequency was monitored by reading the signal generated by four piezoresistors in a Wheatstone bridge configuration placed on the central one. This layout tries to increase thermal isolation and to reduce the cross-talk between heaters and the Wheatstone bridge [3, 4].

1-4244-0396-0/06/$25.00 ©2006 IEEE

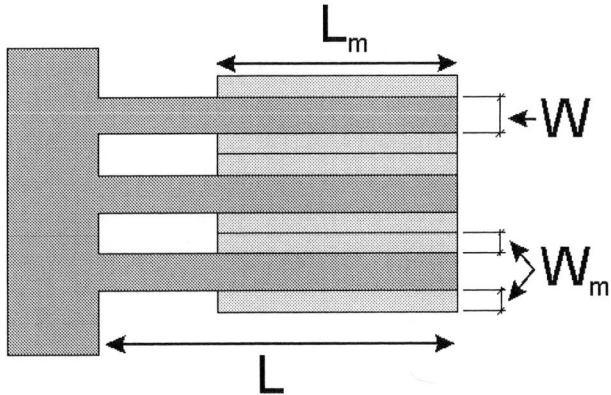

Figure 1. Schema of a cantilever and definition of the parameters.

In this work, two chips were fabricated, both of then containing 12 devices. In this paper, eight different devices are used; these cantilevers were previously selected because their higher values of resonance frequency and quality factor (see Figure 2 and Table 1).

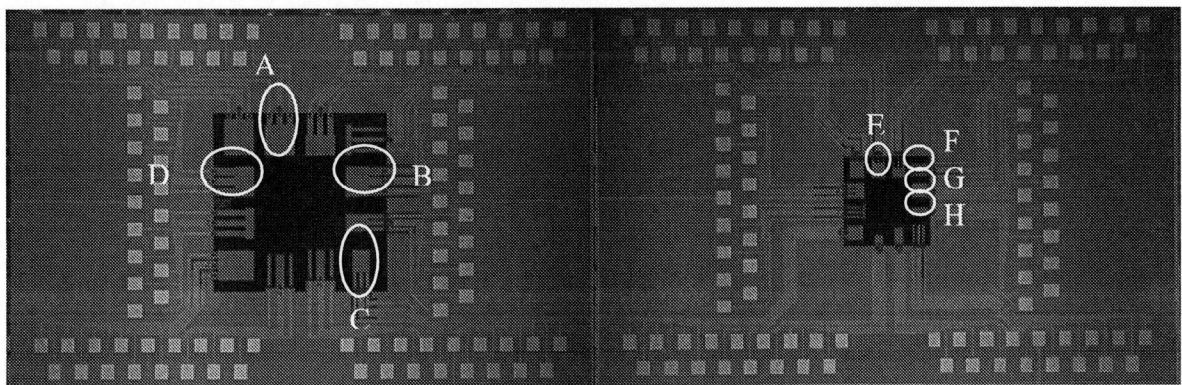

Figure 2.Photograph of the two chips used in this work.

Structure type	Length of the beam L (µm)	Width of the beam W(µm)	Length of the extra mass L_m (µm)	Width of the extra mass W_m(µm)
A	400	64	300	36
B	400	92	235	8
C	400	50	235	8
D	400	84	200	16
E	200	32	150	18
F	200	28	25	22
G*	200	150	0	0
H	200	20/34/20	185	6

Table 1. Dimensions of different structures. For all the beams the thickness is 15µm. *The structure type G was designed as a whole cantilever.

3. Fabrication process

The fabrication process is illustrated in Figure 3. The resonators were fabricated following a 7-mask process, starting with an N-type SOI double side polish wafer. The silicon layer is 15µm thick over a 2µm buried oxide and 450µm of bulk silicon. The process starts with a 180Å grown dry oxide layer and then a 1175Å silicon nitride layer is deposited (Figure

3a). The first level mask is used to define the active zone in the frontside, and then a 10600Å field oxide is grown. After that, with the second mask the backside window is defined (Figure 3b). Next, the silicon nitride on the frontside is removed and a Boron implantation of $1.0*1015at/cm^2$ and 50Kev is performed to define the resistivity of the Wheatstone bridge resistors, subsequently trough the third mask a new implantation fix de resistivity of contacts and heaters. Then a 1.3μm BPTEOS oxide is deposited (Figure 3c). The fourth mask is used to open contacts. Aluminium is deposited and patterned using the fifth mask (Figure 3d) to define metal connections and bonding pads. Afterwards a 0.4μm PECVD oxide and 0.4μm PECVD nitride are deposited (passivation layer) and patterned with the sixth mask (Figure 3e). The seventh mask is used to define the motif on the frontside, after that using the nitride mask on the backside the silicon substrate is etched using a KOH bath. Finally the 15μm membrane is etched by reactive ion etching and the cantilever is released (Figure 3f).

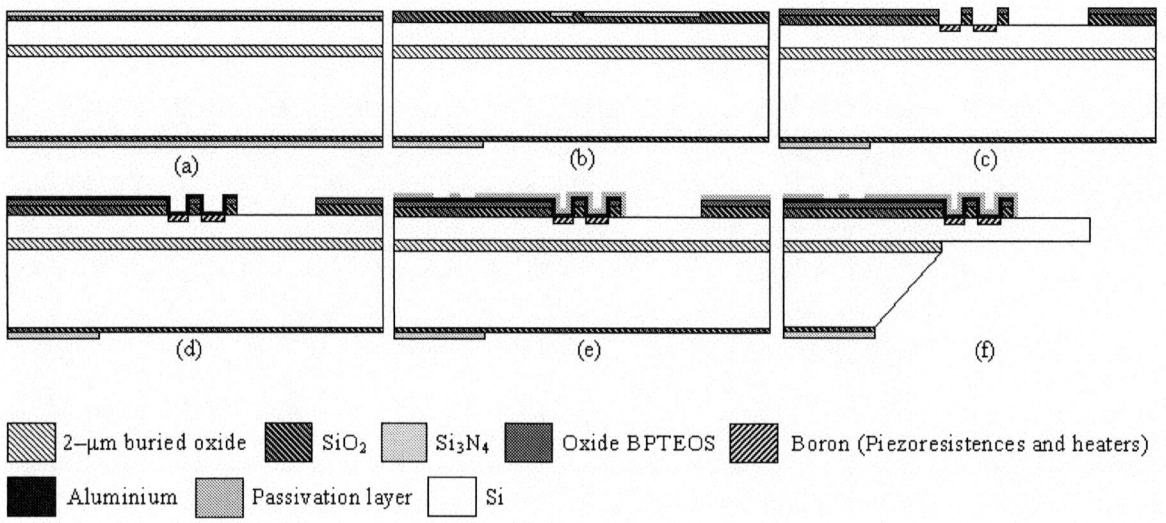

Figure 3. Fabrication process flow.

4.   Characterization and results

The fundamental resonance frequency has been obtained by measuring the amplified output voltage of the Wheatstone bridge. The excitation voltage is a harmonic ac voltage with offset applied to the heaters. The Table 2 shows the resonance frequency and quality factors obtained.

When comparing experimental and simulated values of the resonance frequency, approximately a 15% of error is obtained. This could be because the difference between the theoretical and experimental values of the materials constants and because the damping was not included in simulations.

In previous work, similar structures (type A, B, C and D) have been studied [6]. Comparing these results, the quality factor remains almost the same and a small improvement in resonance frequency (~10 KHz) is obtained.

Structure type	fr (Hz)	Q
A	92046	873
B	107336	771
C	101918	863
D	102990	856
E	333541	1196
F	402268	935
G	439131	931
H	434475	1108

Table 2. Results for the experimental resonance frequency (fr) and quality factor (Q).

5. Conclusions and outlook

In this paper resonant silicon cantilevers was designed, fabricated and examined. Resonators were designed as a structure formed by 3 cantilevers that are hold together by means of an extra, square mass because this approach is most accurate (error ~15%) than the classical one (error ~30%). The experimental results show small improvement in structures type (A, B, C and D) and considerable enhancement in structures type (E, F, G and H), obtaining a resonance frequency and a quality factor up to 434475 Hz and 1108.

Mechanical simulations have been performed in ANSYS showing good agreement with experimental results; however the simulation model must be enhanced including some factors such as: damping, stress (due to fabrication process) and more realistic material constants values. An extension of this work, deposit polymer on the cantilever, could be used to study the cantilever sensitivity and its application as a sensor.

Acknowledgement

This work has been financed by Spanish CICYT project n° TIC2001-0554-C03-02.

References

[1] A. T. Ferguson *et all*, "Modelling and design of composite free-free beam piezoelectric resonators", Sensors and Actuators A 118, 2005, pp.63-69.

[2] H. Jianquing *et all*, "Dependence of the resonance frequency of thermally excited microcantilevers resonators on temperature", Sensors and Actuators A 101, 2002, pp.37-41.

[3] H. Baltes *et all*, "Micromachined thermally based CMOS microsensors", Proceedings of the IEEE, 1998, vol.86, No.8, pp.1660-1678.

[4] M. Morata *et all*, "Thermal and mechanical simulation of bulk resonators", DTIP of MEMS/MOEMS, May 2003, Cannes (France), pp.208-213.

[5] E. Figueras *et all*, "Micro-cantilevers for gas sensing", Spanish Conference on Electron Devices, February 2005, Tarragona (Spain), pp.565-567.

[6] E. Figueras *et all*, "Mechanical characterization of micro-resonators structures". Spanish Conference on Electron Devices, February 2005, Tarragona (Spain), pp.57-60.

[7] R. Candler *et all*, "Investigation of energy loss mechanisms in micromechanical resonators", Actuators and Microsystems, June 2003, pp.332-335.

Electric energy harvesting inside self powered microsystem

JANÍČEK VLADIMÍR, HUSÁK MIROSLAV

Department of Microelectronics, FEE CTU Prague
Technicka 2, 16627 Prague 6, Czech Republic
e-mail: janicev@fel.cvut.cz and husak@fel.cvut.cz

Abstract: - Integration of the power supply unit into today's electronic devices gives the possibility to build autonomous electronic devices structure totally separated from the outside world. As the storage of generated energy conventional batteries or polymer capacitors can be used. A vibration-powered micro-generator, based on a polymer piezoelectric material, is proposed to be used as an energy generator for this purpose.

1. Introduction

The interest in application of all kinds of electronic devices and everyday's demand on implementation of microelectromechanical systems in the last decade has produced rapid progress in the efforts of miniaturizing sensors and actuators. With this demand on making everything smaller there is the same situation in the field of sensors and their wide range of applications such as process control units, medical implants, embedded sensors etc.

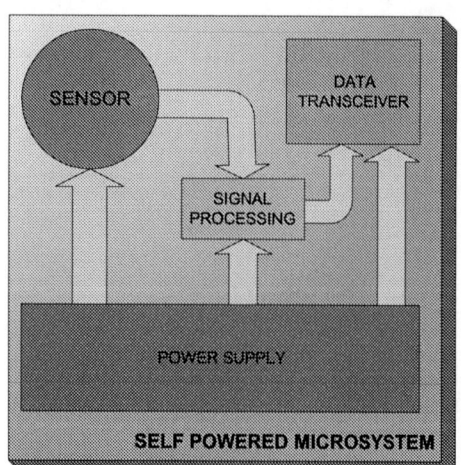

One specific aspect is the problem of supplying electrical power to such a device. Conventional power supplies could be used. But there are many applications, however, that require sensors to be completely embedded in the structure with no physical contact to the rest of the world. Supplying energy to such a system is difficult and the only possible solution is to make them so-called self-powered microsystems (See Fig.1). With integration of power supply unit on the same chip there are several advantages associated like noise reduction, elimination of crosstalks and reduction of power delivery control system complexity [1,2,3].

Fig. 1: Self-powered microsystem principle

The application of this kind of systems can be however expanded for several other applications such as:

- Applications that make it impossible to be hard wired to the base station. For example in machines [4], if there is a possibility of wires to become twisted due to the rotation of the tested object.
- Applications in contained area, such as in extreme heat, cold, humidity or chemical reactive conditions.
- Applications in which long distances are to be bridged or if there are several distributed components, like as it is in so-called smart wireless homes.
- Mobile and wireless applications, e.g. monitoring environmental conditions or positions of mobile goods.

2. Self power microsystem description

Three main units of such integrated self powered microsystem are: power supply unit with an energy generator, energy storage block and signal data processing and transmission unit.

2.1 Power supply unit with energy generator

An unconventional solution is to design a micro generator to convert energy from an existing ambient energy into electrical energy – here piezoelectric effect. But today's most used piezoelectric material PZT was substituted by piezoelectric polymers. It's a good candidate because of their ideally suitable characteristics - lightweight, flexibility and reasonably good force/displacement. The micro-mechanical generator will basically be constructed by depositing metal based electrodes onto thin film Polyvinylidene fluoride (PVDF).

2.2 Generated energy storage

Ideally an energy storage mechanism that can guarantee high power requirements and energy densities with the smallest size possible is required. The candidates are:

- Electrochemical capacitors or so-called super capacitors
- Secondary type (rechargeable) batteries

In general, battery technologies provide high energy density but lack of sufficient power density and capacitor technologies provide high power density but they are limited by energy density. Some new combination of irradiating PVDF and coating it with a thin, amorphous acrylate polymer film has been done. This produces a single-layer capacitor of power density on the order of 5.0 joules per cubic centimeter (J/cm^2). Commercial capacitors on the market today provide about 2.0 to 2.5 J/cm^2. According to Sigma [5], a single-layer capacitor with power density on the order of 20 J/cm^2 can be produced. Coating PVDF with acrylate polymer improves the melting point of the PVDF to 300°C, enabling the capacitor to handle higher voltage. It also acts as a skin that helps a capacitor self-heal. There has been a capacitor-battery combined system investigated. Such electrochemical system can result a smaller and more efficient system (See Fig.2).

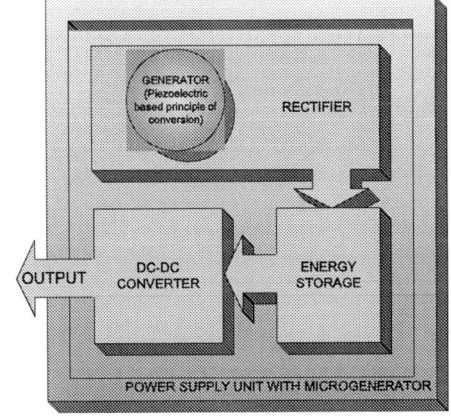

Fig.2: Power supply unit structure

2.3 Output signal data processing and transmission unit

The system must be able to transmit the measured data to the outside world and it must be done while the sensor system is physically isolated. The objective is to use as small energy as possible to transmit the data. Then the power can be reduced to a low level by limiting the data rate.

3. PVDF as piezoelectric material

Piezoelectric materials create electrical charge when mechanically stressed. Unfortunately, the most widely used zirconate titanate (PZT) is fragile and it is very difficult to produce it in large sizes. Piezoelectric polymers are increasingly considered as favorable materials for MEMS applications due to their fast response, low operating voltages and greater efficiencies of operation. The properties of polymers are different in comparison to inorganics (Table 1) that they are uniquely qualified to fill areas where single crystals and ceramics can not be used effectively. As shown in Table 1, the polymer piezoelectric strain constant (d_{31}) is lower than ceramic. However, piezoelectric polymers have much higher piezoelectric stress constants (g_{31}) indicating that they are much better sensors than ceramics. Polymers also exhibit high strength and high impact resistance. Polymers have low dielectric constant, low elastic stiffness, and low density, which result in a high voltage sensitivity (excellent sensor characteristic), and low acoustic and mechanical impedance (crucial for medical and underwater applications).

Table 1

Property comparison for standard piezoelectric polymer and ceramic materials

Property	Units	PVDF	PZT
Density	$\dfrac{g}{cm^3}$	1.78	7.6
Relative permittivity	$\dfrac{\varepsilon}{\varepsilon_0}$	12	1200
Elastic modulus	$\dfrac{10^{10}\,N}{m}$	0.3	4.9
Piezoelectric strain constant	$\dfrac{10^{-12}\,C}{N}$	$d_{31}=23$ $d_{33}=-33$	$d_{31}=-171$ $d_{33}=374$
Coupling constant	$\dfrac{CV}{Nm}$	$k_{31}=0.12$ $k_{33}=0.15$	$k_{31}=0.34$ $k_{33}=0.69$
Piezoelectric stress constant	$\dfrac{10^{-3}\,Vm}{N}$	$g_{31}=216$ $g_{33}=-330$	$g_{31}=-11$ $g_{33}=25$

Polymers also typically possess a high dielectric breakdown and high operating field strength, so they can resist much higher driving fields. Polymers offer the ability to pattern electrodes on the film surface, and pole only selected regions. The coupling constant shown in Table 1 represents the mechanical to electrical energy conversion efficiency of the material. The subscripts of the constants indicate the direction or mode of the mechanical and electrical interactions. "31 mode" indicates that strain is caused to axis 1 by electrical charge applied to axis 3.

Conversely, strain on axis 1 will produce an electrical charge along axis 3. Bending elements, made by an expanding upper layer and a contracting bottom layer, are made to exploit this mode in industry. In practice, such bending elements have an effective coupling constant of 75 storage of mechanical energy.

PVDF is very flexible. The cost, however, is that PVDF's coupling constant is significantly lower than PZT's. Also, shaping PVDF can reduce the effective coupling of mechanical and electrical energies due to edge effects. In addition, the material's efficiency degrades depending on the operating temperature and the number of plies used. There has been already published PVDF based solutions using energy from ocean waves [6].

4. Fabrication of PVDF cantilever

PVDF is material which can be shaped and formed into required shape and size. The most commonly used manufacturing process is laser micromachining, electroplating and punching (microembossing).

5. Layout description

The proposed electromechanical microgenerator is essentially a resonant mechanical structure based on cantilever modifications. The PVDF is coated on both planar sides with a conductive metal layer. These layers act as electrodes. Serpentine cantilever (See Fig.3) has been designed to achieve a low resonant frequency structure as well as a low damping effect when

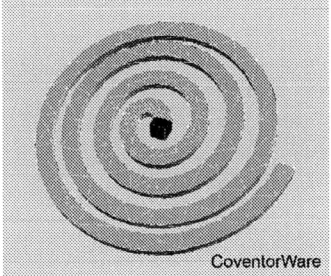

it resonates. A small mass is attached to the free end of the beam. There has been already published concept of microgenerators [7], which use a magnet and coil arrangement to generate the electrical power. One of the factors affecting output power delivered is the resonant frequency of the beam; the higher the frequency the more power will be available. The application areas envisaged for our system do not generally exceed a few hundred Hertz, and are often restricted to below 100 Hz.

Fig.3: Layout of the proposed serpentine cantilever with mass in the centre of the cantilever.

As already described most important aspect is the power generation efficiency to surface ratio. To maximize the microsystem area and to optimize the energy efficiency of the layout there has been designed a field of hexagonal shaped serpentine cantilever (See Figure 4).

Fig.4: Layout of the hexagonal shaped serpentine cantilever field with elongated proof mass

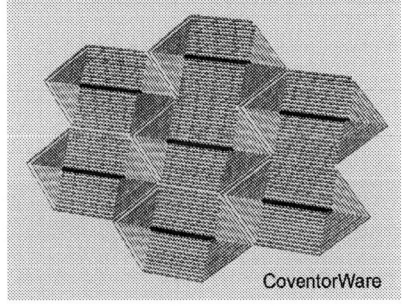

6. Conclusion

This paper describes concept of self powered microsystem with micro generator based on piezoelectric polymer material. The device is not optimized yet and significant improvements are envisaged in the future.

References
[1] M. Klein, H. Haspeklo, H. Wunderlich: "Sensor Systems using wireless Signal and Power Transmission: Applications and future requirements", *Proc. of Sensor '99*, p. 153-156, 1999
[2] H. Wunderlich, G. Hettich, M. Klein, R.Schraub, J. Schrenk: "Concepts and Steps in the Development of Wireless Sensors and Actuators for Automotive Applications", *Proc. of Sensor '99*,p. 157-162, 1999
[3] R. Puers et al: "A Telemetry System for the Detection of Hip Prosthesis Loosening by Vibration Analysis", Proc. of Eurosensors XIII, p. 757-760, 1999, The Hague
[4] J. D. Turner, L. Austin: "Sensors for automotive telematics", *Measurement Science and Technology*, Vol. 11 (2000), p. 58–79.
[5] A.Yializis, „Ultra-High-Energy Density Polymer Film Capacitors", *www.sigmalabs.com* , 2003.
[6] C.B. Carroll. 5,814,921: "Frequency multiplying piezoelectric generators:, *US Patent*, 1998.
[7] P. Glynne-Jones, M. J. Tudor, S. P. Beeby, and N. M. White, „An electromagnetic, vibration-powered generator for intelligent sensor systems", *Sensors and Actuators, A: Physical*, 110(1-3):p. 344-349, February 2004.

Monitoring of Psychosomatic Properties of Human Body by Skin Conductivity Measurements using Thin Film Microelectrode Arrays

E. Vavrinský, V. Stopjaková, L. Majer, V. Tvarožek, M. Weis, P. Marman

Department of Microelectronics, Slovak University of Technology,
Ilkovičova 3, 812 19 Bratislava, Slovakia
e-mail: erik.vavrinsky@stuba.sk

This paper describes a new modification of non-invasive biomedical monitoring of psychosomatic processes. The proposed method is based on skin conductivity measurements by interdigitated array (IDA) of microelectrodes, which allows continual monitoring and analyzing of complicated physiological, pathophysiological and therapeutic processes. Main goal is to monitor psycho-galvanic reflex of the human skin in medical and psychological experiments.

1. Introduction

Measurements of the electrical conductivity, resistance or impedance of skin surface have a 120 year history but the way how physiological changes in living tissue are reflected in electrical impedance parameters is still not very clear [1]. Technical realization of such measurements is very simple but in practice, there is a problem with reproducibility and each other comparison. The psycho-galvanic reflex is the main detection parameter of stress, excitement stimuli or shock, and it is characterised with immediate changing of the skin conductivity [2]. First, it was assumed that increase of skin conductivity during a stress stimulus is only caused by skin perspiration but a very important factor here is the potential barrier near the stratum lucidum layer, which thickness changes due to the nervous system [3].

2. Physiological principle of PGR monitoring

Several conditions can occur if electrodes are applied on human skin. In case of using some macroelectrodes, when the distance between coupled electrodes is greater than the thickness of the stratum corneum with potential barrier, the vector intensity lines of the electric field are enclosed across the planar skin structures (Fig. 1 a, b). If microelectrode pairs are utilized, when the distance between the electrodes is less than the electric thickness of the skin (stratum corneum with potential barrier), then the lines of electric field are enclosed in longitudinal circuit relative to laminar skin structures of epidermis (in stratum corneum). From deeper layers of skin, the electric field intensity lines are embossed to the surface (to the area with a lower conductivity) by the instrumentality of the potential barrier (Fig. 1c). However, under a stress stimulus the potential barrier narrows down and the electric field can reach deeper layers of human skin with higher conductivity, and therefore, the total conductivity is increasing (Figure 1d). Such a configuration is therefore ideal for the analysis of electrophysiological processes in human skin under stress.

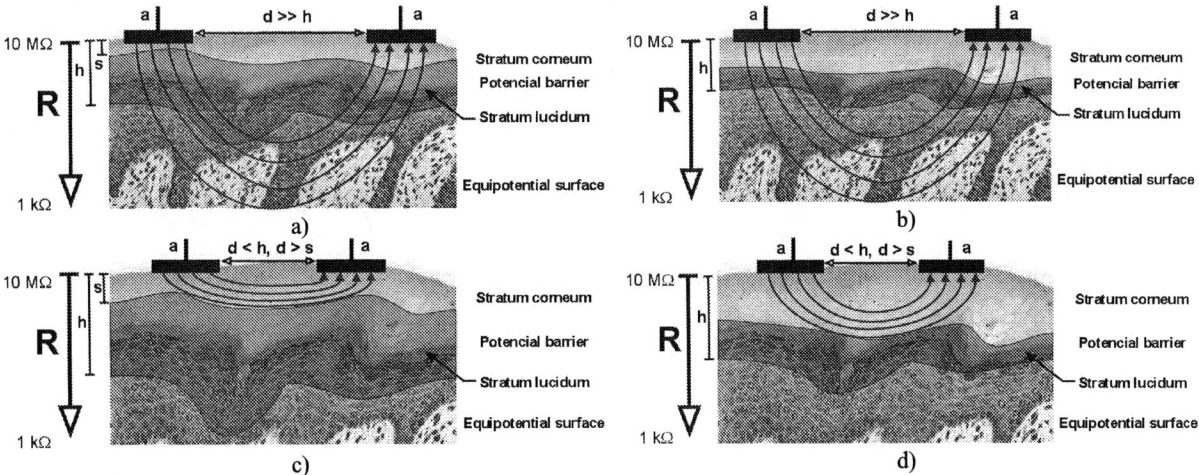

Fig. 1: The vector intensity lines of the electric field in human skin by using:
- macroelectrodes: a) in relaxation time, b) under stress stimulus
- microelectrodes: c) in relaxation time, d) under stress stimulus

3. Experiments

In first, experiments on the electrodermal response to stress stimuli were performed by variations of IDA microelectrodes conductivity ΔG (relative change of current ΔI at constant supply voltage U and frequency f). The IDA system was placed on forefinger of non-dominant hand. During conductivity measurements a drift of output signals occurs due to polarization effects in skin – electrodermal phenomena (EDF) (Fig. 2a). We minimized this effect by software using exponential function $G(t) = A + B(1-e^{-t/C})$, where G is skin conductivity, t is time and A, B, C are constants (Fig. 2 b). This experiment led to very important result: the microelectrode probes are able to monitor simultaneously the electrodermal response as well as the hearth pulses (Fig. 2b).

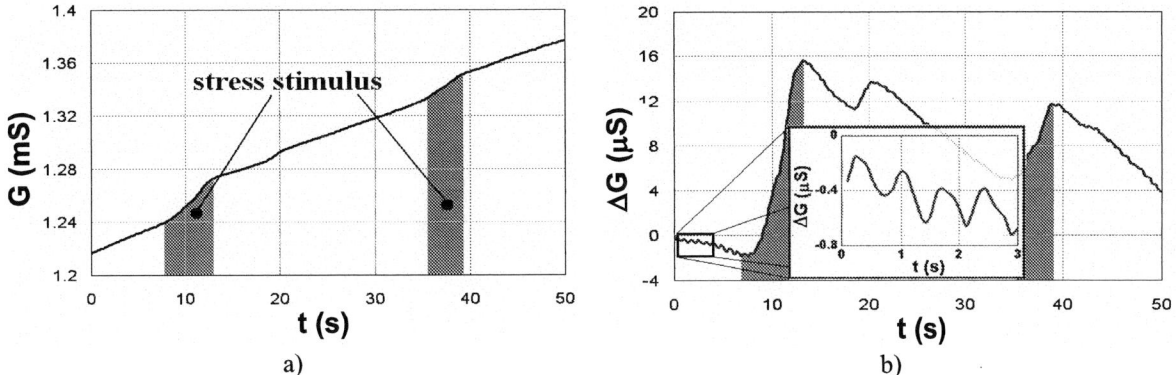

Fig. 2: Typical time responses of EDR (uncorrected and corrected signal with the heart-pulses)

In many next experiments, the influence of different factors such as microelectrodes dimensions, supply voltage parameters, skin hydration (sweating), and the microelectrode placement on output signal have been investigated. The achieved results show that even though the absolute values of conductivity differ, the conductance time responses are very similar. Optimal input signal amplitude and frequency of 3V and 1 kHz have been found, respectively. A proper IDA microelectrode dimension is: 200μm/200μm (finger/gap ratio). Experiments also prove thesis about the potential barrier thickness changing and its influence on the electric field distribution and the penetration depth into the different skin layers.

At last, a comparison of our microelectrodes method to the commercial macroelectrode approach based on galvanic skin response (GRS) (used on Faculty of Philosophy, Comenius University) was carried out. Standard psychotests show that the responses were similar but the microelectrode signals are more stable with a shorter response time (Fig. 3).

After these experiments our next step was to develop integrated monitoring device.

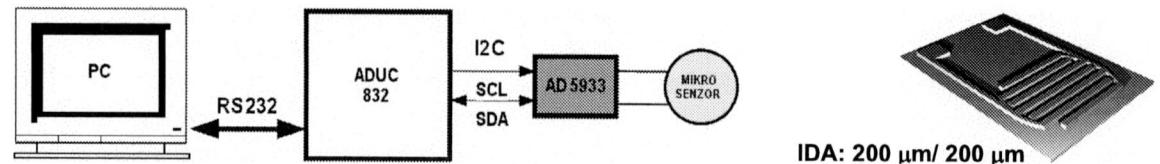

Fig. 3: Comparison of our microelectrode system and classical macroelectrode GSR method

4. Integrated monitoring system

Finally, complex measurement environment for continuous and non-invasive monitoring of the skin impedance has been developed. The proposed measurement system (Fig 4.) consists of the microsensor described above, integrated circuit AD5933, microprocessor ADuC832, and a personal computer.

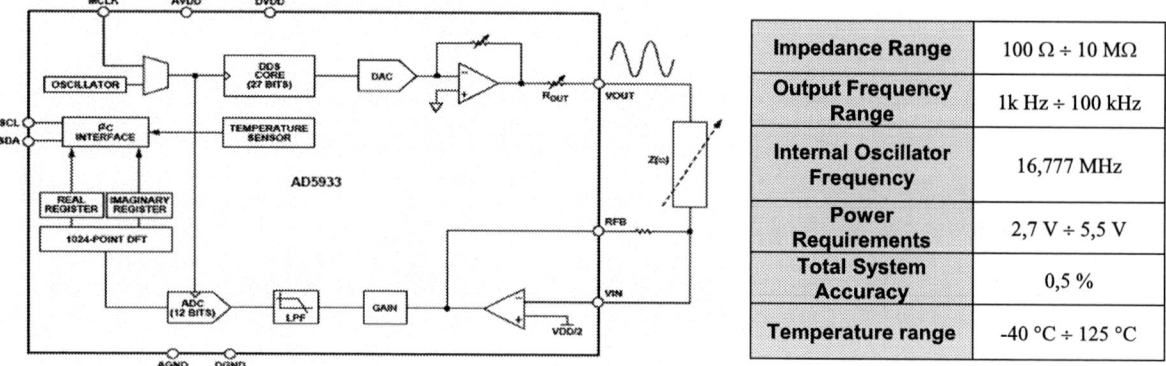

Fig. 4: Block diagram of the measurement system

Impedance Range	$100\ \Omega \div 10\ M\Omega$
Output Frequency Range	$1k\ Hz \div 100\ kHz$
Internal Oscillator Frequency	$16,777\ MHz$
Power Requirements	$2,7\ V \div 5,5\ V$
Total System Accuracy	$0,5\ \%$
Temperature range	$-40\ ^{\circ}C \div 125\ ^{\circ}C$

Fig. 5: Integrated circuit AD5933

The integrated circuit AD5933 (Fig. 5) by Analog Devices [4] provides measurement of the human skin impedance sensed by the microsensor system. The measurement is controlled by the microprocessor ADuC832 [5] via I2C interface. Using a serial interface RS232, the microprocessor then sends the measured data to a personal computer providing data storage. Additionally, the microcontroller also provides an initial configuration for the integrated circuit AD5933 needed at the measurement start. This configuration includes mainly frequency and amplitude of the input signal used for measurement of an unknown impedance. The controller also controls time slots during which the measurements are performed. After the measurement in the time slot is done, the microprocessor reads the data from AD5933 circuit and sends them to a PC, where the data are stored and further processed. The AD5933 circuit is composed of the following parts: an input signal generator, a 12-bit A/D converter, a DFT (Discrete Fourier Transform) circuit, a thermal sensor, and I2C interface. The generator provides a sine wave input signal of certain frequency and amplitude at the output VOUT. Unknown impedance is connected between VOUT and VIN terminals. Thus, the magnitude and phase of the current flowing through a load depend on its impedance. This current is then transformed to voltage that is converted into a digital signal by the D/A converter. Finally, the DFT circuit provides discrete Fourier transform of the converted signal. Resulting data give information about the values of the real and imaginary parts of a load admittance.

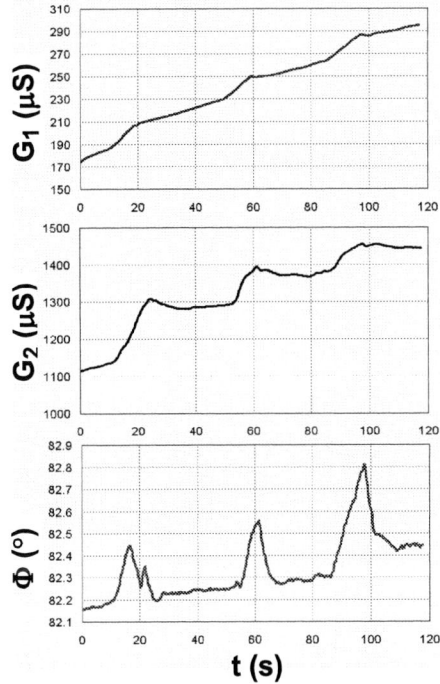

Fig. 6: Comparison of the achieved results

As the last experiment, we compared the results measured using the developed integrated monitoring system ($G2$ and Φ – admittance amplitude and phase, respectively) to the results obtained by the former measurement method ($G1$) (Fig. 6). This experiment shows that the phase of the skin impedance may offer more sensitive monitoring of psychogalvanic response since it significantly reflects admittance changes.

5.   Conclusions

An integrated measurement system for long time monitoring of the psychoglavanic reflex for human stress detection was developed. Experimental results show that the developed microelectrode probes are able to monitor the electrodermal response as well as the blood pulse caused by the heart rhythm, simultaneously. Moreover, thesis about the influence of the potential barrier and its thickness changing due to nervous system has been proven. Interesting outcome was observed – the psychogalvanic reflex might be much more accurately sensed by the admittance phase since this parameter reflects the human skin conductivity changes very precisely. The achieved accuracy, voltage and frequency ranges are suitable not from human stress monitoring point of view but also from the system integration and miniaturization requirements. The possibility to measure also heart rhythm with one IDA microsensor might be appealing and innovative aspect for psychological and medical research.

Acknowledgement

The presented work has been supported by the Ministry of Education of the Slovak Republic under project No. VTP1013/2003.

References

[1]   Olmar S., Bioelectrochemistry and Bioenergetics, 1998, vol. 45, no. 2, pp. 157-160 (4)
[2]   Qubit systems – Human electrophysiology, http://www.qubitsystems.com/electro.html
[3]   Weis, M., Danilla, T., Matay, L., Hrkút, P., Kákoš, J.: Noninvasive Biomedical Sensors on the Biology - Interface of Human Skin., 1995, pp.89-91 (3)
[4]   AD5933 – Datasheet, http://www.analog.com/
[5]   ADuC832 – Datasheet, http://www.analog.com/

PREPARATION AND CHARACTERIZATION OF MICROHOTPLATE FOR GAS SENSORS

A. Reháková[1], D. Tengeri[1], I. Hotový[1], T. Lalinský[1], V. Řeháček[1], L. Spiess[2], H. Romanus[2], Š. Haščík[3]

[1]*Department of Microelectronics, Slovak University of Technology, Ilkovičova 3, 812 19 Bratislava, Slovakia*
[2]*Technical University Ilmenau, PF 100565, D-98684 Ilmenau, Germany*
[3]*Institute of Electrical Engineering, Slovak Academy of Sciences, Dubravska cesta 9, 842 39 Bratislava, Slovakia*

E-mail: andrea.rehakova@stuba.sk

In this work, we describe preparation and characterization of microhotplate for gas sensor structures. The microhotplate was prepared by DC magnetron sputtering using a TiN/Pt doublelayer. Measuring of electrical parameters demonstrated the structure to have linear I-V characteristic and resistance of 63Ω. The electro-thermal characterization showed good linearity of the microhotplate at temperatures below 220°C. Temperature dependent measuring of different samples displayed influence of membrane thickness on selfheating process. According to these measurements, we could deduce some optimizations in design and fabrication process of the sensor structure.

1. Introduction

In recent years, the importance of gas sensing structures strongly expanded because of increasing number of applications from health care and safety to quality control in industry. The importance of sensors leads to continuous research including the optimization of sensing materials, device design and the fabrication process. The structure described in this work is a sensor structure based on semiconducting gas sensing materials. These structures use often metal oxide thin layers as gas sensitive element. Metal oxides can detect specific gases by increasing or decreasing their resistance (according to their oxidizing or reducing character) caused by chemical reactions. To achieve higher effectivity of chemical reactions between metal oxide and molecules of detected gas, the sensitive layer has to be warmed up.

Our research is focused on development, fabrication and optimization of microheater for semiconductor gas sensors. Described here is the heating meander fabrication, physical and electro – thermal characterization.

2. Construction and fabrication process

The sensor structure was designed and fabricated on GaAs substrate (thickness 300 μm), which was chosen as substrate material for its electrical, mechanical and thermal properties and better compatibility with integration processes. The structure is shown on Fig.1. The NiO layer lies on interdigitated electrodes fabricated for resistance measurements. Because of relatively high operating temperatures, the structure has to be warmed up. The main part of such sensor structure is the microhotplate, often constructed as heating meander or double spiral element.

Our sensor structure was made by several fabrication steps as it is shown on Fig. 2.

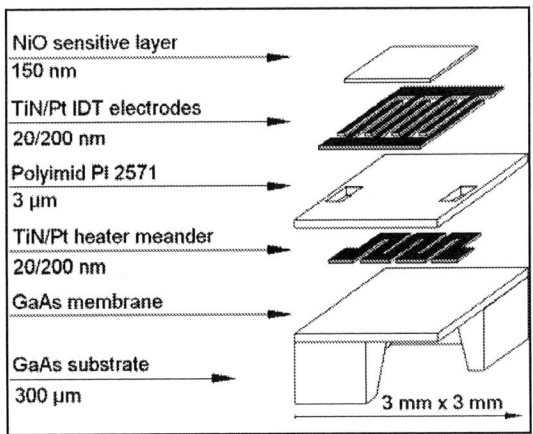

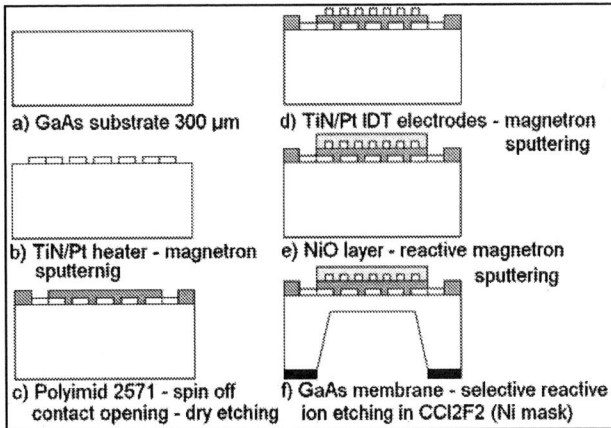

Fig. 1: Gas sensor structure

Fig. 2: Fabrication process

The microhotplate meander structure and interdigitated electrodes based on TiN/Pt thin films were prepared by DC magnetron sputtering using the TiN thin film as adhesive layer. A polyimide layer is used to insulate interdigitated electrodes from the microhotplate. This layer has an electrically insulating and thermal conducting function. The gas sensitive NiO layer was prepared using DC reactive magnetron sputtering in a mixture of O_2 and Ar. In one post processing step – selective reactive ion etching of GaAs substrate in CCl_2F_2, a closed membrane structure was obtained. A Ni mask was used.

3. Microscopic observation

Sensor chips were characterized by scanning electron microscopy and optical microscopy. Sensor structure elements, fabrication process accuracy and mask alignment were observed. On Fig.3 is a view of sensor chip from optical microscope displaying single elements overlay and contact pads of microheater and interdigitated electrodes.

A backside view of sensor structure chips from SEM observations is on Fig. 4. Single backside openings made by etching process can be observed.

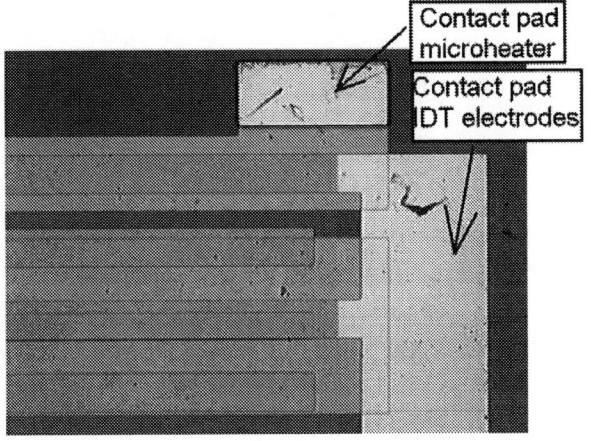

Fig. 3: Gas sensor structure
Optical microscope view

Fig. 4: Sensor chips
SEM image

4. Electro thermal characterization of sensor structure

The measurement set-up consisted of two devices KEITHLEY 238 and 237 operating as current and voltage sources and digital multimeters. These devices combined with a PC and workplace with optical microscope DM23 completed our set-up.

Temperature dependent measurements were realized in oven Eurotherm 2408 with automatically controlled temperature connected to our workplace.

To achieve optimal heating parameters we realized electro-thermal characterization of microheater structure. All samples exhibit good linearity and thermal stability at temperatures below 220°C. Resistance of microheater at room temperature is 63Ω. We also measured I-V characteristics of NiO sensing layer, which could be used as reference temperature sensor element. Room temperature resistance of NiO layer is 17.5 GΩ calculated at 2V.

Several measurements realized at temperatures up to 300°C were made to investigate the temperature dependence of heater resistance. The resistance values increase with temperature up to 250°C and then fall down (Fig. 5.). This phenomenon could be assigned to the interdiffusion effect of thin layers on GaAs substrate [2] and has to be inspected in future investigations. From these measurements we can define an important microheater parameter – thermal coefficient of resistance. Calculated from dependence on Fig. 5, TCR of the Pt heater is 1080×10^{-6}/K. On Fig.6 a power-temperature conversion characteristic of microhotplate is displayed.

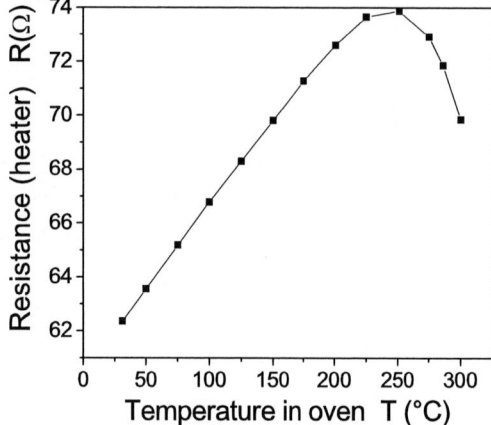

Fig. 5: Temperature dependence of microheater resistance

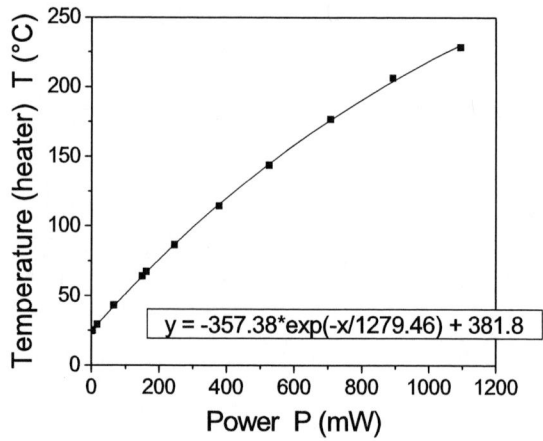

Fig 6: P-T curve of microheater

5. Effect of etch depth on selfheating process

For complex characterization we examined temperature dependence of NiO layer resistance. The resistance of NiO layer decreases with increasing temperature. The curve has an exponential character decreasing from values of over 300 MΩ down to values lower than 1MΩ. The next point of our measurement was the NiO layer resistance depending on current through the heating element. Current values from 10 mA to 130 mA were applied on the microhotplate and after 3 minutes of warming up the I-V characteristic were measured. Then, the resistance was calculated.

To avoid heating losses in the substrate we studied the influence of substrate thickness under the microheater structure on its electro-thermal properties. Two sensor chips with different membrane thickness (etch depth) were measured (chip 03 with depth 100 nm and chip 06 with etch depth 364 nm).

Following these measurements, the NiO layer temperature depending on heater current was calculated. The curve is shown on Fig. 7. Temperature of NiO layer represents surface temperature of sensor structure achieved by microheater. The thin membrane structure achieves higher temperatures than the thicker one at heating power higher than 300 mW. For lower values, the influence of membrane thickness is obscure.

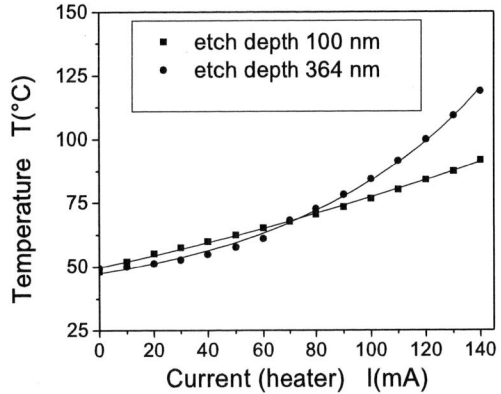

Fig. 7: Temperature rise caused by microhotplate current

6. Results and conclusions

Electrical characterization of sensor structure defined several structure parameters. The microheater resistance is 63 Ω and calculated TCR of Pt microhotplate is 1080×10^{-6}/K. The resistance of NiO layer is 17.5 GΩ at room temperature. According to our electro-thermal measurements, we can conclude, that heat losses in substrate can not be neglected. There is a significant influence of substrate thickness under the active area on heating effectivity. This effect enforces at heating power higher than 300 mW. The solution could be another construction of whole sensor structure, where the active area of sensor would be connected with substrate through microbeams with low thermal conductivity. Applying these optimization steps and experiences we can expect the reduction of power dissipation of sensors and the improvement of gas sensor parameters.

Acknowledgement

This work was supported by the Scientific Grant Agency of the Ministry of Education of the Slovak Republic and the Slovak Academy of Sciences, No.1/3095/06, by Science and Technology Assistance Agency under the contract No. APVT-20-021004, and by a Grant of DAAD.

References

[1] I. Hotový, T. Lalinský, V. Řeháček, L. Spiess, M. Gubisch, S. Haščík, *Fabrication and characterization of microheater on GaAsfor gas senzore*, IWK 2006, Ilmenau, Germany, 2006

[2] C. C. Chang, S. P. Murarka, V. Kumar, and G. Quintana, *Interdiffusions in thin-film Au on Pt on GaAs (100) studied with Auger spectroscopy,* Journal of Applied Physics, Volume 46, Issue 10, October 1975, pp.4237-4243

[3] Jiang S.R.1; Feng B.X.; Yan P.X.; Cai X.M.; Lu S.Y., *The effect of annealing on the electrochromic properties of microcrystalline NiOx films prepared by reactive magnetron rf sputtering*, Applied Surface Science, Volume 174, Number 2, 16 April 2001, pp. 125-131(7)

Reduced-order modeling of capacitive MEMS microphones using mixed-level simulation

M. Niessner[1], W. Bedyk[1], G. Schrag[1], G. Wachutka[1], B. Margesin[2], and A. Faes[2]

[1]Institute for Physics of Electrotechnology, Munich University of Technology,
Arcisstr. 21, 80290 Munich Germany, email: niessner@tep.ei.tum.de
[2]ITC-irst, Microsystems Division, via Sommarive 18, 38050 Povo (TN), Italy

We demonstrate a reduced-order modeling methodology based on the mixed-level simulation approach that allows for the rapid evaluation of microphone design variations with arbitrary arrangement and shape of the acoustic holes. The resulting reduced-order models take multiply coupled energy domains into account and are capable of describing the effects caused by non-linear fluidic damping. The methodology is easily approachable, as the generation of the reduced-order models is automated by the use of a MATLAB toolbox.

1. Introduction

Virtual prototyping of capacitive MEMS microphones requires the ability to quickly predict the characteristics of the device and, thus, the requirements on the circuitry. Models that allow for predictive simulation have to take into account the deformation of the diaphragm, the gas film damping, which is strongly influenced by acoustic holes, and the electric field between the diaphragm and the back electrode. As several energy domains have to be coupled and the acoustic holes must adequately be resolved by the mesh, FEM simulation proves to be computationally highly expensive or even prohibitive. Hence, reduced-order models are applied for the virtual prototyping. In the following we present a reduced-order modeling method based on the mixed-level simulation approach [8] and identify its capabilities by comparison with modeling via reduced-order equivalent circuits [1, 2, 3]. The demonstrator for the comparison is the microphone presented in [6, 7] and depicted in fig. 1.

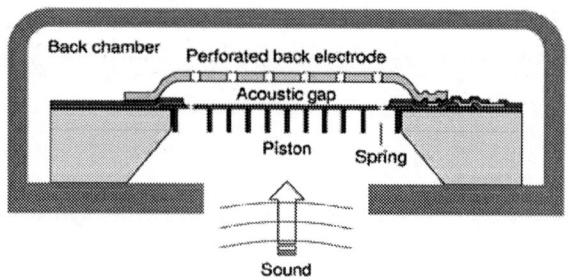

Fig. 1: *Schematic view of the capacitive microphone with its principal functional components: multi-layered and stiffened membrane, suspension springs, acoustic air gap, and perforated back electrode.*

2. Reduced-order equivalent circuits with lumped elements

In the modeling approach via equivalent circuits, the structure is decomposed into subsystems that are first represented by a set of lumped elements and then form a Kirchhoffian network. A network suitable for the modeling of the microphone under investigation is shown in Fig. 2. Here, the radiation impedance $Z_r=R_r+j\omega M_r$ [4] accounts for the air in front of the device that is displaced by the vibrating diaphragm. The mass M_d and the lumped compliance C_d approximate the undamped mechanical behavior of the diaphragm. The gas damping due to the air in the gap between the diaphragm and the perforated back electrode is modeled by the air gap compliance C_g and the fluidic resistances R_g and R_h of the air gap and the acoustic holes in the back plate [1, 2, 3].

1-4244-0396-0/06/$25.00 ©2006 IEEE

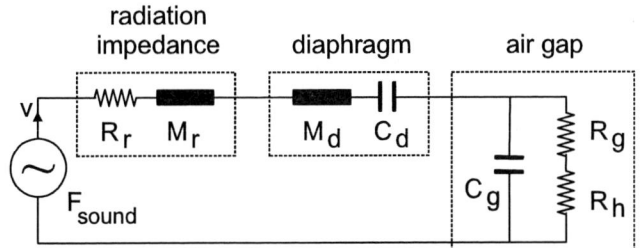

Fig. 2: *Equivalent circuit of the capacitive microphone. The variables governing this mechanical equivalent circuit are force ("across quantity") and velocity ("through quantity").*

The proper employment of this modeling strategy requires an in-depth understanding of the lumped element models and parameters. For instance, the model underlying R_g, which was derived in [5] under the assumption that all air in the gap flows concentrically towards the acoustic holes, does neither account for the air leaving via the edge nor for non-linear damping effects that might occur at small gap heights, high sound pressures or high frequencies.

Consequently, employing this approach without reflecting the assumptions underlying lumped models can easily lead to incorrect predictions in certain regions of the operating area.

3. Reduced-order mixed-level model

In order to generate a physical model that allows for predictive simulation even in regions where equivalent circuits yield no longer reliable results, the mixed-level simulation method [8] proves to be an efficient approach. The framework for it is provided by generalized Kirchhoffian network theory. Following this methodology, the microsystem is decomposed into subsystems that constitute single functional blocks or whole energy domains. These subsystems are then represented by physically based reduced-order models which are derived in terms of conjugate variables ("across"- and "through" variables) and can thus be linked together to a generalized Kirchhoffian network that, by preserving Kirchhoff's conservation laws, governs the exchange of energy and other quantities between the subsystems. In the case of the microphone considered, we decompose it in its mechanical, electric and fluidic domain. The generation of the model starts from a discretized FEM model of the diaphragm.

Reduced-order modeling in the mechanical domain is achieved by employing the Galerkin method in combination with modal superposition: the discretized deformation $z(t)$ of the diaphragm, represented as a vector of length n (= the number of nodes of the FEM model), is approximated by an affine linear combination of m discretized eigenmode shapes Θ_i added to the displacement z_0 in equilibrium:

$$z(t) \approx z_0 + \sum_{i=1}^{m} \Theta_i \cdot q_i(t) \quad (2)$$

Order reduction is achieved, if $m \ll n$. It can be shown that the merely mechanical (i.e. uncoupled) dynamical behavior of the diaphragm, represented by the m modal amplitudes $q_i(t)$, is governed by the following decoupled set of modal equations of motion:

$$\ddot{q}_i + \omega_i^2 q_i = 0 \quad (3)$$

Here, ω_i denotes the eigenfrequency of the i-th mode. For the microphone considered, $m=2$ eigenmodes have been selected from the spectral decomposition as obtained from the FEM modal analysis. According to Lagrange's formalism, the conjugate variables in the network representation of the mechanical domain are the modal amplitudes q_i as "across-" and the modal moments $p_i = \dfrac{\partial \mathcal{L}}{\partial \dot{q}_i}$ as "through-variables", where $\mathcal{L}(\boldsymbol{q}, \dot{\boldsymbol{q}})$ denotes the Lagrangian functional.

Fig. 3: *Fully meshed model of the diaphragm (left) and the first eigenmode shape at 38 kHz (right). A pseudo-material with averaged Young's modulus and internal stress is used instead of the poly-Si/Si$_3$N$_4$/SiO$_2$ sandwich with stiffeners building the real structure.*

To include the electric energy domain, a lumped plate capacitor element $C(q)$ is derived from the device geometry and plugged into the equations of motion by means of the Lagrangian energy functional (cf. eq. 4). Here, the electric voltage and the electric current are chosen as conjugate variables.

In the fluidic energy domain, the air in front of the diaphragm is modeled by the radiation impedance given in [4]. The pressure p_{eff}, which exerts a surface force on the diaphragm, is coupled to the mechanical domain by linking each modal moment to an area-weighted portion of this surface force, $p_{eff,i} \cdot A_{eff,i}$, through the conversion factors $\Theta_{eff,i}$ (cf. eq. 4). The implementation of the coupling in terms of network elements is illustrated in Fig. 5. The damping force exerted by the air in the gap between the diaphragm and the perforated back electrode is calculated by solving Reynolds' equation. To this end, the FEM-nodes from the discretized geometry are converted into network nodes and connected to build a finite fluidic network – again a Kirchhoffian network – that is governed by pressure differences and mass flows as "through-" and "across-variables" (see Fig. 4). Each of the network nodes is coupled with the mechanical domain through the local gap between the diaphragm and the back electrode, which is a function of the modal shapes and amplitudes. The fluid flow between two nodes is controlled by network elements which are automatically deduced from the local discretization of Reynolds' equation according to the geometry specification. Topological boundary effects, caused by the acoustic holes in the back electrode and the edges and corners of the diaphragm, are taken into account by attaching lumped compact models to the network where necessary.

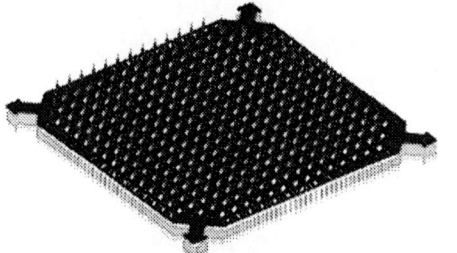

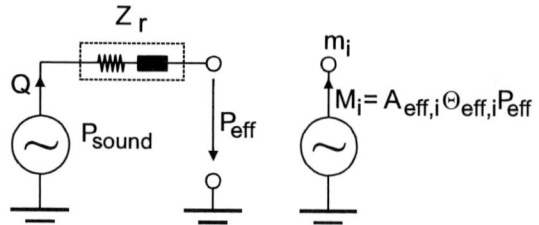

Fig. 4: *Visualization of the finite network used for solving Reynolds' equation, supplemented by compact models accounting for edge effects and acoustic holes.*

Fig. 5: *Circuit with voltage-controlled current source describing the coupling between the fluidic and the mechanical domain*

The nodal damping forces $F_{damp,j}(q)$, acting on the surface elements of the diaphragm as calculated from the finite fluidic network, are coupled with the modal equations of motion by weighting them with the local modal shape factor Θ_{ij}:

$$\ddot{q}_i + \omega_i^2 q_i = \frac{1}{2}\frac{\partial C(q)}{\partial q_i}V^2 + \sum_j \Theta_{ij}F_{damp,j}(q) + \Theta_{eff,i}A_{eff}p_{eff} \qquad (4)$$

The resulting mixed-level model, which consists of $m=2$ equations of the form (4) and the finite network, is coded in VHDL-AMS and, hence, amenable to a straightforward implementation in existing standard circuit simulators.

4. Simulation results

We assumed the following set of design parameters and plugged them in the mixed-level model described above.

Bias voltage	1.0V
Gap diaphragm – substrate	2.7 μm
Square acoustic holes with length of a side	30 μm
Spacing of acoustic holes	20 μm

The automated model generation was performed using the MATLAB toolbox, as network simulator we used SPECTRE. In a virtual experiment, the microphone was excited by a sinusoidal sound pressure wave at a frequency of 5 kHz. Our simulations predict that the damping of the diaphragm's vibrations will be linear up to a pressure of 450 Pa, but that non-linear damping effects will distort the vibration at higher sound pressures (see Fig. 6 and 7). This finding could not be made with the reduced-order equivalent circuits mentioned in section 2.

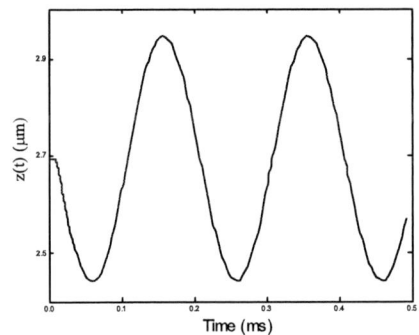

Fig. 6: *Linearly damped vibration of the diaphragm up to 450 Pa of sound pressure.*

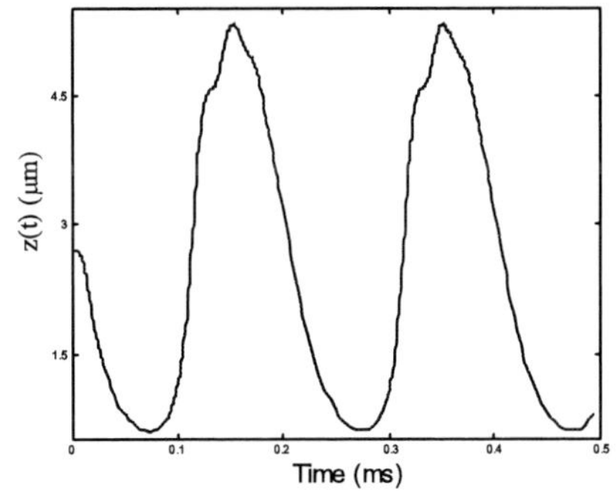

Fig. 7: *Non-linearly damped vibration of the diaphragm above 450 Pa of sound pressure.*

5. Conclusion

We demonstrated an automated reduced-order modeling methodology, based on the mixed-level approach, with reference to capacitive MEMS microphones. In contrast to reduced-order equivalent circuits, the resulting mixed-level model is able to describe the distortion of harmonic vibration due to non-linear damping at high sound pressure. Because the model generation process is automated, microphone designs with arbitrary geometric pattern and shape of the acoustic holes can be quickly and accurately analyzed.

References

[1] J. Bergqvist, and F. Rudolf, *Sensors and Actuators A* **45**, 1994, p. 115-124.

[2] W. Kuhnel, and G. Hess, *Sensors and Actuators A* **30**, 1992, p. 251-258.

[3] P.-C. Hsu, C. H. Mastrangelo, and K. D. Wise, *Proceedings of the MEMS'98 Conference,* Heidelberg, Germany, 1998.

[4] P. M. Morse, *Vibration and Sound,* McGraw-Hill, New York, 1948.

[5] Z. Skvor, Acustica, 19, 1967, p. 295-299.

[6] A. Faes, F. Giacomozzi, B. Margesin and M. Zen, in *Proceedings of the AISEM'04 Conference,* 2004.

[7] B. Margesin, A. Faes, F. Giacomozzi, A. Bagolini and M. Zen, in *Proceedings of the AISEM'03 Conference,* 2003.

[8] G. Schrag, and G. Wachutka, *Sensors and Actuators A* **97-98**, 2002, p. 193-200.

Recent Advances in Organic Electronic

J. Kováč, J. Jakabovič, L. Peternai, O. Lengyel and M. Kytka

Department of Microelectronics, Slovak University of Technology
Ilkovičova 3, 812 19 Bratislava, Slovak Republic
e-mail: jaroslav.kovac@stuba.sk

The development of integrated organic electronic devices is a gateway for a variety of applications, and is of great relevance for the general purpose of achieving highly integrated optoelectronic systems. For example, the combination of organic light emitting diodes (OLEDs) and organic thin film transistors (OFETs) is needed for the development of all-organic active matrix display technology. This article is concerned with the recent advances of organic devices made of organic materials, as electrically and optically active components, in devices ranging from simple single-component OLEDs, through double- and multi-layer OLEDs to organic displays and OFET's.

1. Introduction

Organic electronic received a great deal of attention in last years for electronic applications due to their several advantages over the conventional semiconductors. Non-traditional materials such as conjugated organic molecules, short-chain oligomers, longer-chain polymers, and organic–inorganic composites are being developed that emit light, conduct current, and act as semiconductors. The ability of these materials to transport charge (holes and electrons) due to the p-orbital overlap of neighbouring molecules provides their semiconducting and conducting properties. The self-assembling or ordering of these organic and hybrid materials enhances this p-orbital overlap and is key to improvements in carrier mobility. The recombination of the charge carriers under an applied field can lead to the formation of an exciton that decays radiatively to produce light emission [1]. The first electroluminescence (EL) from the organic solid material was observed by Helfrich et al. in 1964 on antracene crystal with thickness few millimetres [2]. There was needed to have high applied voltages (> 1000 V) to excitate EL, which didn't make it possible or practical using. By reduction of the organic layer thickness below 1 μm allowed to achieve electrical fields comparable to those applied to single crystals but now at considerable lower voltage. The development of organic low-molecular weight multilayer structures considerably improved the efficiency of light emission by Tang and Van Slyke from Eastman Kodak [3] with the potential in lightning and display applications [4]. In addition, since the discovery of electroluminescence in conjugated polymers by the Cambridge group [5] these materials have also been widely examined with equally prospects as the low-molecular weight materials. Since the first presentation of working organic thin film transistor (OTFT) in 1987 [6], there has been a growing interest in research and fabrication of thin film transistors based on organic materials. The performance of the best OTFTs is similar to the inorganic transistors based on hydrogenated amorphous silicon (a-Si:H), that are widely used in the active matrix liquid crystal displays. The existing achievements in device area have greatly stimulated the studies in the related aspects of research in semiconductor organic materials structure, electrical and optical properties as well as new physical phenomena, deposition techniques and device processing and development.

1-4244-0396-0/06/$25.00 ©2006 IEEE   208

2. Organic light emitting diodes (OLED's)

Organic light emitting diodes have been undergone dramatic improvements in performance in the last five years. OLED's contains thin layers of electron- and hole-injecting layers, and confinement of the injected carriers is used to increase the radiative recombination. The principle of operation of OLEDs is similar to that of inorganic light emitting diodes (LEDs). Holes and electrons are injected from opposite contacts into the organic layer sequence and transported to the emitter layer. Recombination leads to the formation of singlet excitons that decay radiatively. In more detail, electrons are injected from a low work function metal contact (cathode), e.g. Ca or Mg. The last one is chosen for reasons of stability. A wide-gap transparent indium-tin-oxide (ITO) thin film is used for hole injection (anode). The efficiency of electron-hole recombination leading to the creation of singlet excitons is mainly influenced by the overlap of electron and hole densities that originate from carrier injection into the emitter layer.

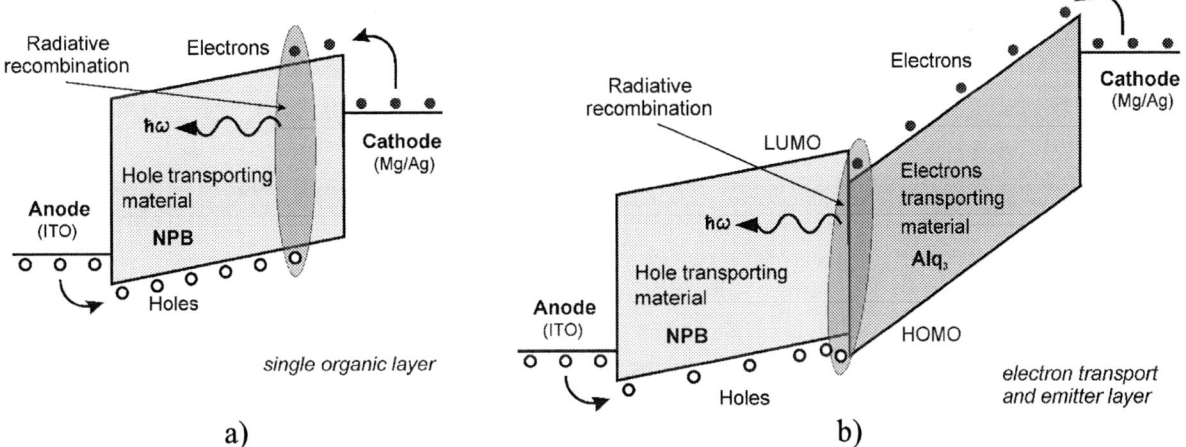

Fig.1 Energy diagrams for OLEDs with (a) single layer, (b) two layers - single heterostructure

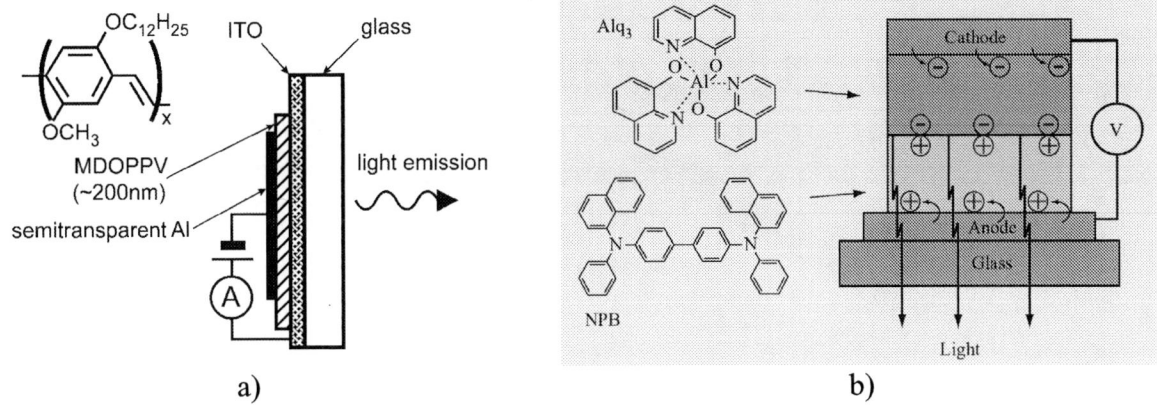

Fig.2 Schematics of OLED: a) MDOPPV polymer single layer structure [7]
b) NPB/Alq₃ single heterostructure [8]

Efficient device operation not only depends on the choice of molecules with appropriate electronic and optical properties, but also on the design of the device structure. Electroluminescence is already achieved with a simple single layer device (Fig. 1a), however the performance is poor since electrons and holes reach the opposite contact and excitons are quenched at the electrodes. The two-layer or single heterostructure device (Fig. 1b) introduces a separate hole transport layer. Holes are injected into the combined emitter and electron transport layer and recombine with electrons near the interface. An optimum thickness is found for the combined layer as a result of sufficient distance of the interface to the metal contact and maximum thickness for a given operating voltage.

Schematic of typical organic light-emitting diode structures which uses polymer single layer alkoxy derivatives of PPV (Fig. 2a) and single heterostructure organic low-molecular weight (small-molecule) Alq_3, tris (8-hydroxyquinolinato) aluminum, as the electron transport and emitting layer, and NPB, N, N'-di (naphthalene-1-yl)-N, N'diphenylbenzidine, as the hole transport layer is provided in Fig. 2b. The turn-on voltage of about 4 V is mainly determined by the total layer thickness (60 nm NPB, 60 nm Alq_3) of green single heterostructure devices. Two main technologies for OLED's have emerged in the last decade, either based on conjugated polymers, or sublimed films of small molecules. With a commercial history of just seven years, OLED manufacturing remains at an early stage, both in terms of technique and equipment. Small-molecule OLEDs are made using vapour deposition techniques, such as evaporation through a shadow or integrated shadow mask [9], because OLED materials are too delicate for photolithography. Polymer (PLEDs) are made by solution processing, either spin-on techniques (for monochrome) or inkjet printing (for colour), although the latter has not yet been commercialized.

The enhancements in performance have been seeing in the development of organic light-emitting diodes. Fig. 3 shows the dramatic increase in performance of light-emitting molecular solids and polymers compared to typical inorganic LEDs over a 15- year time scale [4]. PHOLED refers to molecular phosphorescent OLEDs. Currently, the highest observed luminous efficiencies of derivatives of these materials exceed that of incandescent light bulbs, thus eliminating the need for the backlight that is used in AMLCDs. Extremely rapid advances in OLED efficiencies have been made since the early 1990s, with peak efficiencies of 70 lm/W in the green for molecular PHOLEDs.

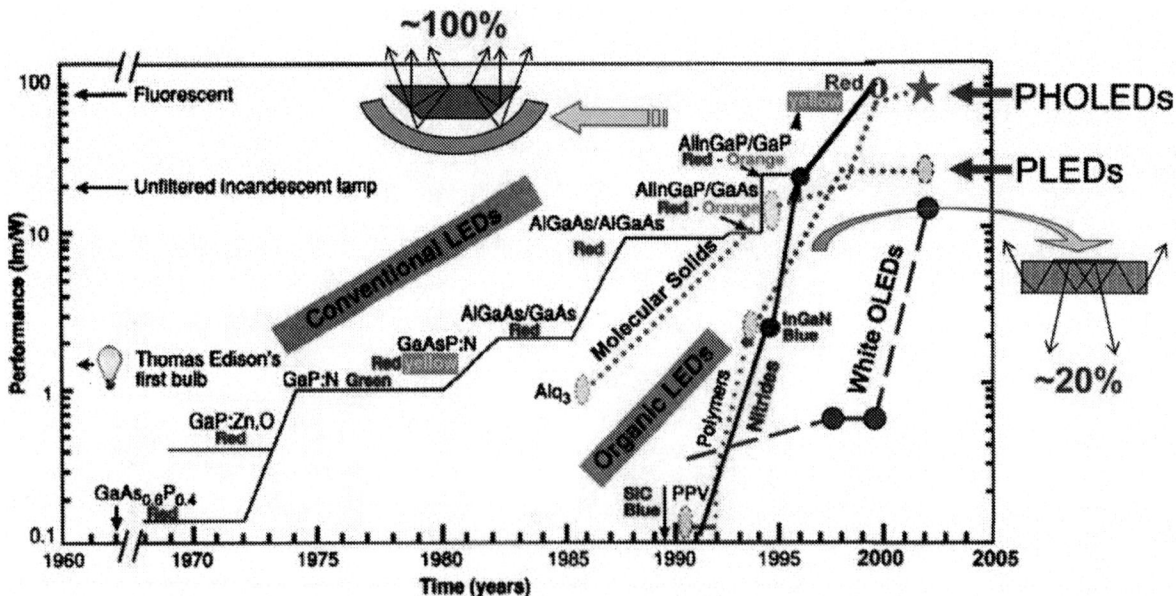

Fig. 3 Evolution in LED's and OLED's performance [4]

In contrast, the highest polymer OLED (PLED) efficiencies are ~20 lm/W. Note that the OLED efficiencies are measured for devices on flat glass substrates, where the total outcoupling is only ~20%. This is compared to ultrahigh brightness inorganic AlInGaP red LEDs used in, among other applications, traffic signals, where nearly all emitted light is projected into the viewing direction, leading to external efficiencies approximately equal to their internal efficiencies. Clearly, a remaining significant challenge for OLEDs is to device schemes that improve the light coupling efficiency.

If this were achieved by a simple and low cost method, OLEDs would unquestionably be the most efficient light source available. Indeed, this suggests that OLEDs may also soon compete with incandescence, and eventually even fluorescent lighting, as low cost, conformable white illumination sources and next generation of flat panel displays. OLEDs are a new and attractive class of solid-state light sources, which are opening up completely new applications in large area illumination. OLEDs are flat light sources and could be made in future on flexible substrates. OLEDs could also be used in lighting systems with controllable colour, allowing users to customize their light atmosphere at home. Furthermore, as a high efficient light source, the technology has the potential of achieving substantial energy savings.

3. OLED displays

The most utilization of OLED technology is expected in displays. They allow more design flexibility than inorganic LED's and thus lead to the high-resolution displays. Liquid-crystal displays (LCDs) are ubiquitous in everyday technology, from mobile phones and laptops to car stereos and coffee machines. But the OLED display is emerging as a credible flat-panel alternative thanks to some important advantages over LCDs. OLED displays fall into two categories: passive matrix and active matrix. Active matrix means that every pixel is individually switched, as opposed to a passive matrix arrangement, where row and column electrodes are used to control the pixel at a given intersection. Unfortunately for manufacturers, OLED driving schemes tend to be more complicated than LCD devices. The reason behind this is that OLEDs are current-driven and are sensitive to slight fluctuations in current. LCDs on the other hand are voltage-driven. Instead of needing one thin film transistor (TFT) per pixel in an active matrix scheme, OLEDs need between two to five, arranged in a compensation circuit. Different organic materials emit red, green, blue or other wavelengths of light. Because they are emissive, OLEDs also have an excellent viewing angle, good contrast and high brightness. Unlike an LCD, an OLED does not need a backlight, which means that the display panel can be thinner. This is an important advantage in mobile devices. OLEDs also offer the potential for lower power consumption compared with an LCD, which is always constrained by the power consumption of the backlight. OLEDs supply power only to the pixels illuminated in a given image. OLED materials and device structures are becoming so efficient that an active-matrix OLED (AMOLED) a few inches in diagonal, showing video (on average 30% of full brightness), consumes less power than an equivalent LCD. In addition, because OLEDs can operate at high speed - around 100 times faster than an LCD - devices can support video rates without blurring. However, the biggest hurdle facing OLED developers is short lifetime. Although OLED materials and device structures have improved greatly over the past few years, manufacturers can still only guarantee between 5000 and 15,000 h of operation before the brightness of the panel is reduced to half of its initial value. This performance is sufficient for mobile phones and other consumer electronics, but inadequate for television and more sophisticated products. Despite the challenges involved, OLEDs have already reached the market in several key applications. The first commercial OLED product was a small-molecule, passive matrix monochrome car stereo display from Pioneer in 1999. Sold as an aftermarket device, the display was blue-green to resemble vacuum-fluorescent versions commonly in use. Since then, OLEDs have moved into mobile phones, MP3 players, a Kodak digital camera, various industrial and medical devices, and a few other consumer electronics. The worldwide market for OLED panels was valued at $520 million (€400 million) in 2005, and is expected to reach $743 million in 2006, rising to $3.5 billion in 2012. This represents a compound annual growth rate of 29% from 2006 to 2012 [10]. Looking at the detail, the growing importance of portable media applications and mobile phone main displays is clear.

A comparison of the efficiencies of several different display and light sources is shown in Fig. 4 [4]. It is apparent that OLEDs provide the highest efficiency of any emissive display source. In particular, active matrix liquid crystal displays (AMLCDs) have efficiencies of ~2 lm/W, compared to ~10 lm/W for molecular organic phosphorescence (or PHOLED) based displays. Note that emissive display pixels are only turned on when needed, whereas LCD backlights must be fully on during use. Given that only ~25% of the pixels need to be illuminated when displaying a ''typical'' image, this alone provides for a significant power saving over LCDs. Then, given the very high efficiencies of PHOLEDs, even a greater advantage in power saving is realized.

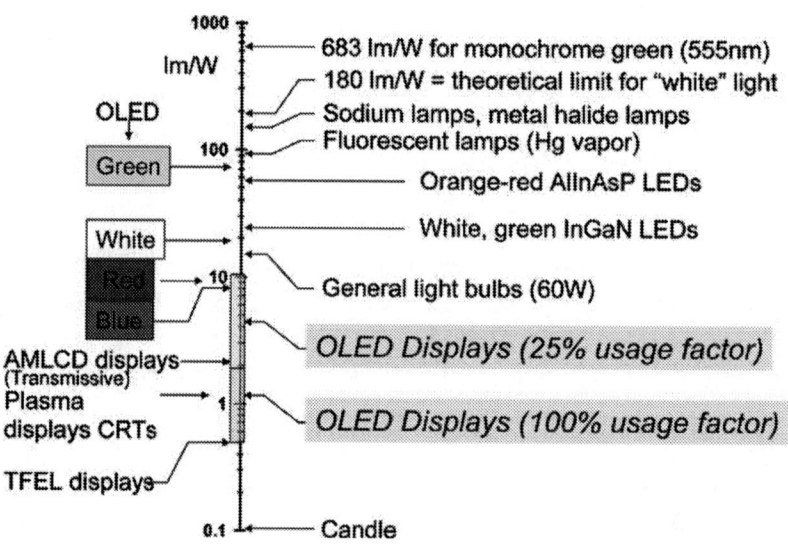

Fig.4 Comparison of the efficiencies of several different display and light sources [4]

The growth of the OLED market depends heavily on the success of AMOLED. The first commercial AMOLED reached the market in April 2003 and was used in a Kodak EasyShare LS633 digital camera back display. The key application for AMOLED is the mobile phone main display. It offers the largest total available market, and is well suited to the OLED's attractive image, low power consumption and thin profile. In addition, the increasing use of video on mobile devices also favours the OLED's fast speed. The ultimate dream of many OLED panel makers is to serve the PC monitors and large-screen television market. OLED is well suited to TV- it has fast speed, good colour, excellent viewing angle and high contrast ratio. The first world's largest full colour OLED display (13 inches diagonally with a resolution of 800x600 pixels) was developed by Sony in 2001 [11] followed by IDTech with 20-inch full-colour OLED display in 2003 [12] and Seiko Epson with 40-inch diagonal full-colour OLED display prototype in 2004 [13]. For this display Seiko Epson developed an original inkjet process for depositing organic layers and plans to start selling products using the technology in 2007. Samsung Electronics announced in 2005 that it developed the largest single-panel active matrix-based (AMOLED) display for TVs [14]. At 21-inches, this OLED features the highest resolution at 6.22 million pixels (WUXGA: wide ultra-extended graphics array). In addition, the company adopted active matrix based technology for its low power consumption and high-resolution qualities. Samsung's new OLED offers brightness of 400 nit, contrast ratio of 5000:1, colour gamut of 75 percent and fast response times, making the product ideal for viewing HD-resolution video images.

4.   Organic field effect transistor (OFET)

For more than a decade now, organic thin-film transistors (OTFTs) based on conjugated polymers, oligomers, or other molecules have been envisioned as a viable alternative to more traditional, mainstream thin-film transistors (TFTs) based on inorganic materials. Because of the relatively low mobility of the organic semiconductor layers, OTFTs cannot rival the performance transistors based on single-crystalline inorganic semiconductors, such as Si and Ge, which have charge carrier mobilities about three orders of magnitude higher [15, 16]. Consequently, OTFTs are not suitable for use in applications requiring very high switching speeds. However, the processing characteristics and demonstrated performance of OTFTs suggest that they can be competitive for existing or novel thin-film-transistor applications requiring large-area coverage, structural flexibility, low-temperature processing, and, especially, low cost. Such applications include switching devices for active-matrix flat-panel displays (AMFPDs) based on either liquid crystal pixels (AMLCDs) or organic light-emitting diodes (AMOLED's). Other applications of OTFTs include low-end smart cards and electronic identification tags. Organic thin film transistor can contain either a molecular

or polymeric channel connecting the source and drain contacts. The gate electrode is first deposited onto an insulating substrate such as glass or plastic, followed by deposition of the gate insulator, which can consist of either an organic or inorganic dielectric film (Fig. 6). Standard photolithographic techniques can be employed to result in gate lengths in the range of 50 – 5 µm. Device configurations of OTFT can be either top- contact where source and drain electrodes are evaporated onto the organic semiconducting layer (Fig.6) or bottom-contact device, with the organic semiconductor deposited onto prefabricated source and drain

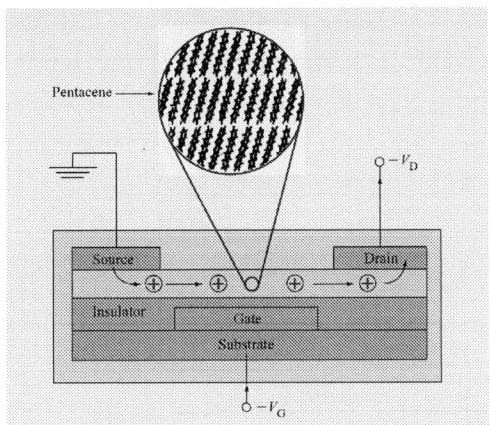

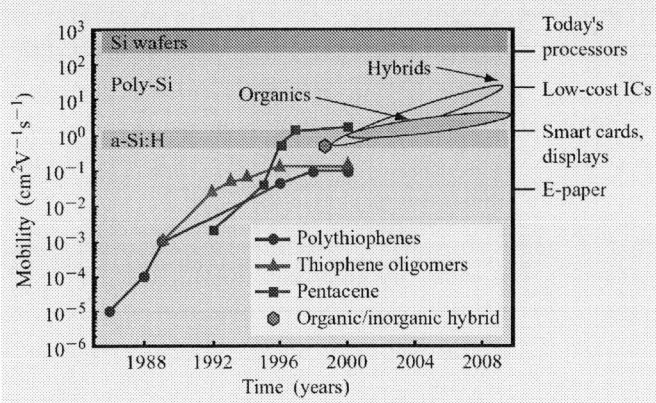

Fig. 6 Schematic of organic thin-film transistor with top contacts [8]

Fig. 7 Performance of organic and hybrid semiconductors [8]

electrodes. In disordered organic semiconductors the carrier transport takes place by hopping between localized states and carriers are scattered at every step. Hopping is assisted by phonons and mobility increases with temperature. The boundary between the band transport and hopping is defined by materials having mobilities between 0.1 and 1 cm^2 V^{-1} s^{-1} [15]. Highly ordered organic semiconductors, such as several members of the acene series including anthracene and pentacene, have room temperature mobilities in this intermediate range, and in some cases temperature-independent mobility has been observed, even in polycrystalline thin films of pentacene. Reported mobilities of representative classes of organic materials are seen in Fig. 7. Evaporated films of pentacene have achieved mobilities comparable to that of the amorphous silicon used to fabricate the thin-film transistors (TFTs) that drive the liquid crystal pixels in AMLCD flat-panel displays. Measurements on single organic crystals of p-type pentacene and an n-type perylene which mark the upper boundary of performance show mobilities of 2.7 cm^{2}V^{-1}s^{-1} and 5.5 cm^{2}V^{-1}s^{-1} orders of magnitude lower mobility than single-crystal silicon. A new class of materials, organic–inorganic perovskites, has also achieved the mobility of amorphous silicon [8]. While these carrier mobilities are now useful for applications that do not require high switching speeds, all of the previously reported materials operated at high voltages. Alternative methods such as injection moulding or direct printing have also been successfully employed to generate potentially very low cost circuits.

Recently, organic light emitting transistors (OLETs) have been demonstrated, which combine in a single device the electrical switching functionality of a field-effect transistor and the capability of light generation [17, 18, 19]. OLETs represent a novel class of organic devices, and could pave the way towards nanoscale light sources and highly integrated organic optoelectronics. In addition, it is worth mentioning that OLETs offer an ideal structure for improving the lifetime and efficiency of organic light emitting materials due to different driving conditions with respect to standard OLED architectures, and to optimised charge-carrier balances.

5.   Conclusions

Overall, in this paper we review the recent very rapid progress made over the last few years in developing organic semiconductor devices. Organic light-emitting diodes (OLED), OLED displays, organic field effect transistors (OFET), photodiodes and solar cells are the applications under intense study and first products are already in commercial use. Recent improvements have taken OLED's to luminous efficiency higher than 20 lm/W. OLEDs currently compare favourably in efficiency with incandescent sources like light bulbs, especially when colored. However, in the future, OLEDs offer the potential to become as efficient as energy-saving bulbs or even more so. Once efficient OLEDs are available at reasonable cost, they can be used for a multitude of applications in lighting. However, in contrast to conventional LED's, OLED's share many of the properties associated with other organic substances and polymers.

Similarly, there has been tremendous progress in OTFT performance during the last decade. Organic semiconductors such as pentacene, deposited by vacuum sublimation, remain the best performers because of their very well ordered structures, resulting from the use of this highly controllable deposition method. However, substantial improvements have taken place in solution-processes organic semiconductors, and their mobilities are currently only one order of magnitude lower than those of vapour deposited pentacene TFTs. There is a potentially important cost advantage associated with the solution processing of organic

TFTs, because it eliminates the need for expensive vacuum chambers and lengthy pump-down cycles .The electronic and optical properties of these "active" organic materials are now suitable for some low performance, low-cost electronic products that can address the needs for lightweight portable devices for the 21st century and should be dominant in OLEDs flat panel displays, and perhaps in solid state illuminations.

There will be continued growth in the field of organic electronics, fuelled by the promise of the new products and applications that can be derived from electronically and optically active organic and hybrid materials. These include low-cost and perhaps even flexible displays for e-newspapers and advertising, and low-cost memory and logic devices. Long-term research efforts and innovation are needed to provide new organic semiconductors, organic light-emitters, and conducting polymers with improved performance, process ability, and environmental stability to oxygen and moisture.

Acknowledgement

This work was supported by the Slovak Grant Agency projects

References

[1.] W. Brüttig, S. Berleb, A.G.Mückl, *Organic Electronics,* **2,** 1, 2001

[2.] W. Helfrich, W. G. Schneider, *Phys. ReviewLetters,* **14,** 229, 1965

[3.] C. W. Tang and S. A. Van Slyke, *Appl. Phys. Lett.,* **51,** 913, 1987

[4.] S. R. Forrest, *Organic Electronics,* **4,** 45, 2003

[5.] J. H. Burrowghes, D. D. C. Bradley, A. R. Brown, R. N. Marks, K. Mackay, R. H. Friend, P. R. Burn, A.B. Holmes, *Nature,* **347,** 539, 1990

[6.] H. Koezuka, A. Tsumura and T. Ando, *Synthetic Metals,* **18,** 699, 1987

[7.] J. Kalinowski, *J. Phys. D: Appl. Phys.,* **32,** R179, 1999

[8.] J. M. Shaw, P. M. Seidler, *IBM J. Res. & Dev.,* **45,** No. 1, 3, 2001

[9.] Ch. Py, D. Roth, I. LeÂvesque, J. Stapledon, A. Donat-Bouillud, *Synthetic Metals,* **122,** 225, 2001

[10.] Optics.org, June 16, 2006:
http://optics.org/optics/Articles/ViewArticle.do?channel=business&articleId=25148

[11.] Sony Press Releases, February 7, 2001:
http://www.sony.net/SonyInfo/News/Press_Archive/200102/01-007aE/

[12.] IDTech Press Releases, March 12, 2003:
http://www.idtech.co.jp/en/news/press/20030312.html

[13.] Epson Corporate news, May 18, 2004:
http://www.epson.co.jp/e/newsroom/news_2004_05_18.htm

[14.] Samsung Press Release, January 4, 2005: http://www.samsung.com/PressCenter/
PressRelease/PressRelease.asp?seq=20050104_0000089670#

[15.] C. D. Dimitrakopoulos, D. J. Mascaro, *IBM J. Res. & Dev.,* **45,** No. 1, 11 2001

[16.] G. Horowitz, *Adv. Mater.,* **10,** No. 5, 365, 1998

[17.] A. Hepp, , H. Heil,W. Weise, M. Ahles, R. Schmechel, H. von Seggern, *Phys. Rev. Lett.,* **91,** 157406, 2003

[18.] C. Rost, S. Karg, W. Riess, M. A. Loi, M. Murgia, M. Muccini, *Appl. Phys. Lett.,* **85,** 1613, 2004

[19.] M. Muccini, *Nature materials,* **5,** 605, 2006

Energy Band Diagram of the Ru/Hf$_{0.75}$Si$_{0.25}$O$_y$/Si Gate Stack

K. Fröhlich[1], J. P. Espinos[2], M. Ťapajna[1,3], K. Hušeková[1] and R. Lupták[1]

[1] Institute of Electrical Engineering, Centre of Excellence CENG, SAS, Dúbravská 9, 841 04 Bratislava, Slovak Republic

[2] Instituto de Ciencia de Materiales de Sevilla, CSIC, Avda. Americo Vespucio s/n., 410 92 Sevilla, Spain

[3] Faculty of Electrical Engineering and Information Technology, STU, Ilkovičova 3, 812 19 Bratislava, Slovak Republic

e-mail: karol.frohlich@savba.sk

We have studied advanced MOS structure containing Ru gate electrode, Hf$_{0.75}$Si$_{0.25}$O$_y$ dielectric and Si substrate by means of capacitance-voltage characteristics (C-V), X-ray photoelectron spectroscopy (XPS) and reflection electron energy loss spectroscopy (REELS). Using experimental values we have constructed energy band diagram of the Ru/Hf$_{0.75}$Si$_{0.25}$O$_y$/Si gate stack.

1. Introduction

Due to aggressive scaling of silicon based metal-oxide-semiconductor field effect transistor (MOSFET) thickness of the SiO$_2$ (or SiON) gate dielectric should decrease below 2 nm. Consequently, nanoscale MOSFET suffers from high gate leakage current caused by direct tunnelling through the gate dielectric. Replacing of SiO$_2$ gate dielectric by thicker high dielectric constant (high-κ) gate oxide film circumvents excessive gate leakage current. It is commonly agreed, that metal gate electrodes should be used in nanoscale MOSFET in combination with high-κ gate oxide films.

High-κ gate stack of the nanoscale MOSFET is a great technological challenge [1]. The most important properties of the MOSFET gate is high capacitance and low leakage currents. To achieve low leakage current, the gate stack should exhibit sufficient barriers against emission of carriers into oxide bands. Conduction and valence band offsets greater than 1 eV are considered as satisfactory barrier heights to suppress leakage currents caused by the emission.

Calculation of the conduction band offsets for most of candidates for gate oxide films were performed recently by Robertson [2]. Unfortunately, experimental values of the band offsets are scarce in the literature. In this work we have analysed Ru/Hf$_{0.75}$Si$_{0.25}$O$_y$/Si gate stack by X-ray photoelectron spectroscopy, XPS. In combination with capacitance-voltage measurements and reflection electron energy loss spectroscopy, REELS, we determined valence and conduction band offsets and consequently, energy band diagram of the Ru/Hf$_{0.75}$Si$_{0.25}$O$_y$/Si gate stack.

2. Experimental

The Hf$_x$Si$_{1-x}$O$_y$ layers were grown by a metal organic chemical vapour deposition, MOCVD, based technique - AVD® (Atomic Vapour Deposition®) in an AIXTRON Tricent® system [3]. The Hf$_x$Si$_{1-x}$O$_y$ layers Hf/Si with nominal composition of 75/25 (Hf$_{0.75}$Si$_{0.25}$O$_y$), thickness 7 nm and dielectric constant 18 were used in this study.

The Ru films were deposited by MOCVD at 290 °C using bis(2,2,6,6-tetramethyl-3,5-heptandionato) (1,5-cyclooctadiene) ruthenium, Ru(thd)$_2$(cod)) dissolved in toluene. Thickness of the Ru film was 30 nm.

Ru/Hf$_x$Si$_{1-x}$O$_y$/Si MOS capacitors for electrical characterization were prepared by optical lithography and Ar ion milling. The electrical properties of the structures were determined by capacitance-voltage measurements at frequency 500 kHz using an Agilent 4284A LCR meter. The Ru/Hf$_{0.75}$Si$_{0.25}$O$_y$/Si structures were exposed to forming gas (FGA, 10% H$_2$ + 90% N$_2$) annealing at temperature 430 °C for 30 min.

XPS analysis was performed with a VG ESCALAB 210 instrument, in an UHV chamber with a base pressure of $2*10^{-10}$ mbar, using unmonochromatised Mg K$_\alpha$ radiation (XPS). Depth profilings of the films were done by Ar ion bombardment with 3 keV kinetic ion energy. Binding energy scale in XPS experiments was calibrated against the Ru 3d5/2 peak position at 280.86 eV.

The band gap energy was determined from the reflection electron energy loss spectrum (REELS). For the REELS spectra, the hemispherical analyzer was operated at a constant pass energy of 20 eV, and primary electrons with 1500 eV of kinetic energy. The incidence angle was 60 ° from the surface normal and the takeoff angle was 0 °. The full width at half maximum of the elastic peak was 0.7 eV.

3. Results and discussion

Work function Φ_M of the gate electrode can be extracted from flat band voltage shift V_{FB} capacitance-voltage characteristics of MOS structures with different dielectric thickness. If we assume, that the fixed oxide charge in the dielectric, qN_{ox}, is located near the dielectric/Si interface, the gate electrode work function Φ_M can be expressed as

$$\Phi_M = \Phi_{Si} + qV_{FB} + \frac{qN_{ox}}{\kappa_{ox}}t_{ox} \qquad (1)$$

where Φ_{Si} is the work function of the Si substrate, q is the electronic charge, κ_{ox} is the dielectric constant and t_{ox} is thickness of dielectric. Extrapolation of V_{FB} plot versus EOT (equivalent oxide thickness) to zero gives work function of the gate electrode.

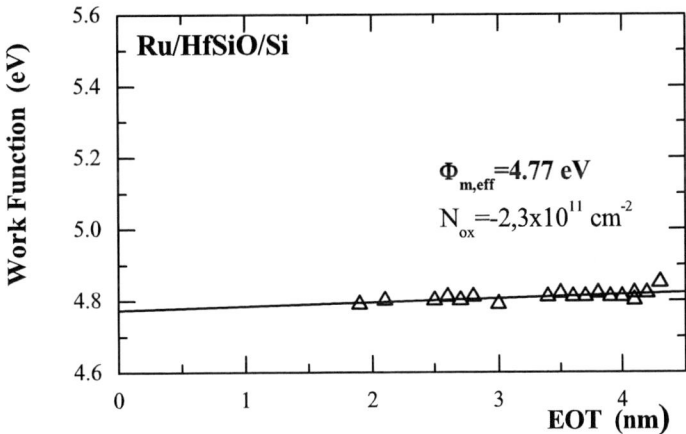

Fig. 1. Work function determination from capacitance-voltage measurements.

Effective work function of the Ru gate electrode was extracted from the C-V measurements on wet etched $Hf_{0.75}Si_{0.25}O_y$ sample (slanted sample). The slanted sample was prepared by gradual etching of the $Hf_{0.75}Si_{0.25}O_y$ film in $HF/HCl/H_2O$, [HF]=0.03 M, pH=1 solution [4]. The technique of the work function extraction from C-V characteristics on slanted dielectric gives precise results because oxide charges distribution is constant within the sample. Figure 1 shows plot of the flat band voltage vs. equivalent oxide thickness (EOT). Extracted value of the work function for the Ru gate electrode is 4.77 eV.

To determine $Ru/Hf_{0.75}Si_{0.25}O_y$ valence band offset we analysed the $Ru/Hf_{0.75}Si_{0.25}O_y$ interface by XPS. To reach the $Ru/Hf_{0.75}Si_{0.25}O_y$ interface, upper part of the Ru film was removed by Ar milling until Hf $4f_{7/2}$ core level peak emerged in the XPS spectrum. Figure 2 displays the spectra of the $Ru/Hf_{0.75}Si_{0.25}O_y$ interface in the vicinity of the valence band and Hf $4f_{7/2}$ core level peak. The spectra were taken after sequential sputtering and were calibrated against Ru $3d_{5/2}$ core level peak positioned at 280.86 eV. The onset of the spectrum (0 eV binding energy) corresponds to Fermi level of the Ru electrode. The gradual shift of the Hf $4f_{7/2}$ core level peak with the depth we ascribe to charges in the $Hf_{0.75}Si_{0.25}O_y$ film.

We have determined valence band alignment from XPS spectra following Kraut et al. [5]. The method is based on assumption, that all energy bands (conduction band, valence band, core levels) are bended at the semiconductor/metal interface by the same amount. The valence band offset $\Delta E_v^{Ru-HfSiO}$ at the $Ru/Hf_{0.75}Si_{0.25}O_y$ interface is then calculated as

$$\Delta E_v^{Ru-HfSiO} = (E_{Hf4f7/2}^{HfSiO} - E_V^{HfSiO}) - (E_{Ru3d5/2}^{Ru} - E_V^{Ru}) - (E_{Hf4f7/2}^{Ru/HfSiO} - E_{Ru3d5/2}^{Ru/HfSiO}) \qquad (2)$$

where subscripts *Hf4f7/2* and *Ru3d5/2* denote corresponding Ru and HfSiO core level peaks and *V* denotes top of the Ru and HfSiO valence band. The valence band offset $\Delta E_v^{Ru-HfSiO}$ was determined as 3.1 eV.

Band gap of the $Hf_{0.75}Si_{0.25}O_y$ was obtained from its reflection electron energy loss spectrum for O 1s photoelectrons by extrapoling the segment of negative slope to the background level of the first electron loss signal, ref. [6, 7], Fig. 3. The band gap of the $Hf_{0.75}Si_{0.25}O_y$ was determined as 5.5 eV.

Knowing $Ru/Hf_{0.75}Si_{0.25}O_y$ valence band offset we calculated barrier height between Fermi level of the Ru electrode and conduction band of $Hf_{0.75}Si_{0.25}O_y$ (2.4 eV) and affinity of the $Hf_{0.75}Si_{0.25}O_y$ (2.37 eV). Taking into account affinity of Si (4.05 eV) and Si band gap (1.12 eV) we determined conduction band offset between $Hf_{0.75}Si_{0.25}O_y$ and Si as 1.68 eV.

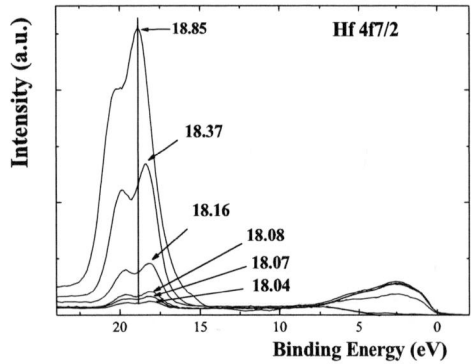

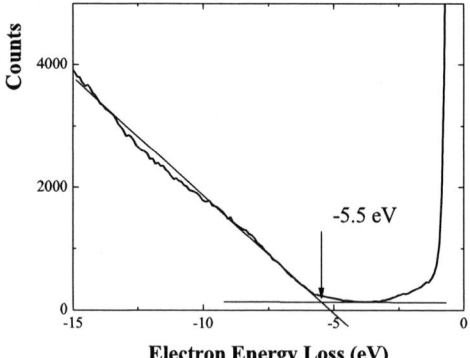

Fig. 2. Hf $4f_{7/2}$ and valence band XPS spectra for $Ru/Hf_{0.75}Si_{0.25}O_y/Si$ stack.

Fig. 3. O 1s reflection electron energy loss spectrum for the $Hf_{0.75}Si_{0.25}O_y$ film.

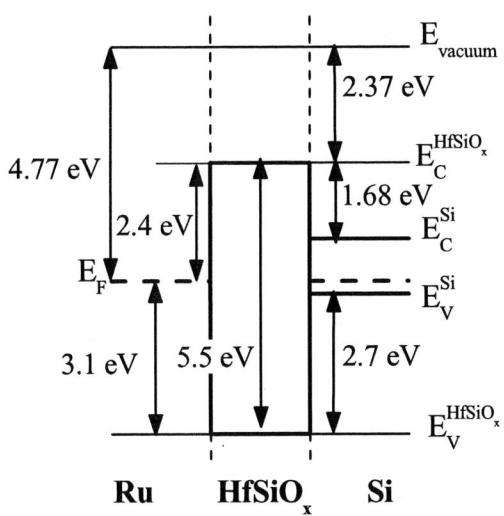

Fig. 4. Energy band diagram of the Ru/Hf$_{0.75}$Si$_{0.25}$O$_y$/Si gate stack.

Complete energy band diagram of the Ru/Hf$_x$Si$_{1-x}$O$_y$/Si gate stack is displayed in the Fig. 4.

4. Conclusions

We have constructed energy band diagram of the Ru/Hf$_x$Si$_{1-x}$O$_y$/Si gate stack using capacitance-voltage measurements, XPS and REELS analysis. We found that barrier heights (conduction and valence band offsets) are sufficient to suppress leakage currents caused by emission of electrons and holes.

Acknowledgment

This work was supported in part by the Slovak grant agency APVT (project APVT-51-017004) and VEGA (project 2/5130/26). The authors acknowledge AIXTRON AG for providing Hf$_{0.75}$Si$_{0.25}$O$_y$ films.

References:

[1] G. D. Wilk, R. M. Wallace, J. M. Anthony, J. Appl. Phys. **89**, 5243, 2001.
[2] J. Robertson, *J. Non-Crystal. Solids*, **303**, 94, 2002.
[3] S. Van Elshocht, U. Weber, T. Conard, V. Kaushik, M. Houssa, S. Hyun, B. Seitzinger, P. Lehnen, M. Schumacher, J. Lindner, M. Caymax, S. De Gendt, and M. Heyns, *J. Electrochem. Soc.* **152**, F185, 2005.
[4] D. Shamiryan, V. Paraschiv, M. Claes, W. Boullart, in *Defects in High-κ Gate Dielectric Stack,* edited by E. Gusev, NATO Science Series, Springer, Netherlands, 2006, p. 331.
[5] E. A. Kraut, R. W. Grant, J. R. Waldrop, and S. P. Kowalczyk, *Phys. Rev. Lett.* **44**, 1620, 1980.
[6] L.A.J.Garvie, P.Rez, J.R.Alvarez, P.R.Buseck, *Solid State Commun.* **106**, 303, 1998.
[7] S. Miyazaki, *J. Vac. Sci. Technol. B* **19**, 2212, 2001.

Towards a Microtechnology based 4-channel infrared detector unit for a miniaturised NDIR system

L. Fonseca, J. Santander, R. Rubio, N. Sabaté, P. Ivanov, E. Figueras, I. Gràcia, C. Cané

Centro Nacional de Microelectrónica, Campus de la Universidad Autónoma de
Barcelona, 08193 Bellaterra, Barcelona, Spain
e-mail: luis.fonseca@cnm.es

An infrared detector unit to be included in a miniaturised Non Dispersive Infrared system (NDIR) is being developed with the aid of different microtechnologies: CMOS technology, silicon bulk micromachining and flip-chip techniques. The chosen application for this optical gas sensing approach is the storage ambient control of climacteric fruit such as apples. For this application the detection of different gases is of interest: ammonia (as a result of leaks of the cooling system), ethylene (as a monitor of the ripeness state of the fruit), and acetaldehyde or ethanol (as a measure of fruit stress).

1. Introduction

Most gases exhibit fingerprint absorptions in the infrared region. For this reason infrared spectroscopy is well established as a laboratory method for the analysis of various gases and features a high sensitivity and selectivity of the gas measurement. The main elements of such an equipment are an infrared emitter, a gas cell (most conveniently a multipath gas cell in order to increase the optical path) and an infrared detector. In the case of applications in field, cheaper and more compact photometers are needed and in this minaturisation microtechnologies are of great help. In this work the attention is focused in the development of such a miniaturised infrared detector

2. A detector for a miniaturised NDIR system

A prototype of detection unit with four thermopile detectors, capable of the detection of the appropriate wavelengths according to the gases involved in the intended application: ethylene, ethanol, ammonia is being constructed in a first approximation. A fourth channel is also added as a reference (using a filter of a wavelength free of any gas absorption). Referring the other channels outputs to this one helps cancel out any effect not directly linked to gas effects (optical windows loss of transparency, emitter aging...). The general architecture of the detection unit will be that shown in Figure 1. It consists of a substrate chip, where the thermopiles are fabricated that is attached to the base of the package. Flip-chipped on it will be the filters. Attached on the top of the package will be a Fresnel multilens, intended to divide the total IR radiation reaching the prototype into four parts and focus each one of the parts into the corresponding absorber zone of each one of the thermopiles.

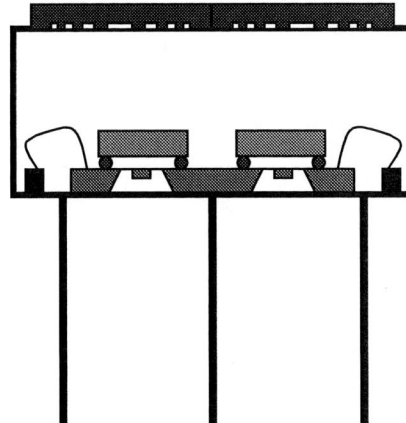

Figure 1. Schematic view of the proposed infrared detector module consisting of an array of 2x2 bulk micromachined thermopiles with different commercial infrared filters attached onto it by flip-chip techniques, wire-bonded to a TO5 house in whose lid a Fresnel lens will be physically attached.

2.1 Thermopiles

Generally speaking a thermal detector consist of a sensitive element that exhibits a variable measurable property linked to its change of temperature, located in an thermally insulated region connected to a heat sink by a high heat resistance path. The better the thermal insulation, the higher the reached temperature under a given infrared flux; the more responsive the sensitive material, the larger the variation of the measurable property for a given temperature increase [1].

Thermally insulated regions are easily defined by micromachining processes well linked to silicon technologies. There are material pairs within silicon technologies, like aluminum-polysilicon pairs, that although not optimum, exhibit quite acceptable Seebek coefficients. This possibility opens the way to the fabrication of monolithic CMOS compatible thermopiles detectors (or arrays) with unequalled expectatives of low cost, size, and mass production.

A four element bulk micromachined thermopile array detector is proposed. Each thermopile chip consists of thermocouples made of n-doped polysilicon and aluminium that extend from the centre of a free-standing thin nitride membrane, where an absorber element is placed, to a silicon rim. The membrane is made of a thin SiO2/Si3N4 layer and defined after performing an anisotropic backside etching of the silicon substrate. The hot contacts of the thermocouples are formed on the thermally insulating membrane over the absorber and the cold contacts are placed on the rim of the structure, which acts as a heat sink. The absorber, which is defined by heavily doping the silicon with a boron source, defines the sensitive area. The high thermal resistance membrane assures the temperature isolation between junctions once the silicon underneath is removed, allowing a temperature difference to develop when IR radiation reaches the absorber layer.

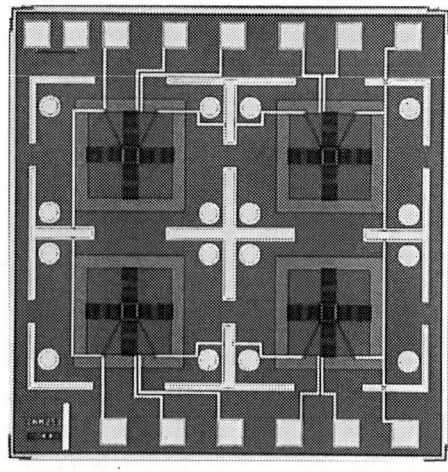

Figure 2. Micromachined 4 element thermopile array.

2.2 Optical filters

The filters will be of commercial origin and will be selected according to the desired wavelengths [2]. They will be attached to the thermopile substrate by way of flip-chip techniques. Flip-chip techniques are usually related to the world of MCMs (multichip modules) where several commercial chips can be mounted face to face to a common connecting substrate through conductive solder bumps. Aluminium pads are not directly solderable, therefore a wettable metal (a Ti/Ni/Au stack) has to be defined on top of the connecting pads. When performing the flip chip procedure, cylinders of solder paste are deposited on each bump pad of the substrate wafer by dispensing the paste through the round holes of a metal stencil mask. Next, a pick-and-place machine takes the chips to be mounted on top and arranges them so that the bumped solder pads of the substrate face their corresponding pad on the chip. Once the chips are placed onto the substrate, the wafers undergo a short inert thermal process using a standard reflow ramp procedure that reaches a maximum temperature of 220°C. The solder paste melts and partially dissolves the wettable metal forming a strong contact between the pads on the substrate and the ones on the flipped chips. A flip-chip approach is a promising way to integrate a stack of two optical devices. It provides the necessary optical alignment and it has been proved compatible to the handling of fragile micromachined devices, such as thermopiles

According to the previous procedure, the commercial filters have been conditioned depositing and defining the corresponding Ti/Ni/Au bump pads. After this operation they have been diced of the appropriate size and attached onto the thermopile chip.

Figure 3. Three different diced narrow bandgap filters flip-chiped onto the thermopile array and wire-bonded onto a TO5 package

2.3 Fresnel lenses

An optical system is needed to focus the beam to increase the radiation intensity falling on the detector. For such purpose diffractive lenses, so called Fresnel lenses, are being developed. In a conventional lens the curvature defines its focal, but the bulk material has no effect to the focusing properties. In a diffractive lens the bulk material is removed, and the surface is a set of grooves with tilted surfaces that correspond to the original surface. The ideal surface is being approximated using eight quantisation levels (using only three masks). With this approach the optical efficiency will remain quite good (about 95%) [3]. Each etching has a different depth. The etching depth is *h* for the first mask, *2h* for the second one and *4h* for the third mask. Using this codification, it is possible to get a stairs-like profile, which is a good alternative to approximate the ideal shape. Such 8 level Fresnel lenses have been frabricated directly on silicon using Reactive Ion Etching (RIE) as shown in Figure 4, and a sharp focus spot and focal depth have been found for them.

Figure 4. SEM image of an 8-level silicon Fresnel lens designed for a 10micron wavelength.

Acknowledgement

This work has been funded by the European Commision under the program FP6-IST-1-508774-IP. The authors would like also to thank the teams of University of Barcelona and the Fraunhofer Institute Physical Measurement Techniques for their assistance in the thermopile and Fresnel lenses optical characterization.

References

[1] P.W. Kruse, D.D. Skatrud, Uncooled infrared imaging arrays and systems.Semiconductors and Semimetals, Vol. 47. Academic Press, New York, 1997.

[2] C.J. Pouchert: "The Aldrich Library of FT-IR Spectra, Vapor Phase, Volume 3", Aldrich chemical company, Inc.,

[3] J. Jahns, S. Walker, "Two-dimensional array of diffractive microlenses fabricated by thin film deposition", Appl. Optics Vol. 29, no.7).

Nano-structures for light management in optoelectronic devices

M. Zeman[1] and J. Krc[2]

[1]Delft University of Technology – DIMES, P.O. Box 5053, 2600 GB Delft, The Netherlands
[2]Fac. of Electrical Eng., University of Ljubljana, Trzaska 25, SI-1000 Ljubljana, Slovenia
e-mail: m.zeman@tudelft.nl, janez.krc@fe.uni-lj.si

Two types of nano-structures are presented for light manipulation in semiconductor devices such as solar cells: (i) photonic-crystal-like structures for wavelength-selective manipulation of high reflectance or transmittance of light and (ii) diffraction gratings for angle-selective management of light scattering. Using optical simulations photonic-crystal-like structures with a high reflectance in the long-wavelength region are designed for a dielectric back reflector in solar cells. Structures formed by thin amorphous silicon and silicon dioxide layers are presented and the important role of refractive indexes and thicknesses of the individual layers is demonstrated. Haze and angular distribution function of scattered light are measured for aluminum based rectangular periodic gratings with different period and height of the rectangles.

1. Introduction

Manipulation of light propagation in optoelectronic devices such as light emitting diodes, waveguides, and solar cells is of great importance in order to further improve their performance. In this article we introduce two thin-film based nano-structures, which have a large potential for the manipulation of light propagation in thin-film silicon (TF Si) solar cells: (i) one-dimensional (1-D) photonic-crystal-like (PC) structures and (ii) diffraction gratings. PC structures represent specially designed *optical interface* which is capable of a **wavelength-selective** management of (high) reflection or transmission of light. Diffraction gratings represent *rough interfaces* with specially designed surface texture that scatter light into (pre) selected angles enabling the **angle-selective** management of scattered (diffused) light.

The PC structures, which are nowadays also thoroughly investigated in the telecommunication research [1,2], have been indicated as future candidates for the realization of a novel type of highly-reflective dielectric back reflectors in solar cells [3]. In principle, the PC structures can provide almost ideal reflection in a limited wavelength region. The selective reflection and transmission properties of 1-D PC structures are also important for application of these structures as intermediate reflectors (interlayers) in multi-junction TF Si solar cells [4-6]. Periodic gratings have already been tested in TF Si solar cells [7-9]. Despite high-scattering ability of the gratings [10], the published results have not indicated yet significant improvements regarding their implementation in TF Si solar cells, compared to the results obtained with optimized randomly-textured surfaces. Therefore, a further investigation resulting in a better understanding and optimization of the gratings for their application in TF Si solar cell is of great importance.

In this article we investigate a potential of the novel light-trapping approaches in TF Si solar cells using the PC structures and diffraction gratings. Utilizing an optical simulator SunShine [11], 1-D PC structures with wavelength-selective high reflectance are investigated for their potential use as a back reflector in TF Si solar cells. Different PC structures consisting of intrinsic hydrogenated amorphous silicon (i-*a*-Si:H) and silicon dioxide (SiO$_2$) are investigated as candidates for the back reflector. Different rectangular periodic gratings have been fabricated and their scattering properties characterized.

2. 1-D photonic-crystal-like structures in TF Si solar cells

Two possible applications of 1-D PC structures in TF Si solar cells are: highly reflective back reflector and wavelength-selective interlayer in multi-junction solar cells. In this article the design of 1-D PC structures suitable for back reflectors is presented. In case of a back reflector it is important to obtain a very high reflectance (close to 100 %) for the wavelengths that reach the back reflector in TF Si solar cells (i.e. from ~ 650 nm to 1100 nm).

The optical simulator SunShine was used to design the 1-D PC structures with required optical properties. The key point for obtaining the required wavelength-selective properties of the 1-D PC structure is to select the materials with proper optical properties for the layers, the design of thicknesses of the individual layers and the number of periods (two-layer stacks) in the PC structure. For verification and calibration purposes a few samples were deposited on glass, consisting of magnetron sputtered ZnO:Al layers and n-*a*-Si:H layers deposited by PECVD. The simulation and experimental results for three simple structures: (i) a single ZnO layer (130 nm), (ii) n-*a*-Si:H (25 nm)/ZnO (130 nm) stack, and (iii) ZnO (130 nm)/n-*a*-Si:H (25 nm)/ZnO (130 nm) deposited on glass substrate are presented in figure 1. Evaluating the reflectivity properties of these three structures, the ZnO/n/ZnO stack can already be assigned to a simple PC with one period $a = d_1 + d_2$ and a half (3 layers). Although several periods are usually required to form a typical PC with high reflectance and wavelength selectivity, in this case the simple structure with three layers exhibits already a pronounced already a selective reflectance around $\lambda = 500$ nm, whereas at $\lambda = 750$ nm it is noticeably suppressed.

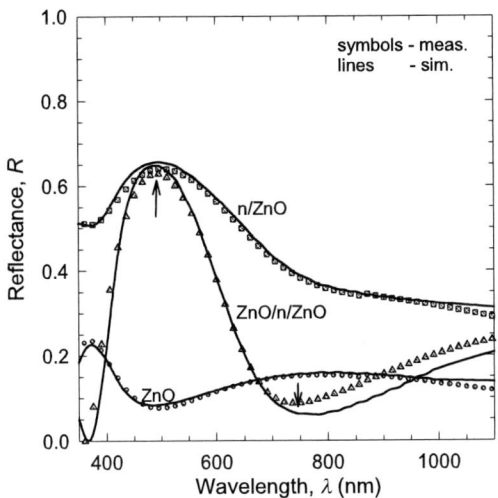

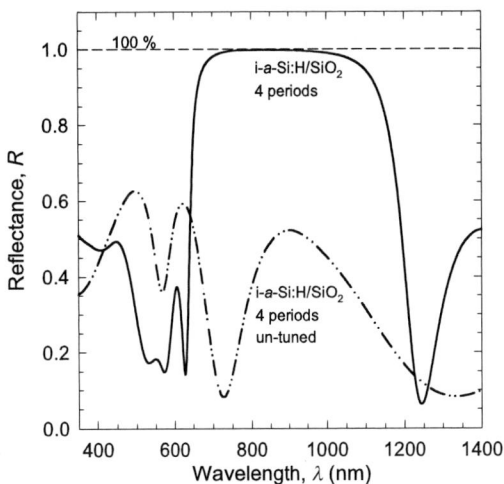

Figure 1: Measured and predicted (simulated) reflectance of the specified simple 1-D PC structures. The thickness of ZnO and n (n-*a*-Si:H) layer are 130 nm and 25 nm, respectively.

Figure 2: Simulated reflectance of a 1-D PC back reflector: i-*a*-Si:H (55 nm)/SiO$_2$ (135 nm) with 4 periods A simulation of un-tuned i-*a*-Si:H/ SiO$_2$ PC structure ($d_1 = d_2 = 40$ nm) is shown as well.

Based on a good agreement between predicted simulation curves and experimental data for the above described structures the optical simulator was used to design more complex 1-D PC structures for back reflectors. Simulations showed that for high reflectance in broad wavelength region it is important to utilise a material with low refractive index, n (e.g. SiO_2 with $n = 1.53$) and a material with a high n (e.g. i-a-Si:H with $n \approx 4$ at $\lambda = 650$ nm). Figure 2 shows the reflectance of i-a-Si:H (55 nm)/SiO_2 (135 nm) structure including 4 periods (8 layers). Four periods are already enough to obtain the high reflectance (close to 100 %) in the required wavelength region when the thicknesses are properly optimized. To demonstrate the important role of layer thickness in a PC structure, a simulation result for a stack with not optimized thicknesses of both layers, i-a-Si:H (30 nm) and SiO_2 (30 nm) is shown in figure 2.

3. Periodic gratings for selective (large) angle light scattering

Periodic diffraction gratings have a potential for efficient light scattering to large angles [10]. They are based on periodic perturbations of the surface morphology. The period of the perturbations, P, should be in the range of the light wavelength in order to achieve desired diffraction effects. Different shapes of the surface features can be used; in this article we focus on rectangular gratings only. To evaluate the scattering properties of a diffraction grating, we introduce two descriptive scattering parameters; haze, H, and angular distribution function, ADF, of scattered light. H describes how much of light is scattered outside of specular direction, whereas ADF gives information about scattering directions (angles).

Two periodic gratings were fabricated using Al deposited on glass substrates. A period P of 6.5 µm was determined by AFM. The height h of the rectangles was around 450 nm and 90 nm, respectively. H for reflected light was determined (figure 3) as a ratio between the diffused and total reflectance, which were measured using an integrating sphere. One can observe that for the sample with $h \approx 450$ nm, H approaches unity (no specular reflection) around $\lambda = 1800$ nm. This phenomena is of great importance for TF Si solar cells where such behavior can significantly enhance the absorption of the long-wavelength light (650 nm $< \lambda$ <1100 nm). To achieve such high H at specific wavelengths, the condition $h \approx \lambda/4$ [3] has to be fulfilled. For the Al grating with $h \approx 450$ nm this is just around $\lambda = 1800$ nm whereas for $h \approx 90$ nm the maximum of H is at $\lambda = 360$ nm, In this case the maximum is not observed (figure 3), since for $\lambda = 360$ nm the first diffraction orders are located closely to the specular direction and cannot be directly separated from it.

Besides the amount of scattered light described by H, the important parameter is also the ADF. By scattering into sufficiently large angles at the back of the cell, the total reflection at the front side of the solar cell can be achieved for light propagating from the back. The total reflection of this light at front interfaces can be considered as a key issue enabling ideal light trapping. The ADF of the Al gratings was determined by using Angular Resolved Scattering measurements. All the results shown in this article correspond to a normal incidence of light beam ($\varphi_{inc} = 0$). Figure 4 shows the normalized ADF (the highest component was normalized to unity) of the sample with $P = 6.5$ µm and $h \approx 450$ nm. The measurement was carried out using a red He-Ne laser with $\lambda = 633$ nm. Pronounced scattering in selective angles (diffraction orders) can be clearly observed. The scattering angles for all possible diffraction orders have been calculated using the grating equation [10] and compared to the positions of the measured ADF peaks. A good agreement is obtained between the measured and calculated (vertical lines) angles for all 9 diffraction orders (see figure 4). Since P of the analyzed sample is rather large compared to the wavelength of the laser beam, the first diffraction order ($m = 1$) is located relatively close ($\varphi_{scatt} = 6.2^0$) to the specular beam ($m = 0$, $\varphi_{scatt} = 0^0$).

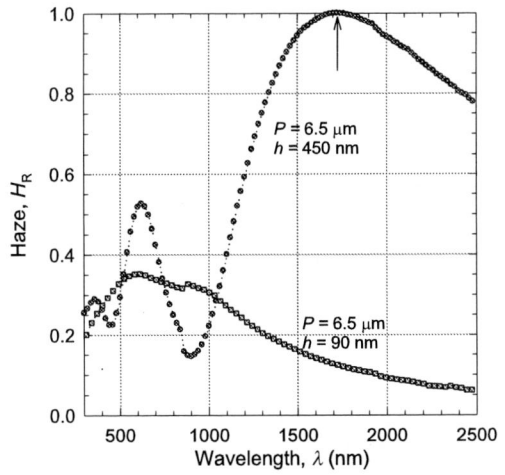

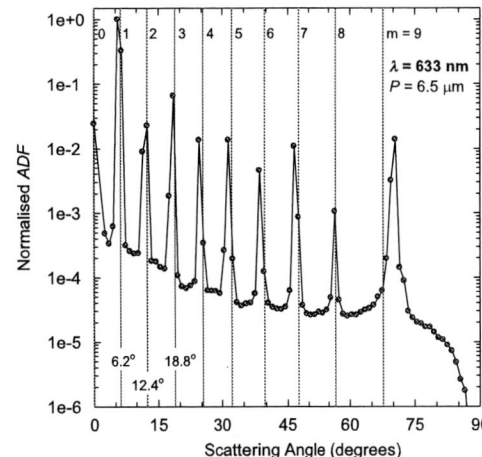

Figure 3: Measured haze of reflected light for two Al gratings with different h.

Figure 4: Normalized *ADF* of the Al grating with $P = 6.5$ mm and $h \approx 450$ nm measured with a red laser ($\lambda = 633$ nm). Calculated scattering angles are plotted as vertical lines.

4. Conclusions

Two novel approaches of light management in TF Si solar cells using nano-structures were investigated: 1-D PC structures wavelength-selective manipulation of high reflection or transmission in a solar cell and periodic diffraction gratings for angle-selective management of light scattering inside a solar cell. Both investigated nano-structures show promising potential for light manipulation in semiconductor devices.

Acknowledgement

The authors thank Martijn Tijssen for the preparation of 1-D PC samples.

References

[1] http://ab-initio.mit.edu/photons/tutorial/
[2] H-Y. Lee and T. Yao, *J. Appl. Phys.* **93** (2), 812, 2003.
[3] L. Zeng, Y. Yi, C. Hong, X. Duan and L. C. Kimerling, *Mater. Res. Soc. Symp. Proc.* **862**, A12.3.1, 2005.
[4] D. Fischer *et al.*, *Proceedings of the 25th IEEE PVSC*, Washington, DC, 1996, p. 1053.
[5] K. Yamamoto et al., *Solar Energy Mat. & Solar Cells* **74**, 449, 2002.
[6] B. Rech *et al.*, *Proceedings of the WCPEC-3,* Osaka, Japan, 2003, p. 2783.
[7] C. Eisele, C. E. Nebel and M. Stutzman, *J. Appl. Phys.,* **89** (12), 7722, 2001.
[8] H. Stiebig, N. Senoussaoui, C. Zahren, C. Hasse and J. Mueller, *Prog. in Photovoltaics*, **14**, 13, 2006.
[9] V. Terrazzoni *et al.*, *Proceedings of the WCPEC-3,* Osaka, Japan, 2003, p. 1596.
[10] P. Beckmann and A. Spizzichino, *The scattering of electromagnetic waves from rough surfaces,* Pergamon Press, 1963.
[11] J. Krc, F. Smole, M. Topic, *Prog. in Photovolt: Res. Appl.* **11**, 15, 2003.

Preparation of p-type ZnO thin films by RF diode sputtering

K. Shtereva[1], I. Novotny[2], V. Tvarozek[2], R. Srnanek[2], J. Kovac[2] and P. Sutta[3]

[1] Department of Electronics, University of Rousse, Studentska 8, BG-7017 Rousse, Bulgaria, e-mail: kshtereva@ecs.ru.acad.bg

[2] Department of Microelectronics, Slovak University of Technology, Ilkovicova 3, 812 19 Bratislava, Slovakia

[3] West Bohemia University, New technologies - Research Center, Univerzitni 8, 306 14 Plzen, Czech Republic

A p-type ZnO thin films were prepared by RF diode sputtering and nitrogen doping. Deposition in plasma N_2 gas source increases the N solubility and thus the incorporation of N_O acceptor that is responsible for p-type conductivity of the ZnO films. Raman analyses performed in back scattering configuration proved the incorporation of the nitrogen acceptor N_O into ZnO:N. Raman spectra show E_2 mode and two nitrogen related local vibrational modes (LVMs) typical for N-doped ZnO. Minimum resistivity of 790 Ωcm, a Hall mobility of 22 $cm^2 V^{-1} s^{-1}$ and the carrier concentration of 3.6×10^{14} cm^{-3} were obtained at 75 %N_2 in Ar/N_2 sputtering gas. X-ray diffraction measurements (XRD) showed that ZnO:N films had the preferential orientation of (002) plane at 25 %N_2 and of (100) plane for higher N_2 concentrations. The average grain size was from 7 to 42 nm for all Ar/N_2 ratios. ZnO:N films exhibit relatively high microstrains (10×10^{-3}).

1. Introduction

Recently transparent conductive zinc oxide (ZnO) thin films have attracted a great attention because of their good electrical and optical properties in combination with a wide band gap, and great natural abundance and absence of toxicity that make the material a good candidate for short wavelength optoelectronic devices, surface and bulk acoustic wave devices, piezoelectric transducers, chemical and gas sensing and solar cells [1, 2, 3].

Zinc oxide is a group II-VI semiconductor, with direct band gap of 3.37 eV at room temperature, large exiton binding energy of 60 meV, and high refractivity (melting point of 1975 °C) and chemical stability. At present the research interest is focused on p-type ZnO since the operation of many devices is based on the nature of p-n junction. Undoped zinc oxide exhibits intrinsic n-type conductivity and the fabrication of a reproducible and low-resistivity p-type ZnO still remains a challenge. The asymmetry of ZnO p-type doping is explained with high activation energy of acceptors, low solubility of acceptor dopants and self-compensating process on acceptor doping. ZnO can be doped intrinsically, through native defects, and extrinsically, through impurity doping with group III elements (Al, Ga or In), group I elements (Li, Na and K) or group V elements (N, P and As). Nitrogen doping, which has been successful in fabricating p-type ZnSe, is considered an effective method to realize p-type ZnO thin films [4].

Nitrogen doped ZnO thin films can be deposited by variety of methods such as magnetron sputtering, evaporation, metalorganic chemical vapor deposition, (MOCVD), molecular beam epitaxy (MBE), sol-gel process, spray pyrolysis, plasma enhanced chemical vapor deposition (PECVD), pulsed laser deposition (PLD), atomic layer deposition (ALD) and filtered cathodic vacuum arc technique.

1-4244-0396-0/06/$25.00 ©2006 IEEE

To the best of our knowledge p-type conductivity in nitrogen doped ZnO films is reported in few articles and one of the most quoted is Joseph et al. [5]. These studies point that the key elements for successive synthesis of p-type ZnO thin films were N dopant source and the substrate temperature. Novelty of our approach is in use of plasma assisted deposition method - RF diode sputtering - which allows performing direct action of ions, ion complexes and energetic electrons on growing film with the aim to increase the substrate temperature as well as to form suitable nitrogen acceptors.

2. Experimental

Thin films were prepared in a planar RF sputtering diode system Perkin Elmer 2400/8L. ZnO:N films were deposited on Corning glass substrates from ZnO target in Ar/N_2 atmosphere. Pre-sputtering for 10 minutes cleaned the target. The vacuum chamber was evacuated to a base pressure of 2×10^{-5} Pa. High purity nitrogen (N_2) acted as a doping source and the content of N_2 in sputtering gas varied from 0 % to 100 %. The total gas pressure of 1.33 Pa was maintained constant during the sputtering process. The sputtering power was 500 W and the sputtering time varied from 30 to 60 minutes and accordingly thicknesses of ZnO films were in range from 430 nm to 870 nm. The deposited ZnO:N thin were annealed in nitrogen at temperature 600 °C for 10 minutes.

The film thickness was measured by Talystep instrument. The crystal orientation and microstrains of the films were investigated by an automatic powder X-ray diffractometer AXS Bruker D8 with Eulerian cradle and 2D detector (CoKα, $\lambda = 0.179$ nm). Micro-Raman spectra of ZnO:N thin films were obtained at room temperature by a Dilor system working with a He-Ne laser, operating at a wavelength of 632.8 nm in the back scattering geometry on a JobinYvon-Spex spectrometer equipped with an air cooled CCD detector. Hall measurements were carried out at room temperature and at following measurement parameters: currents 10 nA - 200 nA, temperature 300 K, magnetic field 0.385 T.

3. Results and discussion

The XRD diffraction lines of ZnO (100), (002), (101), and (110) were observed for all studied samples. The set of two-dimensional diffraction patterns of ZnO:N films deposited at 75 % N_2 in sputtering gas displays three Debye rings of ZnO Bragg reflections (Fig.1). The white circle is a Laue spot diffracted by substrate material (c-Si) due to the continuous X-ray radiation. The highest intensity of (002) peak for ZnO:N films deposited at 25% N_2 suggests c- axis preferential orientation of (002) plane. The increase of the nitrogen content into the sputtering gas led to the decrease of the intensity of (002) peak and ZnO:N deposited at 50 % N_2 and 75 % N_2 showed the preferential orientation of (100) plane (Fig. 2). The change of the preferential orientation with increase of the dopant concentration agrees with other authors results [6, 7]. ZnO films orientation depends on the substrate temperature and the dopants and co-dopands concentration. If the deposition is carried out at the non-equilibrium conditions such as low temperature or high deposition rate, orientation different than

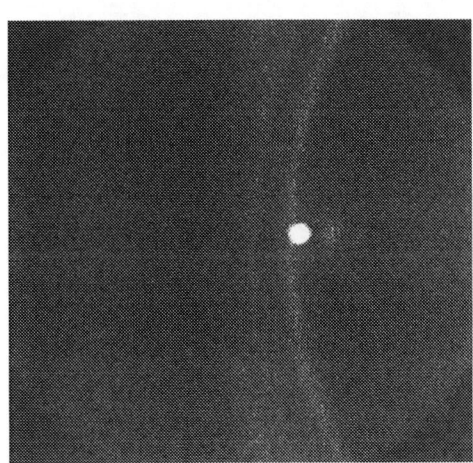

Fig. 1. Two dimensional diffraction patterns of ZnO:N films

(002) can be achieved. The highest intensity of (002) peak at 25% N_2 content may be explained with the increase of the concentration of N_O and the decrease of oxygen vacancies

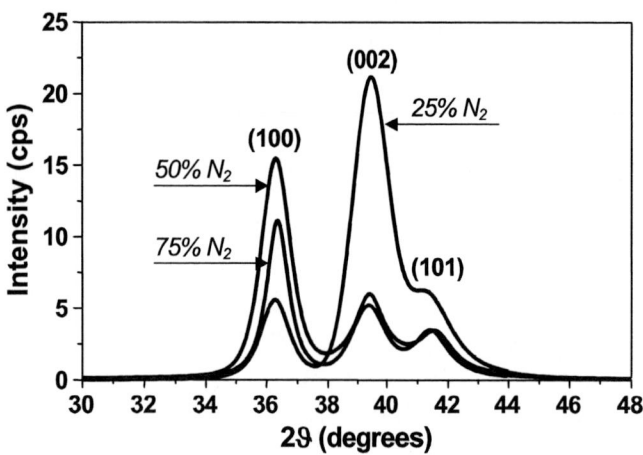

Fig. 2. XRD patterns for ZnO films doped with different percentage of N_2

(V_O), which are detrimental to the crystallinity of ZnO films [8]. The followed decrease of the intensity of (002) peak with increasing nitrogen content from 50 % N_2 and 75 % N_2 is a result of the degradation of the crystallinity possibly because of the direct incorporation of N_2 molecules in ZnO:N film. Corresponding parameters of real structure ZnO:N agree with the above results: the average grain size was in the range $7 - 42$ nm for all Ar/N_2 ratios; average microstrains were relatively high (10×10^{-3}).

The Raman spectra of diode sputtered nitrogen doped ZnO thin films, as grown and annealed, show three obvious peaks related to ZnO and ZnO:N thin films (Fig.3, a) and b)). The broad peak at about 435.75cm^{-1} for as grown and at 437.17 cm^{-1} for annealed samples is assigned to the two non-polar optical phonons E_2 modes of the nitrogen doped ZnO films. The intensity of the peak increases after annealing. A theoretical calculation based on a modified valence-force model shows the value of 272 and 580cm^{-1} for LVMs of nitrogen on a substitutional oxygen site in the ZnO lattice [9,10,11]. Hence, the mechanism

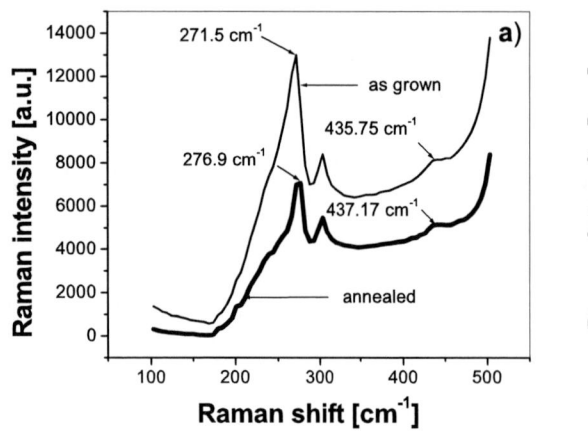

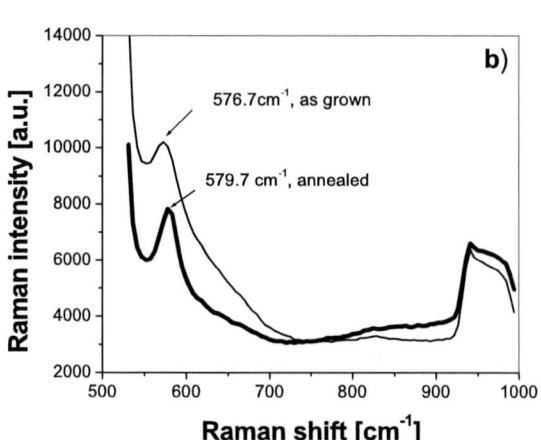

Fig. 3. Raman spectra of ZnO:N films

responsible for the peaks near 272 and 580cm^{-1} modes should be due to the N-related LVM. We find both that Raman peaks, one at 270.7 and the other one at 576.73cm^{-1}, characteristic for nitrogen doped ZnO thin films. The intensity of nitrogen related LVM increases with higher nitrogen concentration, and it agrees with the annealing behavior of our samples. The decrease of the intensity of 270.7cm^{-1} Raman peak after annealing can be explained with the decrease of the nitrogen concentration after annealing treatment.

The electrical properties of N-doped ZnO thin films as a function of nitrogen percentage in Ar/N_2 sputtering gas are shown in Table 1 and Fig. 4a, b.

230

ex. No.	h [nm]	N_2 [%]	ρ [Ωcm]	R_H [cm^3C^{-1}]	μ_H [$cm^2V^{-1}s^{-1}$]	P, N [cm^{-3}]	D [nm]	$\varepsilon \times 10^2$ [-]
8	600	0	---	---	---	---	42	1.01
9	870	10	5.4×10^4	---	---	---	41	1.40
10	840	25	1.5×10^3	2.4×10^3	2	P ~ 2.6×10^{15}	9	0.77
11	710	50	7.0×10^2	4.7×10^3	7	N ~ 1.3×10^{15}	12	0.82
12	660	75	7.9×10^2	1.7×10^4	22	P ~ 3.6×10^{14}	10	0.69
13	580	100	2.8×10^3	3.5×10^3	1	P ~ 1.8×10^{15}	11	0.23

Table 1. Electrical resistivity ρ, Hall constant R_H, Hall mobility μ_H, carrier concentration P or N, grain size D and microstrain ε of as grown ZnO:N films deposited on Corning glass substrate

Results of Hall measurements of films prepared at different dopant concentration show a strong dependence of the electrical properties on dopant concentration. The undoped ZnO thin film (0 % N_2 in sputtering gas) was not measurable and ZnO films deposited in 10 % N in the sputtering gas showed a high resistivity (5.4×10^4 Ωcm). The p-type conductivity was recorded at ZnO:N samples deposited in 25 %, 75 % and 100 % nitrogen in the sputtering gas. The conductivity of the sample deposited at 50 % nitrogen was rather controversial and statistically was determined to be n-type.

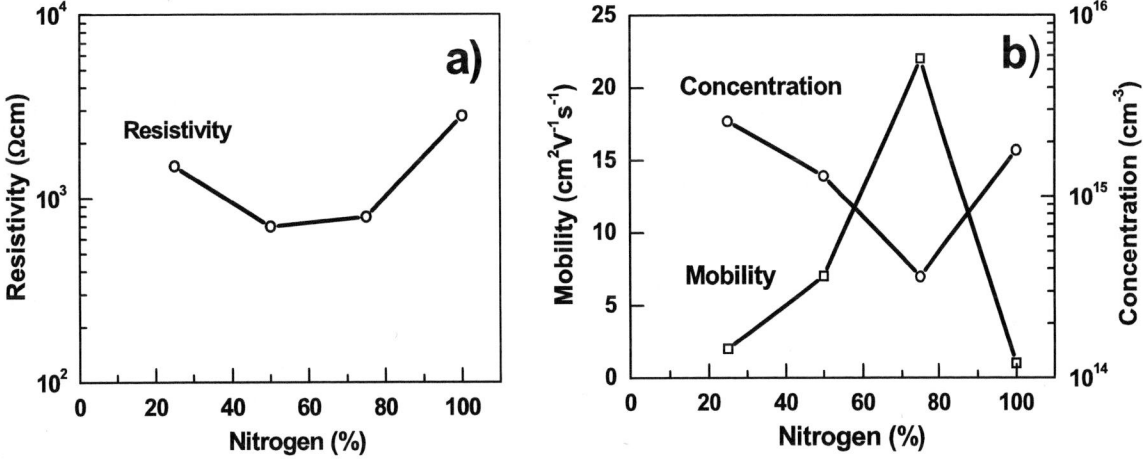

Fig. 4 Dependence of a) resistivity, b) Hall mobility and carrier concentration on nitrogen percentage for as grown ZnO:N films

The film's resistivity decreases with the increase of the nitrogen percentage from 25 % to 75 %. The further increase of nitrogen percentage in the sputtering gas from 75 % to 100 % leads to the increase of the resistivity. p-type ZnO films show the lowest resistivity of 7.9×10^2 Ωcm, the highest Hall mobility of 22 $cm^2V^{-1}s^{-1}$ and the carrier concentration of P ~ 3.6×10^{14} cm^{-3}. The lowest resistivity is a result of the high mobility. The highest carrier concentration P ~ 2.6×10^{15} cm^{-3} is recorded for 25 % N_2 in the sputtering gas. As a comparison to these results Josef et al. reported p-type ZnO:N with a hole concentration of 2×10^{10} cm^{-3} for ZnO:N films [5]. p-type ZnO:N films obtained from Z.-Z Ye et al. [12] had the lowest resistivity of 1.2×10^3 Ωcm, Hall mobility of 84.9 cm^{2}V^{-1}s^{-1} and the carrier concentration of 6.0×10^{13} cm^{-3}.

p- type ZnO can be obtained at appropriate dopant concentration and sputtering conditions. Films with small dopant content exhibit good crystal structure in terms of grain size and grain orientation that is confirmed from the above XRD measurement results but the concentration of active acceptors N_O formed from NO molecules remains low. Hence low carrier mobility and high resistivity were obtained at samples deposited with 25% N_2 in the sputtering gas. The increase of the nitrogen percentage into the sputtering gas from 25% to 75% may lead to the direct incorporation of N_2 molecules in ZnO introducing a large amount of $(N_2)_O$ double shallow donor besides the formation of NO molecules, which generate N_O acceptor [13]. This leads to the decrease of the hole concentration with the increase of the N_2 percentage in the sputtering gas and even to the ambiguous type of conduction at ZnO films grown at 50% nitrogen in the sputtering gas. The increasing carrier concentration in films deposited at 100% N_2 can be a result of the high concentration of NO incorporated into the films.

4.   Conclusion

Using plasma assisted deposition method - RF diode sputtering - and nitrogen as a dopant we fabricate p-type ZnO thin films on Corning glass substrate by. ZnO films were prepared in Ar/N_2 sputtering gas and the percentage of N_2 varied from 0 % to 100 %. The hole carrier concentration, Hall mobility and resistivity are in the range of 10^{14} - 10^{15} cm^{-3}, 1 – 22 cm^{2}V^{-1}s^{-1}, and 7 x 10^2 Ωcm – 2.8 x 10^3 Ωcm. The incorporation of NO molecules, a generator of N_O acceptor that is responsible for p- type features of ZnO:N films, was proved from Raman analysis. In Raman spectra of ZnO:N films, E_2 and two N-related LVM peaks at 270.7 and at 576.73cm^{-1} characteristic for nitrogen doped ZnO thin films appear. The deposition conditions of RF diode sputtering are significantly different from thermodynamic balance. Therefore the properties of p-type ZnO:N films grown by RF diode sputtering will be metastable and highly dependent on nitrogen content in sputtering gas. The improvement on properties of p-type properties ZnO thin films can be achieved by combining of nitrogen doping with other co-dopants.

Acknowledgement

Presented work was supported by the SK grants VEGA of the Slovak Grant Agency and in part by PPP Programme project DAAD 5/2005.

References

1. H. Ohta, and H. Hosono, *Materials Today,* **7**, 42, 2004.
2. I.T. Tang, H.J. Chen, W.C. Hwang, Y.C. Wang, M.P. Houng, and Y.H. Wang, *Journal of Crystal Growth*, **262**, 461, 2004.
3. S. Zwegart, *Siemens & Shell Solar GmbH*, Munich, Germany, Precision Nov. 2001.
4. E.Ch. Lee, Y.S. Kim, Y.G. Jin, and K.J. Chang, *Physica B*, **308-310**, 912, 2001.
5. M. Joseph, H. Tabata, H. Saeki, K. Ueda, and T. Kawai, *Physica B*, **302–303**, 140, 2001.
6. Y.J. Zeng, Z.Z. Ye, J.G. Lu, L.P. Zhu, D.Y. Li, B.H. Zhao, and J.Y. Huang, *Applied Surface Science*, **249**, 203, 2005.
7. C. Zhang, X. Li, J. Bian, W. Yu, and X. Gao, *Solid State Communications*, **132**, 75, 2004.
8. F. Zhuge, L.P. Zhu, Z.Z. Ye, J.G. Lu, B.H. Zhao, J.Y. Huang, L. Wang, Z.H. Zhang, and Z.G. Ji, *Thin Solid Films*, **476**, 272, 2005.

9. R.A. Asmar, J.P.Atanas, M.Ajaka, Y. Zaatar, G. Ferblantier, J.L.Sauvajol, J.Jabbour, S. Juillaget, and A. Foucaran, *Journal of Crystal Growth*, **279**, 394, 2005.

10. X. Wang, S. Yang, J. Wang, M. Li, X. Jiang, G. Du, X. Liu, and R.P.H. Chang, *Journal of Crystal Growth*, **226**, 123, 2001.

11. B. Marí, F.J. Manjón, M. Mollar, J. Cembrero, and R. Gómez, *Applied Surface Science*, **252**, 2826, 2006.

12. Z.Z. Ye, F.Z. Ge, J.G. Lu, Z.H. Zhang, L.P. Zhu, B.H. Zhao, and J.Y. Huang, *Journal of Crystal Growth*, **265**, 127, 2004.

13. L. Zhu, Z. Ye, F. Zhuge, G. Yuan, and J. Lu, *Surface & Coatings Technology,* **198,** 354, 2005.

Coupling Capacitances of Connecting-lead Systems in Integrated Circuits

J. Novak, J. Foit, V. Janicek

Department of Microelectronics
Czech Technical University in Prague
Technicka 2, CZ-166 27 Prague 6, Czech Republic
novakj2@feld.cvut.cz, http://micro.feld.cvut.cz

The development of integrated circuits has reached a situation today that the circuit operating speed is not limited by the parameters of the transistors any more, but rather by the electrical parameters of the interconnections inside the integrated circuit [1]. For this reason, it is necessary to take in account the properties of interconnecting conductors from the start of the circuit design process. Undesirable parasitic electromagnetic couplings appear between the interconnecting lines, causing the transfer of interfering impulses onto the signal-carrying lines. These interfering impulses then result in random errors and/or disturbances in the integrated circuit.

1. Introduction

Equations for finding the primary electrical parameters of interconnecting systems can be defined on the basis of analytical solutions of electromagnetic fields [2]. These equations were defined under certain simplifying conditions and therefore they are only valid for simple configurations of interconnecting systems.

The numerical solution is based on an approximation of the solved area by a network of nodal points and its accuracy depends on the number of nodal points within that area [3]. The better accuracy is required, the larger number of nodal points must be present in the area solved. Equations are set up for each nodal point and the solution of an area of nodal points represents a solution of a set of equations. These equation sets are then solved by some of the existing numerical calculation methods. The size of the system of equations grows with the number of nodal points and, as a result, the calculation speed depends on the computing performance of the computer used.

2. Solving of electromagnetic fields in IC structures

The networks of nodal points are created according to the spatial layout of the interconnecting system. The models represent only certain elements of interconnecting systems, describing the relevant properties of these systems. The networks in interconnecting systems are designed as gradual, with higher nodal point spatial density in the immediate vicinity of the interconnecting leads [4].

A rectangular block was used as the basic spatial element in designing the network. A "modified block" [5] appeared as advantageous in the case of a line-type interconnecting grid where there do not appear geometrical alterations of the system in one dimension. This element type makes possible to create a very complex grid structure in a two-dimensional surface, and the spatial element is created by a simple expansion into the third dimension. An example of such a case for a three-conductor interconnecting system is shown in Fig. 1. Lengthwise, the network consists of blocks with identical dimensions.

1-4244-0396-0/06/$25.00 ©2006 IEEE

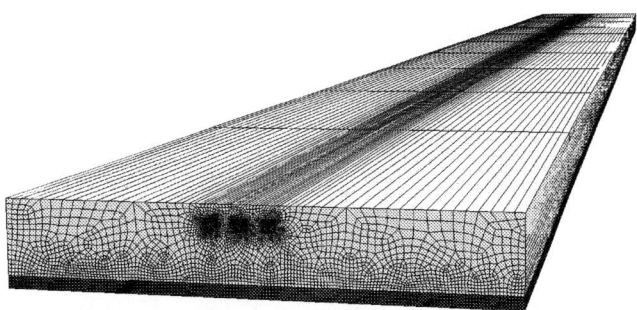

Fig. 1 Example of line network design using the modified block element

2.1 Two-conductor interconnecting system

The most frequent case of unwanted coupling is the case of two parallel lines in the same metallization layer. For this reason we will concentrate our attention to this configuration. The situation is described by a cross-section through the spatial model shown in Fig. 2.

Fig. 2 Cross-section of a two-conductor system

The interconnecting leads are shown in red, silicon dioxide in blue. The dimension to be followed further are shown as numerical values, all other dimensions correspond to the M3 metallization layer in the AMI C05M-A 3M/2P/HR [6] process technology. The segment treated is 10 μm long and the width of the cross-section is $d = 20$ μm. The design rules for the M3 metallization layer set the minimum line width to 1.1 μm and minimum inter-line space s = 1.1μm. The height $h = 3.8$ μm is determined by the M3 metallization layer. The coupling capacitances C_{12} (capacitance between conductors No. 1 and No. 2), C_{10} (capacitance between conductor No. 1 and the ground plane) were calculated for these particular system dimensions. Since we are dealing with a linear system, it is possible to relate these capacitances to unit length. In the following text, these relative capacitances will be branded as C'.

Using parametric simulations, it is possible to follow mutual influencing between conductors No. 1 and No. 2 (Fig. 2). As can be expected, the capacitance C'_{12} between the conductors is inversely proportional to the separation between conductors No. 1 and No. 2, while the capacitance C'_{10} to the ground plane rises slightly. (Fig. 3). The errors of monotonicity of the curves are caused by variations of the nodal points network density.

2.2 Electrical shielding of the conductors

Shielded conductor systems can be divided in systems with direct shielding of the electrical field and systems using indirect shielding. The direct shielding systems use an additional conductor fitted between the interfering and interfered conductor, connected to the ground potential.

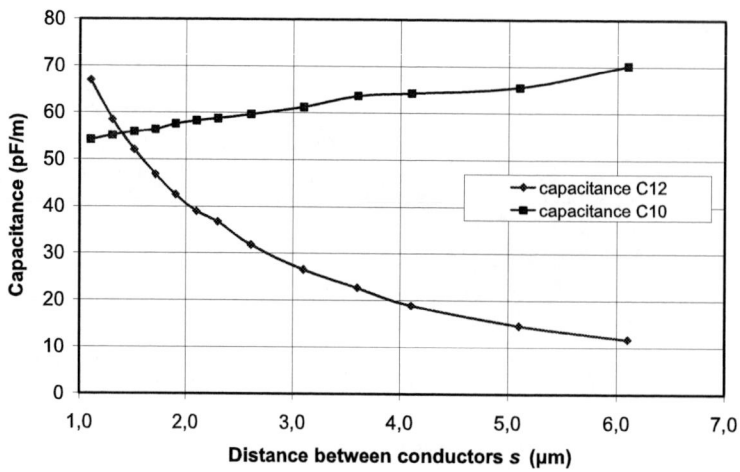

Fig. 3 Coupling capacitances C'_{12} and C'_{10} in a two-conductor system versus conductor spacing

In systems with indirect shielding, the shield conductors are located outside the interacting conductors plane. In an integrated circuit, these shield conductors are located in other metallization layers than the interacting signal conductors. The indirect shielding is based on the principle of deformation of the electromagnetic field due to the shield conductor. The presence of the shield conductor changes the electric induction flux between the interacting conductors in such a way that a part of it is absorbed by the shield conductor.

An example of indirect shielding is shown in Fig. 4. The signal conductors No. 1 and No. 2 are located in the M2 metallization layer, the shield conductor lies in the M3 layer. The shield conductor covers the gap between signal conductors and is connected to the ground potential.

Fig. 4 Cross-section of an indirect shielding system with the shield in the M3 layer

First of all, we have found the capacitances between conductors No. 1 and No. 2 in the absence of the shield conductor No. 3, for system segment length 10 μm. The results show that the coupling capacitance between the signal conductors No. 1 and No. 2 was decreased by 37.5%. At the same time, the total capacitive load of the conductor No. 1 $C'_1 = C'_{10} + C'_{12} + C'_{13}$ (for the shielded case) was increased by only 18.3% as compared to the unshielded case.

Fig. 5 shows a graph of the dependence of the C'_{12} capacitance on the gap size s between the conductors No. 1 and No. 2 (Fig. 4) for an unshielded two-conductor system, as well as for a system using a shield conductor No. 3 in the M3 metallization layer ($w = 1$ μm). By comparison of both graphs (Fig. 5), it is possible to find the efficiency of indirect shielding, as the difference of the coupling capacitance C'_{12} in systems with and without shielding. The largest drop in coupling capacitance was found to appear for conductor spacing $s = 1.5$ μm, the drop amounts to 42%. Further spacing increase is not as effective, but it can help to a substantial decrease of the coupling capacitance.

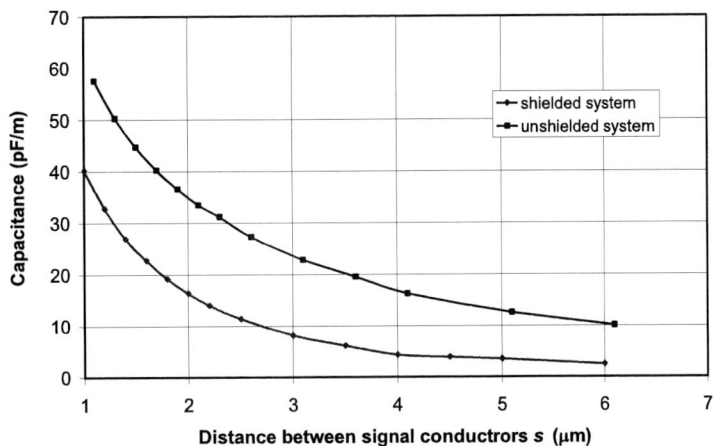

Fig. 5 Coupling capacitance C'_{12} versus conductor spacing s in a shielded (M3) and an unshielded system

3.   Summary

The coupling capacitance between signal conductors can be limited by the application of electric shielding. The shielded system in Fig. 4 shows the application of indirect shielding, using a shield conductor located in another metallization layer. By this type of shielding, the mutual coupling capacitance C'_{12} is cut by 50% as compared to the two-conductor system (Fig. 2), while keeping the area occupied by the interconnecting system at the original size.

The accurate calculation of parasitic capacitances of the signal interconnections helps us to design detailed circuit models of the interconnecting systems [7]. Critical spots in an electronic system, where parasitic couplings cause failures, can be found by simulation of the circuit models. Then, it is possible to take appropriate measures to decrease the parasitic capacitances of signal conductors, like adjusting the integrated circuit layout or by applying shield conductors.

References

[1]   Thomas Kropewnicki, Process Integration Challenges for Copper Dual Damascene Interconnects, *Georgia Tech MiRC Seminar,* 2002

[2]   C. S. Walker, Capacitance, Inductance and Crosstalk Analysis, Norwood, *Artech House,* 1990

[3]   J. M. Min, The finite element method in electromagnetism, New York, J. Wiley 1993

[4]   B. W. Boast, Principles of Electric and Magnetic Fields, New York, *Harper and Brothers*, 1986

[5]   S. Kagami, I. Fukai, Application of BEM to Electromagnetic Problems, *MTT-32*, No. 4, 1984, pp. 455 - 461

[6]   Design rule manual C05M-D, DS13315_V5.0.pdf

[7]   M. Kuhlmann, S. S. Sapatnekar, Exact and Efficient Crosstalk Estimation, *IEEE Trans. Computer-Aided Design*, vol. 20, no.7, pp. 858-866, July 2001

The Effect of Rapid Thermal Annealing on Oxygen Precipitation in Nitrogen Doped Silicon Substrate

Ľ. Stuchlíková[1], L. Harmatha[1], M. Ťapajna[1,2], P. Ballo[1],
P. Písečný[2], M. Benkovič[1], J. Jakabovič[1]

[1] Faculty of Electrical Engineering and Information Technology,
Slovak University of Technology, Ilkovičova 3, 812 19 Bratislava, Slovak Republic
[2] Institute of Electrical Engineering, Slovak Academy of Sciences,
Dúbravská cesta 9, 841 04 Bratislava, Slovak Republic
e-mail: lubica.stuchlikova@stuba.sk, ladislav.harmatha@stuba.sk

Nitrogen introduced into a silicon substrate by the Czochralski method brought about an increase in the density of Si-SiO₂ interface traps and in the density of electrically active defects in Si by an enhanced nucleation of oxygen precipitates. Rapid thermal annealing above 800°C led to decomposition of N_xO_y clusters in the vicinity of the Si-SiO₂ interface and to an increase of the width of the denuded zone.

1. Introduction

The effect of nitrogen introduced into silicon by the float zone (FZ) technology has for a long time been the subject of interest of researchers [1]. Similarly like in the case of doping by oxygen [2], nitrogen doped (NFZ) silicon exhibits better mechanical properties and lower amount of defects created during the production of electron devices.

In the case of the growth of a silicon crystal by the method of Czochralski (CZ) the situation is more complicated because of the presence of a high concentration of oxygen ($\sim 10^{18}$ cm^{-3}) which penetrates into the melt from the surrounding quartz crucible. Nitrogen introduced into Si is captured by interstitial oxygen, which gives rise N_2O [3]. At the same time, the growth of SiO_x precipitates is enhanced. Oxygen precipitates are used in the so-called intrinsic gettering [4] during thermal loading. The result of intrinsic gettering is creation of a low-defect denuded zone in the subsurface region of Si. The problem is how to ensure creation of a uniform denuded zone in the whole silicon wafer. It has been verified that N-doped wafers with large diameters have a more homogeneous denuded zone. This is caused by satisfactory precipitation in all regions of the Si wafer [5, 6].

The issues of formation of electrically active defects and of their identification are addressed with respect to high-temperature loading of the wafers. The depth profile of the defects in the subsurface region of silicon was analyzed by in terms of the lifetime of minority charge carriers retrieved from non-equilibrium capacitance measurements on MOS structures. The results are completed by DLTS measurements and by determining the energy distribution of Si-SiO₂ interface traps from quasistatic low-frequency capacitance measurements.

2. Experiment

Phosphorus doped (n-type) samples were prepared with intention to investigate the properties of NCZ silicon wafers applied in MOS processing. A number of silicon wafers coated on both sides by a 1 μm thick LP CVD grown layer of Si_3N_4 were used as a source of nitrogen. Nitrogen-doped wafers and reference CZ wafers with the same resistivity (between 2 and 5 Ωcm) and thickness 300 μm were used. The SiO₂ gate layer was prepared by thermal oxidation in $O_2 + H_2O$ ambient at 1050°C resulting in oxide thickness 100 nm for n-type wafers. Nitrogen concentration 1.6×10^{15} cm^{-3} was detected by Fourier transform infrared spectroscopy (FTIR). Aluminium gates were vapour deposited and patterned photo-lithographically. After manufacturing the MOS structures, the samples were annealed in the forming gas at 460°C for 20 minutes.

A part of NCZ (NCZ_T MOS) as well as CZ (CZ_T MOS) samples underwent rapid thermal annealing (RTA) in the temperature range 700-1000°C for 10 seconds in N_2 ambient. The temperature was ramped at a rate of 200°C/s.

Electrically active defects in MOS structures prepared on NCZ silicon substrate and on reference CZ Silicon substrate before and after RTA (NCZ MOS, CZ MOS, NCZ_T MOS, CZ_T MOS) were studied by standard Deep Level Transient Spectroscopy (DLTS) and by its modification – Double Deep Level Transient Spectroscopy (DDLTS). The values of electron deep level activation energies, ΔE_T, and capture cross sections, σ_T, are calculated from an Arrhenius diagram using the emission process theory.

All MOS structures mentioned above were investigated by capacitance-voltage (C-V) and capacitance-time (C-t) measurements. From the measured C-V curves we determined the concentration of impurities, N_D, the density of the total defect charge of the MOS structure, N_{eff}, retrieved from the flat-band voltage, V_{FB}. The relaxation time, t_r, generation lifetime of minority charge carriers, τ_g, and the surface generation velocity, S_g, characterizing the electrical behaviour of defects in the MOS structure were measured by a non-equilibrium C-t method [7]. The depth distribution of τ_g was achieved by a modified, constant capacitance cC-t method [8]. Energy distributions of the interface trap density, D_{it}, of Si-SiO$_2$ MOS structures were derived from the low frequency C-V curve measured by a quasistatic method [9].

3. Results

The quality of MOS structures is well characterized by capacitance measurements. Figure 1 shows typical C-t and C-V curves. The results are summarized in Tab. 1.

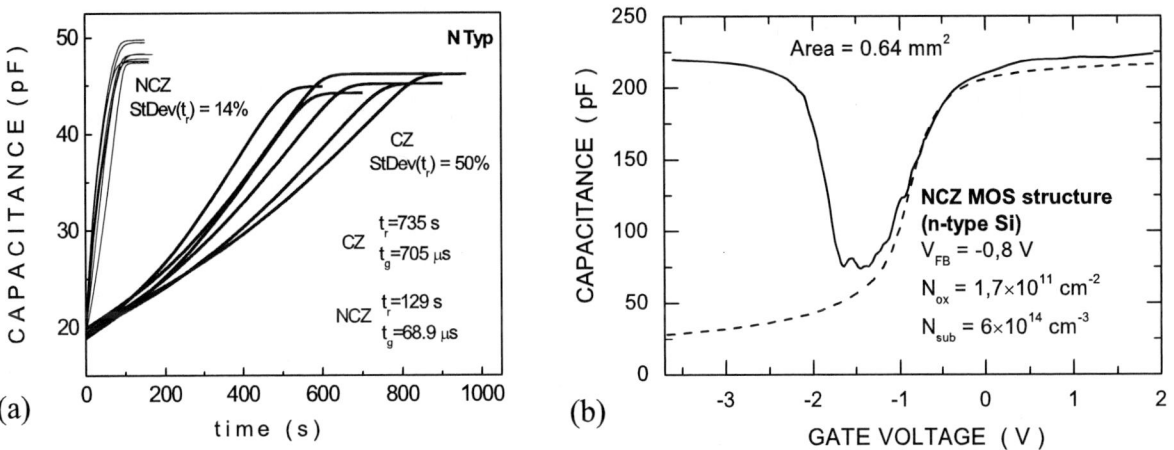

Fig. 1. (a) C-t measured curves of MOS structures with CZ and NCZ Si. (b) Deep-depleted (dashed line) and low frequency (solid line) C-V curves of a MOS structure prepared on a Ni-doped n-type Si substrate.

DLTS measurements were performed at the sensitivity edge of the DLTS apparatus. The spectra measured on the MOS structures exhibit a strong deviation from an exponential dependence. We produced a large set of measured DLTS spectra, the results obtained from DLTS measurements are shown in Tab. 2.

Tab. 1. Summary of experimental data from *C-V* and *C-t* measurements

Structure	Substrate	V_{FB} (V)	N_{eff} (m^{-2})	t_r (s)	τ_g (µs)	S_g (ms^{-1})	D_{it} (cm^{-2} eV^{-1})
MOS n-type Si	CZ	−0.7	1.7×10^{15}	735	705	3.2×10^{-4}	2.6×10^{9}
	NCZ	−0.7	1.7×10^{15}	129	68.9	9.0×10^{-4}	3.4×10^{10}
	CZ_T	−0.8	1.95×10^{15}	245	522.2	9.2×10^{-4}	–
	NCZ_T	−0.8	1.95×10^{15}	215	175.4	11.2×10^{-4}	–

Tab. 2. Summary of experimental data from DLTS measurements

MOS structure	Trap	ΔE_n (eV)	$\sigma_{n,p}$ (cm^2)	Ref. traps in literature
n-type CZ silicon	CZ_ET1	0.546	1×10^{-13}	
n-type NCZ silicon	NCZ_ET1	0.192	8.2×10^{-18}	
CZ_T silicon	CZ_T_HT1	0.476	1.0×10^{-17}	
n-type **NCZ_T** silicon	NCZ_T_HT1	0.254	7.1×10^{-21}	–
	NCZ_T_HT2	0.443	3.8×10^{-18}	NO (0.469 eV)
	NCZ_T_ET1	0.289	7.7×10^{-21}	–
	NCZ_T_ET2	0.381	6.0×10^{-19}	AlO$_2$ (0.385 eV)
	NCZ_T_ET3	0.402	9.7×10^{-19}	N$_2$O
	NCZ_T_ET4	0.480	3.0×10^{-17}	Au (0.510 eV)
	NCZ_T_ET5	0.449	3.6×10^{-18}	–
	NCZ_T_ET6	0.434	1.4×10^{-18}	–

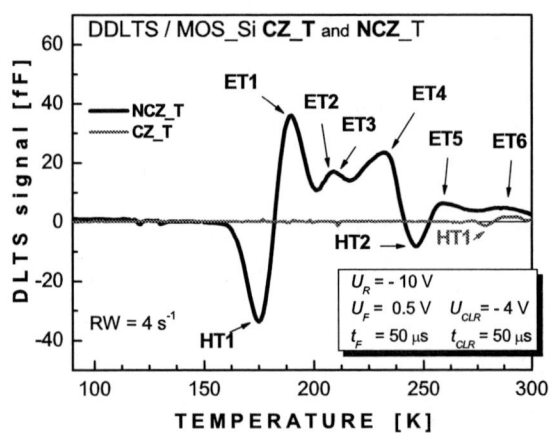

Fig. 2. Typical DDLTS spectra measured on the both type of MOS structures after RTA (CZ_T, NCZ_T)

4. Conclusion

From capacitance measurements a strong change was observed in the effective generation lifetime τ_g (from 705 to 69 µs) after nitrogen doping of CZ silicon. This effect is correlated with an increase of the amount of oxygen precipitates in silicon after nitrogen doping. Nitrogen doping of CZ silicon was found to improve the lateral homogeneity of the relaxation time of MOS structure. All observations, including those from C-t measurements, can be explained by assuming that various N_xO_y complexes were formed in n-type silicon. A significant effect of RTA on oxygen precipitation in NCZ silicon wafer was observed increasing both the generation lifetime τ_g (from 69 to 175 μs) and the relaxation time t_r (from 129 to 215 s). In the case of reference structures decreases were obtained of both the generation lifetime τ_g (from 69 to 175 μs) and the relaxation time t_r (from 705 to 175 μs).

Eight deep energy levels were found in the measured DLTS and DDLTS spectra. The origin of these deep energy levels is still under investigation. Trap CZ_ET1 (0.55 eV) probably corresponds to defects formed by Au. Levels CZ_ET2 and NCZ_ET2 have similar parameters. These levels probably correspond to point defects identified in both types (NCZ MOS and CZ MOS) of structures.

We expected that nitrogen doping caused enhancement of oxygen precipitation in MOS structures prepared on Czochralski-grown nitrogen-doped silicon. We suppose that two of the detected traps correspond to defects formed by nitrogen NCZ_T_HT2 (NO) and NCZ_T_ET3 (N_2O).

Donor and acceptor deep levels were identified from DDLTS spectra measured on MOS structures prepared on Czochralski-grown nitrogen-doped silicon after rapid thermal annealing. Despite this fact the electrical properties of NCZ_T MOS were improved by RTA. The identified deep traps have low concentrations, and their levels have not been resolved until after annealing. We supposed that this effect was caused by decomposition of N_xO_y clusters in the vicinity of the Si-SiO$_2$ interface and by an increase of the defect-free denuded zone in the subsurface region of silicon. The elevated temperature during RTA caused an enhanced growth of bigger precipitates, and destruction of smaller precipitates. The surface region of NCZ Si contains a large amount of smaller precipitates which are destructed during RTA.

Acknowledgement

This work was supported by the Slovak grant agencies VEGA, Project 1/3091/06 and APVT-20-055405.

References

[1] L. Jastrzebski, G.W. Cullen, R. Soydan, G. Harbeke, J. Lagowski, S. Vecrumba, W.N. Henry, *J. Electrochem. Soc.* **134**, 466, 1987.

[2] I. Yonenaga, K. Sumino, K. Hoshi, *J. Appl. Phys.* **56**, 2346, 1984.

[3] P. Wagner, R. Oeder, W. Zulehner, *Appl. Phys.* A **46**, 73, 1988.

[4] T.Y. Tan, E.E. Gardner, W.K. Tice, *Appl. Phys. Lett.* **30**, 175, 1977.

[5] V.D. Akhmetov, H. Richter, O. Lysytskiy, R. Wahlich, T. Müller, *Mater. Sci. Semicond. Process.* **5**, 391, 2003.

[6] L. Harmatha, M. Ťapajna, V. Slugeň, P. Ballo, P. Písečný, J. Šik, G. Kögel, *Microelectron. J.* **37**, 283, 2006.

[7] M. Ťapajna, L. Harmatha, *Solid-State Electron.* **48**, 2339, 2004.

[8] M. Ťapajna, P. Gurnik, L. Harmatha, *J. Electr. Eng.* **53**, 272, 2002.

[9] P. Písečný, M. Ťapajna, L. Harmatha, A. Vrbický, *J. Electr. Eng.* **55**, 95, 2004.

Microstructure of HfO₂ and HfₓSi₁₋ₓOᵧ Dielectric Films Prepared on Si for Advanced CMOS Application

M. Franta[a,b], A. Rosová[a], M. Ťapajna[a,b], E. Dobročka[a], and K. Fröhlich[a]

[a] Institute of Electrical Engineering, Centre of Excellence CENG, SAS, Dúbravská cesta 9, 841 04 Bratislava, Slovak Republic
[b] Faculty of Electrical Engineering and Information Technology, STU, Ilkovičova 3, 812 19 Bratislava, Slovak Republic
e-mail: marekfranta@atlas.sk and Alica.Rosova@savba.sk

We analyzed microstructure of as-deposited and rapid thermal annealed HfO₂ and HfₓSi₁₋ₓOᵧ dielectric films with Ru gate electrode. As-deposited films exhibited dielectric constant 12 and 20 for HfₓSi₁₋ₓOᵧ and HfO₂, respectively. TEM and grazing incidence XRD revealed that as-deposited HfO₂ films have polycrystalline character, while HfₓSi₁₋ₓOᵧ films are amorphous. Rapid thermal annealing makes favourable growth of monoclinic HfO₂ phase in HfO₂ films and tetragonal or orthorhombic phase in HfₓSi₁₋ₓOᵧ films.

1. Introduction

HfO_2 and $Hf_xSi_{1-x}O_y$ (HSO) dielectric films are considered as potential candidates for SiO_2 replacement of in sub-45 nm technology node of complementary metal oxide semiconductor (CMOS) field effect transistors. HfO_2 films exhibit high dielectric constant and therefore films with thickness of several nm offer capacitance equivalent to sub-nanometer thickness of SiO_2. HSO films have lower dielectric constant, depending on Hf/Si ratio, but they remain amorphous up to high temperatures and show higher thermal stability [1-3]. Dielectric properties of HfO_2 and HSO are related to their microstructure and to be used in CMOS devices they must maintain their properties up to 1000°C/10 heat treatment under N_2. To study thermal stability of our samples we have chosen the temperature of 900°C for rapid thermal annealing (RTA). We used Ru layer as electrode and we observed a start of diffusion between hafnium silicate and Ru layers at this temperature by RBS analysis [2, 3].

2. Experimental

HfO_2 and HSO films with nominal thickness range from 3 to 6 nm where prepared by MOCVD at temperatures from 400 to 600°C in IMEC in Leuven. The layers were deposited on (100)-oriented p-type boron doped silicon covered with chemical oxide (0.8 nm for IMEC clean procedure). As an electrode we used Ru layer deposited at 350°C by MOCVD.

Structure of the films with nominal thickness of 6 nm was studied by X-ray diffraction in grazing incidence mode (GI-XRD) on Bruker AXS – D8 Discovery equipment with X-ray tube with rotating anode. The grazing incidence angle of 1.5° was choosen. Transmission electron microscopy (TEM) observations were performed by JEOL JEM 1200 EX microscope with 120 kV accelerating voltage. Specimens for TEM were prepared by mechanical thinning and ion milling of substrate from its bottom. To study thermal stability the stacks were processed by RTA at 900°C for 10 s with the ramp rate 100°C/s in pure nitrogen atmosphere.

Capacitance-voltage (C-V) measurements have been done on the samples annealed in forming gas (90% N_2 + 10% H_2) at 430°C during 30 min (FGA) without RTA. Metal electrodes were patterned by standard optical lithography and Ar ion milling. To provide good

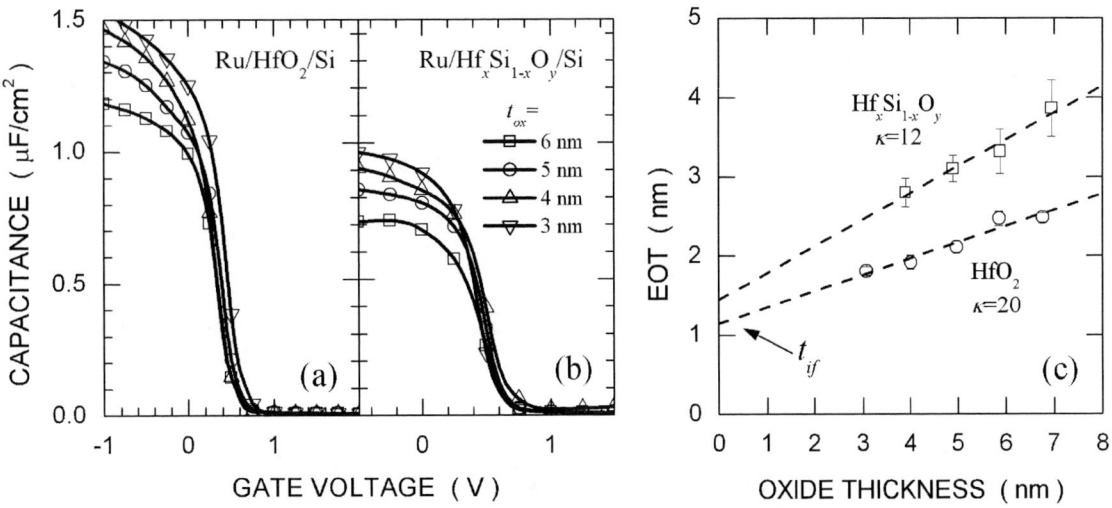

Fig. 1. *C-V* curves of Ru/HfO₂/Si (a) and Ru/HSO/Si (b) MOS capacitors with different dielectric thickness t_{ox}. The symbols correspond to same t_{ox} for both samples. (c) Linear fit to the EOT vs. physical dielectric thickness providing the interfacial layer thickness (intercept) and dielectric constant (inversely proportional to the slope).

bottom contact Al was sputtered to the sample backside. *C-V* curves were measured at 1 MHz and 100 kHz using Agilent 4284A LCR Meter on the gate electrodes with an area of 1.901×10^{-3} cm² and corrected by two-frequency correction [4]. We have used NCSU model for the equivalent oxide thickness, EOT, and the flat band voltage, V_{FB}, extraction [5].

3.   Results and discussions

In order to assess the quality and dielectric constant of the as-deposited HfO₂ and HSO gate dielectrics, we have measured the set of *C-V* curves on MOS structures with various dielectric thickness (fig. 1(a-b)). Dielectric constants κ were determined from the dependence of EOT versus oxide thickness (fig. 1c). EOT was obtained from accumulation part of the *C-V* curves. The HfO₂ and HSO gate dielectrics show κ of 20 and 12, respectively. According to the κ value, one can roughly estimate the composition of $Hf_xSi_{1-x}O_y$ layers from linear interpolation of dielectrics constant for $x = 0$ and $x = 1$, hence pure SiO₂ (3.9) and HfO₂ (20).

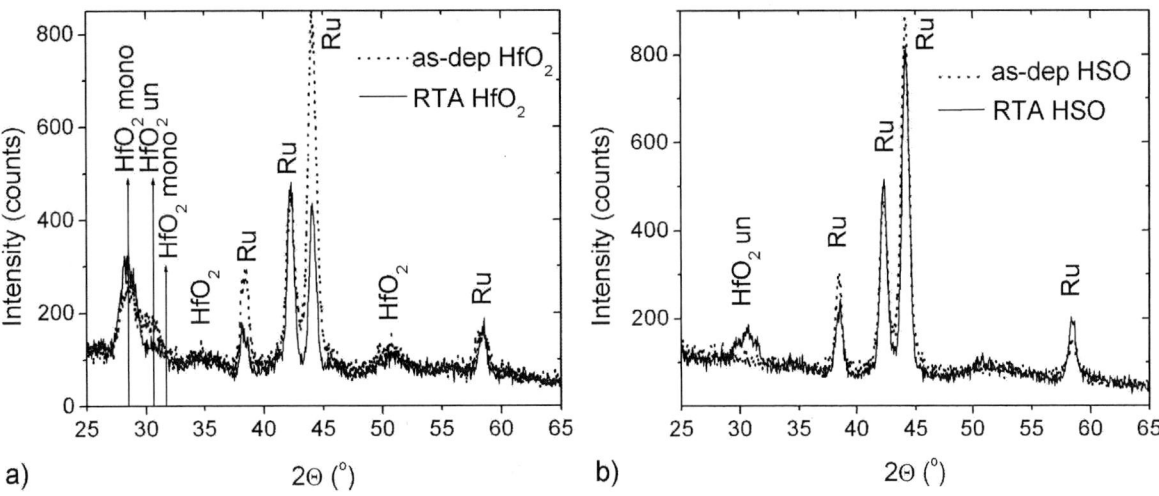

Fig. 2. GI-XRD spectra of Ru/HfO₂/Si (a) and Ru/HSO/Si (b) before and after RTA treatment

243

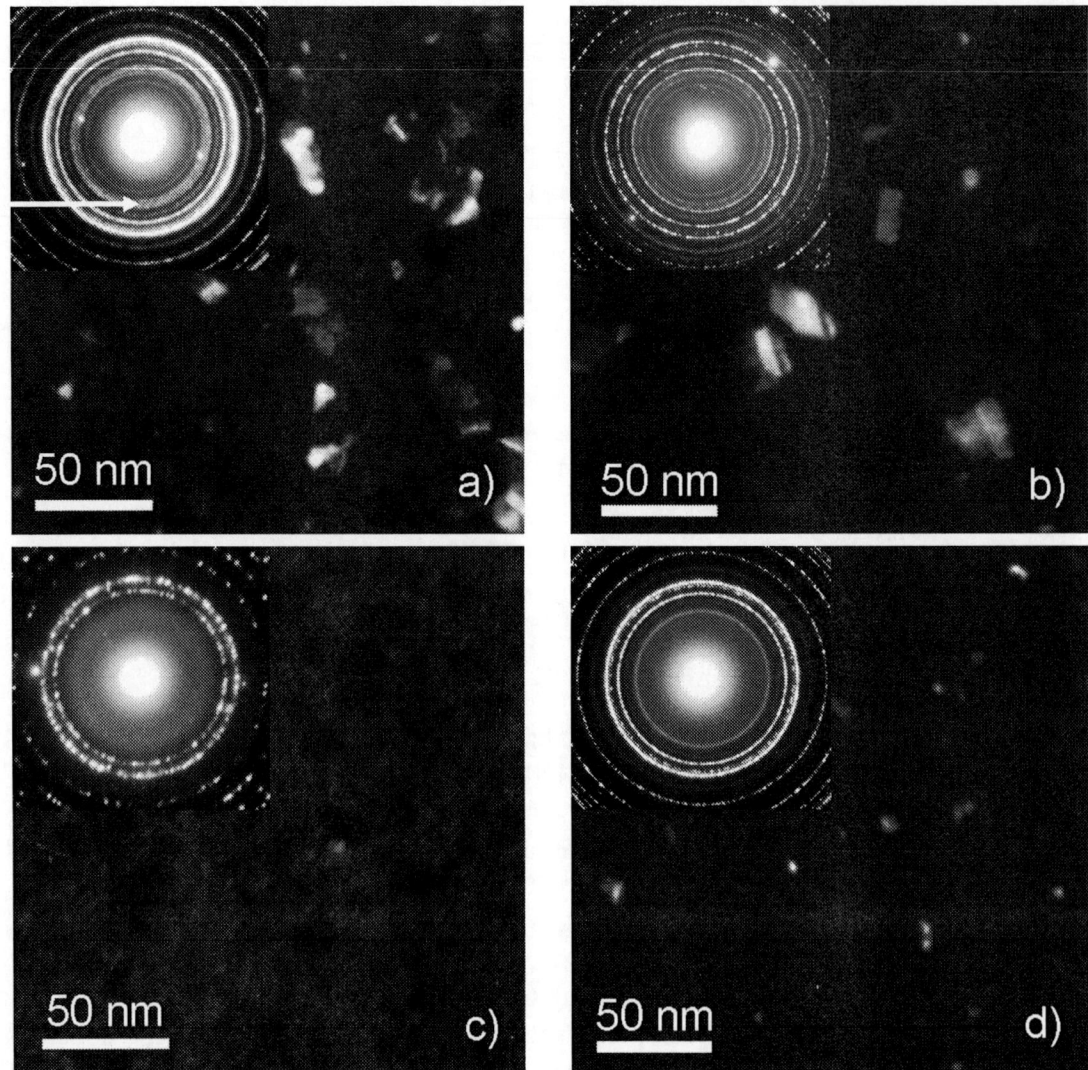

Fig. 3. Dark field images and electron diffraction patterns (insets) of the as-deposited HfO_2 film (a), HfO_2 film after RTA (b), as-deposieted HSO film (c) and HSO film after RTA (d).

We obtain the $x \approx 0.5$. The interfacial layer thickness determined from fig. 1(c) is 1.1 and 1.5 nm for $Ru/HfO_2/Si$ and Ru/HSO/Si MOS capacitor, respectively. Thus its growth during the metal electrode deposition is negligible. From these results, we can conclude good dielectric properties of the MOCVD grown HfO_2 and HSO gate dielectrics.

As-deposited HfO_2 layer has polycrystalline character and HfO_2 is present there as a mixture of major monoclinic phase and second phase marked as "HfO_2 un" on GI-XRD scans (fig. 2a). Diffraction maximum of the second phase was observed at 2Θ values between 30.0 and 30.8°. It is impossible to estimate precisely the value because the peaks are too expanded on GI-XRD spectra and too small due to the low thickness of the film.

On electron diffraction patterns the second phase spots or rings are visible, but they are situated between two stronger rings of monoclinic HfO_2 (all together marked by an arrow on the fig. 3a). From the results we are not able to distinguish if the unknown phase is orthorhombic or tetragonal HfO_2 phase. Both unstable phases were observed as admixture with stable monoclinic phase under certain deposition conditions [6, 7].

In the HfO_2 films RTA makes growth of stable monoclinic phase favourable, (fig. 2a).

As deposited HSO films are amorphous according our GI-XRD measurement. However, TEM shows in these "X-ray-amorphous" layers presence of tetragonal or orthorhombic HfO$_2$ clusters of sub-nanometer or nanometer size as it is confirmed by dark field mode TEM and also by wide diffuse rings on electron diffraction pattern on fig. 3c. During RTA treatment small HfO$_2$ clusters grow up and their unstable "second phase" remains, probably stabilized by small size of grains, amorphous matrix and/or diffusion of Ru. HfO$_2$ nanocrystallites are embedded in amorphous matrix, so they are isolated each other, (fig. 3d). It can also explain the starting interdiffusion between HSO dielectric and Ru electrode [2, 3]. After crystallization and growth of HfO$_2$, Si/Hf ratio in the amorphous matrix increases and the matrix becomes similar to SiO$_2$. Ruthenium interdiffuses with SiO$_2$ [8].

4. Conclusions

We have observed microstructure of hafnium dioxide and hafnium silicate dielectric layers with good dielectric constants (k=20 for HfO$_2$ and k=12 for HSO) and their changes under RTA at 900°C during 10 s in N$_2$ atmosphere. The layers were deposited on Si substrate with thin (~0,8nm) SiO$_2$ layer. Non-equlibrium deposition conditions, presence of SiO$_2$ and small size of growing objects result in appearence of unstable tetragonal or orthorhombic phase. In as-deposited HfO$_2$ layers the unstable phase appeared as a mixture with stable monoclinic HfO$_2$ phase. RTA suppress the unstable phase and favorites the stable monoclinic phase.

As deposited hafnium silicate is „X-ray amorhous", but TEM shows, that even without any additional heat treatment the siliciate is in an early state of phase separation. Clusters of tetragonal or orthorhombic HfO$_2$ are spread in amorphous matrix. The additional RTA promotes crystallisation and grain growth of HfO$_2$ tetragonal or orthorhombic phase.

Acknowledgement

This work was supported by the Slovak grant agency APVT (project APVT-51-017004). We would like to thank to T. Schram and S. De Gendt from IMEC, Leuven for providing HfO$_2$ and HSO films.

References

[1] P. Lysaght, B. Foran, S. Stemmer, G. Bersuker, J. Bennett, R. Tichy, L. Larson and H. R. Huff, *Microelectronic Engineering* **69,** 182, 2003.

[2] M. Ťapajna, K. Hušeková, D. Machajdík, A. P. Kobzev, T. Schram, R. Lupták, L. Harmatha and K. Fröhlich, *Microelectronic Engeneering,* in press.

[3] D. Machajdík, A. P. Kobzev, K. Hušeková, M. Ťapajna, K. Fröhlich and T. Schram, submitted to *Vacuum*.

[4] K. J. Yang, and Ch. Hu, *IEEE Trans. Electron Dev.* **46**, 1500, 1999.

[5] N. Yang, K. Henson, J. Hauser, and J. Wortman, *IEEE Trans. Electron. Dev.* **46**, 1464, 1999.

[6] J. Aarik, J. Sundqvist, A. Aidla, J. Lu, T. Sajavaara, K. Kukli and A. Hårsta, *Thin Solid Films*, **418**, 69, 2002.

[7] J. Q. He, A. Teren, C. L. Jia, P. Ehrhart, K. Urban, R. Waser and R. H. Wang, *J. Crystal Growth*, **262**, 295 2004.

[8] H. - C. Wen, P. Lysaght, H. N. Alshareef, C. Huffman, H. R. Harris, K. Choi, Y. Senzaki, H. Luan, P. Majhi, B. H. Lee, M. J. Campin, B. Foran, G. D. Lian and D. - L. Kwong, *J. Appl. Phys.*, **98**, 43520, 2005.

Deep Defects in MOVPE Grown SiC/AlGaN/GaN Heterostructures

D. Kindl [1], P. Hubík [1], J. Krištofik [1], J. J. Mareš [1], Z. Výborný [1],
M. R. Leys [2], and S. Boeykens [2]

[1] Institute of Physics, Academy of Sciences of the Czech Republic,
Cukrovarnická 10, 162 53 Prague, Czech Republic
[2] IMEC vzw, Kapeldreef 75, B-3001 Leuven, Belgium
e-mail: kindl@fzu.cz and boeyken@imec.be

Deep level transient spectroscopy (DLTS) measurement was carried out on GaN/AlGaN/SiC heterostructures prepared by low-pressure metalorganic vapour phase epitaxy. Two kinds of p-type 4H-SiC substrates were employed for the n-GaN layer growth using an n-AlGaN nucleation layer. Two different aluminium concentrations of 30% and 8% were tested. DLTS spectra of on-axis (0001) grown samples exhibit a peak of majority carrier trap with apparent activation energy of 0.78 eV regardless of nucleation layer composition. As the amplitude of this peak is almost insensitive to the height of filling pulses and the space charge region in AlGaN is negligible, the level may be associated with a hole trap in the substrate. The off-axis grown samples show a majority trap-like DLTS peak only under high positive filling pulses. Therefore, activation energy of 0.66 eV obtained for both nucleation layer compositions is related to an interface defect.

1. Introduction

Wide band-gap Group III-Nitrides have become one of the hottest research topics, owing to their excellent potential for use in high-power electronic and optoelectronic devices. The binary nitrides of aluminium, gallium, and indium as well as their ternary solid solutions are materials with direct band gaps ranging from 1.9 to 6.2 eV [1] and, therefore, they are suitable for photodetectors, lasers and light emitting diodes working in visible and UV emission ranges [2, 3]. Some specific properties of AlGaN, e.g. thermal and radiation stability, chemical inertness, high breakdown voltage, high thermal conductivity and wide band-gap are being utilized in order to fabricate microwave devices designed for high temperature and high frequency operation. Due to recent progress in both MOVPE and MBE technology, Modulation Doped Field Effect Transistors (MODFET) [3], Heterostructure Field Effect Transistors (HFET) [4], and High Electron Mobility Transistors (HEMT) [5] based on GaN/AlGaN are fabricated.

III-Nitrides are primarily grown on sapphire substrates [6-8], while alternative materials such as 4H- and 6H-SiC are increasingly investigated [9-14] because of their higher thermal and electrical conductivity as well as better lattice matching for hetero-epitaxial growth of GaN. Heterojunctions *n*-GaN/*p*-SiC and *n*-AlGaN/*p*-SiC are of interest for use in Heterojunction Bipolar Transistors (HBT). Hence, great effort is paid to the study of band discontinuities [10-12] and deep defects [9, 11] which determine transport properties of these structures.

Recently it has been found [11] that the *n*-GaN/*p*-SiC heterojunctions prepared by hydride vapour phase epitaxy (HVPE) on 4H-SiC substrates are staggered type II. When changing GaN for AlGaN with Al mole fraction of 0.25–0.3 the band alignment becomes normal type I, which theoretically should improve the performance of corresponding HBTs.

1-4244-0396-0/06/$25.00 ©2006 IEEE

Current-voltage (*I-V*) characteristics of both heterojunctions show strong tunnelling via defect states. The tunnelling as well as leakage currents are more pronounced in the AlGaN/SiC heterojunctions [11, 13].

In order to obtain high quality nitride layers on SiC substrates it is necessary to optimize many parameters, especially nucleation layer thickness, composition of this layer, and substrate orientation [14] because of their effect on interface-related properties. In this contribution we present experimental results concerning deep defects in heterostructures GaN/AlGaN/SiC prepared on 4H-SiC substrates. We correlate the appearance of deep levels with Al-content in the nucleation layer and with the substrate orientation.

2. Experiment

All nitride layers were prepared by low-pressure metalorganic vapour phase epitaxy (LP-MOVPE) in a 3×2" vertical close-coupled showerhead reactor. Growth details were published elsewhere [14]. Two different *p*-type 4H-SiC substrates (on-axis orientation, doping level of 2×10^{18} cm^{-3} and 8° off-axis orientation, doping level of 1×10^{18} cm^{-3}) were available for the *n*-type GaN (2×10^{18} cm^{-3}) epitaxial layer growth using an *n*-type AlGaN (10^{19} cm^{-3}) nucleation layer. Two different aluminium concentrations of 30% and 8% were tested. The thickness of all layers was about 300 nm. The samples were processed into mesa structures. A pattern of circles with the diameters ranging from 60 to 700 microns was etched using a Cl$_2$-based RIE etching process. Sidewalls of the mesas were passivated with sputtered SiO$_2$. Ohmic contacts to the *n*-type GaN, i.e. a Ti/Al/Mo/Au metallization, were annealed at 750°C. On the backside of the substrate, a Ti/Al ohmic contact annealed at 1025°C was applied.

The samples were investigated by means of deep level transient spectroscopy (DLTS) using a Polaron S4600 system in order to determine the activation energy as well as the location of deep defects which could be responsible for recombination transport deduced from *I-V* measurements. Besides, an extensive admittance analysis consisting of *C-f*, *G-f*, *C-T*, *G-T*, and *C-V* measurements was performed with the aim to get information about the heterojunction built-in voltage and carrier freezing out below room temperature.

3. Results and discussion

As the DLTS results are almost insensitive to the aluminium content in the nucleation layer, they are presented in two subsections according to the kind of SiC substrate. The results of admittance analysis are summarized for all the samples in the third subsection.

3.1 Deep levels in the on-axis grown samples

Typical DLTS spectra of a GaN/AlGaN/SiC heterostructure prepared on on-axis 4H-SiC substrate are shown in Fig. 1. The plot presents a signal taken under injection biasing conditions (pulse bias equal to +1 V) and different rate windows. A majority carrier peak which dominates the spectrum gives apparent activation energy of 0.78 eV regardless of the nucleation layer composition. Besides, the amplitude of this peak is not sensitive to both the height and the width of filling pulses. As a major part of the space charge region is located in SiC, we deduce that the level is associated with a hole trap in the substrate.

The same deep level was found on the samples with different junction areas. Thus, there is probably no direct relation between this level and excessive recombination currents observed in the case of the large area samples.

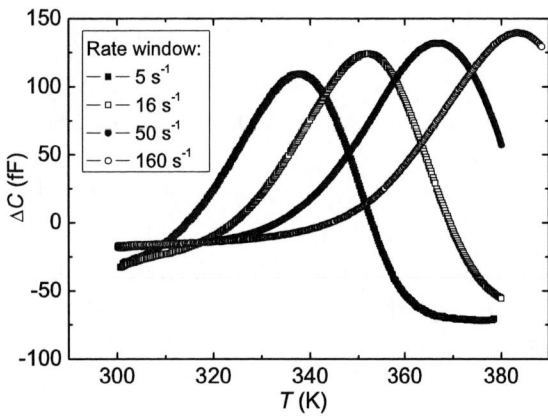

Fig. 1. DLTS spectra of a GaN/AlGaN/SiC heterostructure grown on on-axis 4H-SiC substrate. All the spectra were taken at a reverse bias of -4 V, a pulse bias of +1 V, and a pulse width of 2 ms.

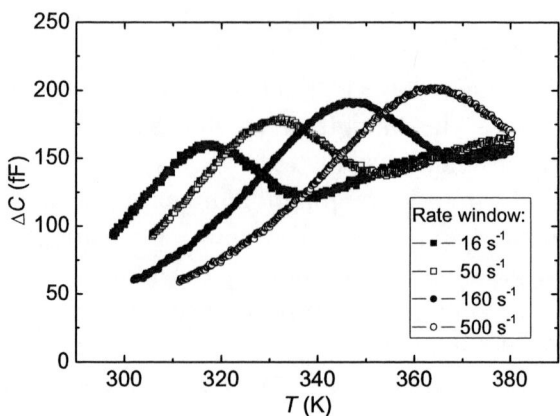

Fig. 2. DLTS spectra of a GaN/AlGaN/SiC heterostructure grown on off-axis 4H-SiC substrate. All the spectra were taken at a reverse bias of -4 V, a pulse bias of +2 V, and a pulse width of 2 ms.

3.2 Deep levels in the off-axis grown samples

Representative DLTS spectra of a GaN/AlGaN/SiC heterostructure prepared on off-axis 4H-SiC substrate are plotted in Fig. 2. In this case, the signal was taken using higher pulse bias of +2 V. Under such conditions we have found a majority carrier trap with energy of 0.66 eV for both nucleation layer compositions. However, the reliability of the energy determination may be affected by the fact that the DLTS signal is very sensitive to the height of filling pulses and contains some temperature dependent background.

We have observed no peaks in DLTS spectra until high forward biasing pulses were applied. This feature indicates an interface states related effect. A similar result has been reported by Polyakov *et al.* [11] for on-axis GaN/4H-SiC heterojunctions grown by HVPE.

3.3 Admittance analysis

Admittance analysis performed in the range of 90-300 K shows carrier freeze out at low temperatures. With decreasing temperature, both the capacitance and the conductance approach zero disabling DLTS scans. The G-f measurements (Fig. 3) at temperatures below 230 K yield activation energy of 0.20 eV for all samples. This value calculated from an Arrhenius plot of the G/ω peak frequencies is connected with the SiC substrate because it coincides very well with the known energy of the Al acceptors in 4H-SiC [11]. At higher temperatures, the frequency dependence of G/ω becomes more complex. This effect may be related to isotype GaN/AlGaN junction or contact series resistance.

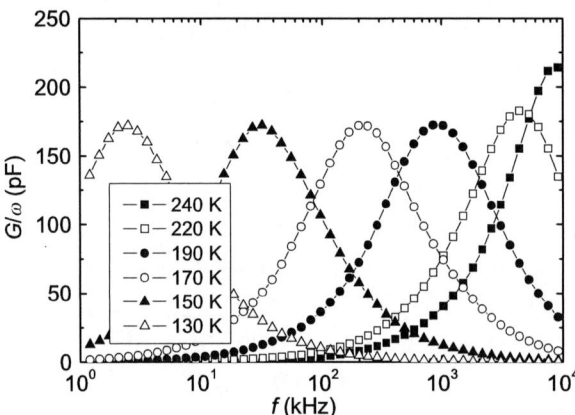

Fig. 3. Conductance/angular frequency as a function of the test frequency. Measurement was performed without dc bias.

Room temperature C-V measurements yield the built-in voltage values in the range of 2.4-2.9 V. The on-axis grown samples exhibit systematically higher built-in voltage than the off-axis ones. This quantity also increases with increasing Al content in the nucleation layer.

4. Conclusion

The DLTS method was used to direct observation of deep defects in GaN/AlGaN/SiC heterostructures grown by LP-MOVPE. The samples prepared on on-axis 4H-SiC substrates exhibit a majority carrier peak in DLTS spectra which gives apparent activation energy of 0.78 eV regardless of nucleation layer composition. This energy may be associated with a hole trap in the substrate. On the other hand, the off-axis grown samples show a DLTS peak only under high forward biasing filling pulses, which indicates an interface related defect. The same deep level with activation energy of 0.66 eV was found for both Al concentrations in the nucleation layer. The use of DLTS is limited by carrier freeze out at low temperatures.

Admittance analysis yields the activation energy of the substrate resistance. Energy of 0.20 eV agrees very well with the known energy of the Al acceptors in 4H-SiC

We believe that the results will help to further improve the technology of optoelectronic and microwave devices based on GaN/AlGaN/SiC heterostructures.

Acknowledgement

This work was partly supported by Grant Agency of AS CR (Contract No. IAA1010404).

References

[1] J. Stejskal and J. Leitner, *Chem. Listy* **96**, 311, 2002.

[2] H. Kim, S. J. Park, and H. Hwang, *IEEE Trans. on Elect. Devices* **49**, 1715, 2002.

[3] R. D. Dupuis, *J. Cryst. Growth* **178**, 56, 1997.

[4] M. A. Khan, M. S. Shur, J. N. Kuznia, J. Burm, and W. Schaff, *Appl. Phys. Lett.* **66**, 1083, 1995.

[5] M. A. Khan, A. Bhattarai, J. N. Kuznia, and D. T. Olson, *Appl. Phys. Lett.* **63**, 1214, 1993.

[6] P. Hacke and H. Okushi, *Appl. Phys. Lett.* **71**, 524, 1997.

[7] W. Götz, N. M. Johnson, D. P. Bour, C. Chen, H. Liu, C. Kuo, and W. Imler, *Mater. Res. Soc. Symp. Proc.* **395**, 443, 1996.

[8] W. Shan, J. W. Ager III, K. M. Yu, W. Walukiewicz, E. E. Haller, M. C. Martin, W. R. McKinney, and W. Yang, *J. Appl. Phys.* **85**, 8505, 1999.

[9] J. T. Torvik, C.-H. Qiu, M. Leksono, and J. I. Pankove, *Appl. Phys. Lett.* **72**, 945, 1998.

[10] J. T. Torvik, M. Leksono, J. I. Pankove, B. van Zeghbroeck, H. M. Ng, and T. D. Moustakas, *Appl. Phys. Lett.* **72**, 1371, 1998.

[11] A. Y. Polyakov, N. B. Smirnov, A. V. Govorkov, E. A. Kozhukhova, B. Luo, J. Kim, R. Mehandru, F. Ren, K. P. Lee, S. J. Pearton, A. V. Osinsky, and P. E. Norris, *Appl. Phys. Lett.* **80**, 3352, 2002.

[12] E. Danielson, C.-M. Zetterling, M. Ostling, A. Nikolaev, I. P. Nikitina, and V. A. Dmitriev, *IEEE Trans. Electron Devices* **48**, 444, 2001.

[13] E. Danielson, C.-M. Zetterling, M. Ostling, D. Tsvetkov, and V. A. Dmitriev, *J. Appl. Phys.* **91**, 2372, 2002.

[14] S. Boeykens, M. R. Leys, M. Germain, R. Belmans, and G. Borghs, *J. Cryst. Growth* **272**, 312, 2004.

Diagnostics of LT GaAs/ InP structures by micro-Raman spectroscopy

R. Srnánek[1], G. Irmer[2], R. Záluský[1], F. Dubecký[3], R. Kúdela[3], A. Vincze[4], I. Novotný[1], J. John[5]

1   Microelectronics Department, Slovak University of Technology, Ilkovičova 3, 812 19 Bratislava, Slovakia, e-mail: rudolf.srnanek@stuba.sk
2   Institute of Theoretical Physics, TU Bergakademie Freiberg, Leipziger-Strasse 23, D-09596 Freiberg, Germany
3   Institute of Electrical Engineering, Slovak Academy of Sciences, Dúbravská cesta 9, 841 04 Bratislava, Slovakia
4   International Laser Centre, Bratislava, Ilkovičova 3, 81219 Bratislava, Slovakia
5   IMEC, Leuven, Belgium

We present diagnostics of LT GaAs/InP structures by micro-Raman spectroscopy, micro-photoluminescence and SIMS methods. Bevelled form of the samples was used for the study. The thickness of the epitaxial LT GaAs layer with presence of high density of antisites As_{GaAs} and As excess was determined. Between InP substrate and LT GaAs layer an interfacial layer composed from InAs, InO and C was detected and studied.

1.   Introduction

Low temperature (LT) GaAs presents an epitaxial layer grown by molecular beam epitaxy (MBE) at lowered growth (T_g) temperature ($T_g < 350$ °C) under As overpressure resulting in incorporation of a large excess concentration of arsenic atoms diluted in a perfect GaAs single crystal. This material performs unique optical and electrical properties such as relatively high electrical resistivity and extremely short carrier lifetimes due to very high concentration of antisite defects As_{Ga} and another defects related to the excess arsenic. These defects significantly influence the electrical and optical properties of device structures based on the LT GaAs. Unique characteristics of the LT GaAs determine the applications such as a buffer or interface layer in high speed FET or directly as active layers in fast opto/photoelectronic devices and THz devices [1, 2, 3]. It has been shown that the electrical and physical properties of the LT GaAs depend on the both, the growth T_g and the annealing - T_{ann} temperatures due to the changes in the size and density of the As precipitates. Through these temperatures the resistivity and the carrier lifetimes can be manipulated. LT GaAs/GaAs structure is for now used in many device applications. In some particular applications InP as a base material is required and its combination with the LT GaAs opens a new technological/application solutions. On the other hand, LT GaAs as an interface/barrier layer could potentially help to overcome problems with formation of a blocking Metal-Semiconductor system on InP. Such system presents from the physical point of view interesting strained pseudomorphic heterojunction. In our best knowledge only a few published results are available up to now

In this work results of diagnostics by micro-Raman spectroscopy (μ-RS), micro-photoluminiscence (μ-PL) and SIMS (secondary ion mass spectrometry) of the LT GaAs grown on bulk SI (semi-insulating) InP substrates are presented. This is due to its potential

1-4244-0396-0/06/$25.00 ©2006 IEEE

utilization as a base structure for fabrication of particle, X- or γ-ray detectors where InP presents the active absorption (stopping) part of the device.

2. Experimental details

LT GaAs epitaxial layers were grown onto 2" bulk SI InP substrate using MBE at 250 °C and BEP = 15 (As beam equivalent pressure) with the final thickness of 500 nm. InP substrate was annealed before growth in the MBE chamber at ambient arsenic vapour at a temperature of 500 °C to reduce an oxide film. As–grown structures were annealed at temperatures 360-640 °C in MOCVD reactor (Aixtron) under As overpresusure. The structures of LT GaAs/InP were bevelled by wet chemical etching. Such samples were used for diagnostics by μ-RS and μ-PL. For the bevel preparation an etchant composed of H_3PO_4, H_2O_2 and H_2O which does not attack InP was used [4]. Prepared bevel angles, determined by Talystep measurements lay in the range of 10^{-4} rad. The depth profile of the sample B124 (annealed at 470 °C) and a cross-section view of studied structure is shown in fig. 1.

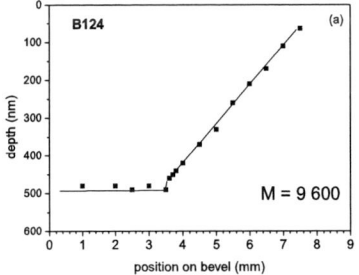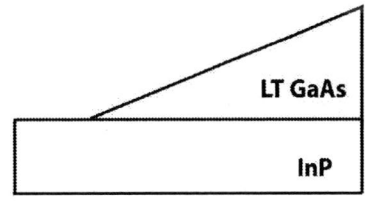

Fig.1. (a) Measured depth-profile of the sample B124 annealed at 470°C
(b) cross-section view of the measured sample.

The μ-RS and μ-PL of the samples were revealed by excitation of YAG: Nd laser generating the second harmonic with wavelength of 532.2 nm. The incident laser beam was focused to a diameter of ~ 2 μm on the sample surface. Time of flight based instrument (Ion-TOF, SIMS IV) with an Au^+ primary ion source for the SIMS measurement was used.

3. Results and discussion

Typical μ-Raman spectra taken from bevelled structure annealed at different temperatures are presented fig. 2a. In the vicinity of frequencies 250-260 cm^{-1} we detected a mode which corresponds to the presence of antisite As_{Ga} defects [5]. The intensity of this mode depends on the annealing temperature; its intensity decreases with the increasing of annealing temperature. Presence and intensity of this mode along the bevelled structures was studied. It is found that the concentration of As_{Ga} point defects in the LT GaAs layer is not constant, however these defects are accumulated near to the InP/GaAs interface. The thickness of this interfacial layer was determined in dependence on the annealing temperature with the values between 13 and 170 nm. The dependence of the mode intensity measured at 250 cm^{-1} on sample B125 (annealed at 560 °C) is shown in fig.2b. Thickness of accumulated layer (69 nm) was determined from this dependence.

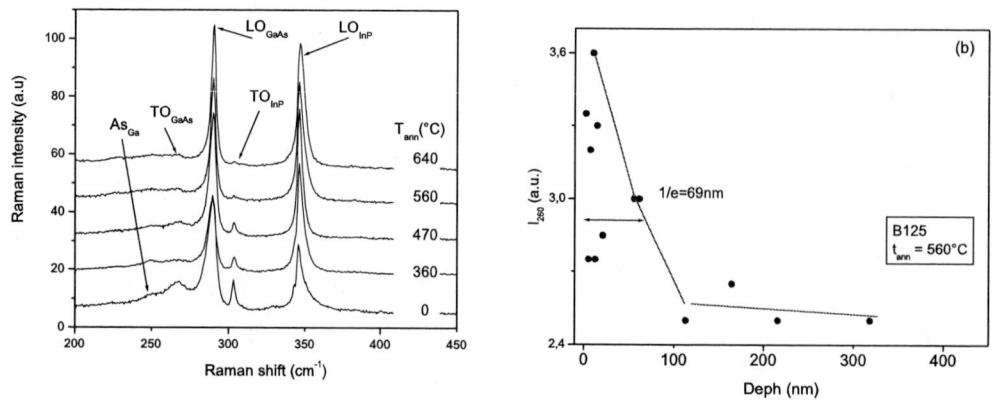

Fig. 2. (a) Typical μ-Raman spectra taken from the bevelled structures. (b) Dependence of Raman intensity at Raman shift 250 cm^{-1} for sample B 125 annealed at 560 °C.

The interfacial strains between the substrate and the epitaxial layer originates from lattice mismatch between the two materials (a_{GaAs} = 0.5653 nm and a_{InP} = 0.58687 nm). This interfacial strain causes a shift of the LO peak in detected Raman spectra. The shift of TO and LO peaks in studied samples is caused also by extension of Ga lattice into the LT GaAs layer due to As$_{Ga}$ antisites. Gant [6] used a model of dielectric continuum to calculate an excess of As atoms from the shifts of TO and LO phonon peaks. We have used this procedure for determination of As excess in studied bevelled samples. The dependence of As excess for bevel B 123 (annealed at 360 °C) is shown in Fig. 3a. We suppose that at the temperature of 640 °C is no As excess in the LT GaAs (precipitates are created from all As atoms). Therefore in this sample the TO and LO shift is caused by the interfacial strain alone. Then the real dependences of As excess in the studied samples were obtained by substraction of the dependence obtained at 640 °C from all others dependencies. The Fig.3b represents the As excess in the position of 50 nm from the interface LT GaAs/InP for different annealing temperatures.

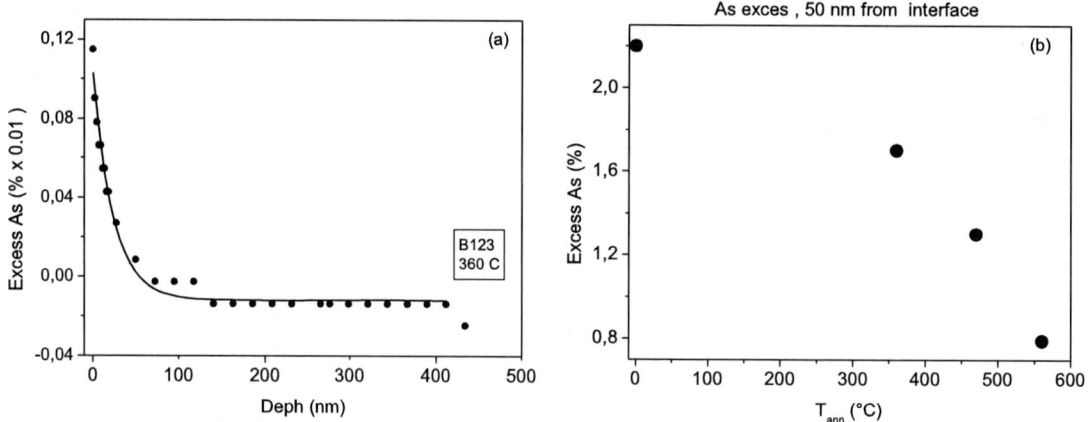

Fig. 3. (a) As excess dependence in LT GaAs layer (b)As excess in position 50 nm from LT GaAs/InP interface at different annealing temperatures

Thin interfacial layer between the LT GaAs layer and InP substrate was observed, explainable as a result of the annealing of InP substrate in ambient As in the MBE chamber prior growth of the epitaxial LT GaAs layer. The observed interfacial layer contains InAs,

InO and C. It was detected and studied by μ-PL and SIMS (see Fig. 4). Its thickness determined from the μ-PL measurement is some few nm.

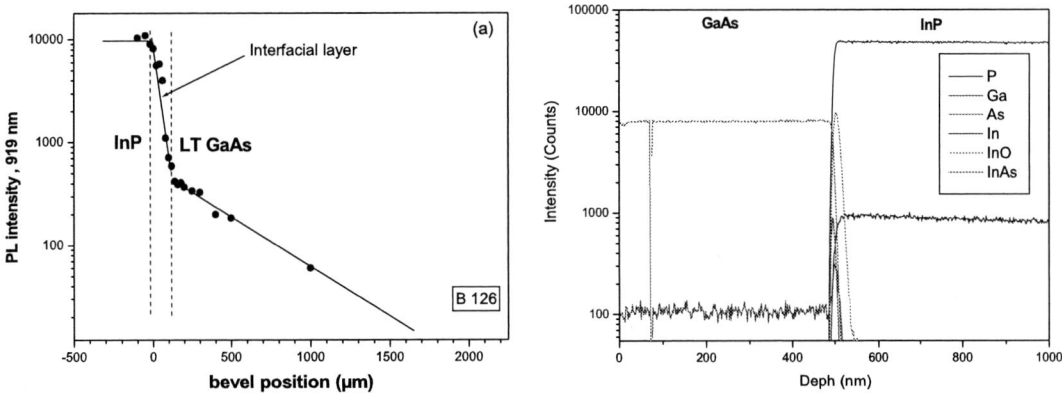

Fig. 4. (a) Detection of interfacial layer between LT GaAs and InP substrate by μ-PL and (b) by SIMS. In the depth of 500 nm was detected presence of InAs, InO and C.

6.   **Conclusion**

A diagnostics of LT GaAs/InP structures convenient for the preparation of X-rays detectors were presented using methods of micro-Raman spectroscopy, micro-photoluminescence and SIMS on bevelled samples prepared by chemical etching. We have detected accumulated layer LT GaAs with the high density of As_{Ga} antisite defects by the means of micro-Raman spectroscopy. The thickness of this layer was determined in dependence on annealing temperature. We determined As excess in LT GaAs from the shift of LO and TO phonon peaks in the LT GaAs using micro-Raman spectra. The interfacial layer between LT GaAs and InP substrate composed from InAs, InO and C was detected and studied.

Acknowledgements

This work was supported by VEGA grants No. 1/3111/06, 1/3108/06, 1/2041/05, 2/4151/26, 1/3098/06 and by the Agency for Promotion Research and Development under the contract No. APVV-99-P06305 and international cooperation Germany and Slovakia.

References

[1]   D. C. Look, *Thin Solid Films* **231**, 61,1993
[2]   D. H. Auston, K. P. Cheung, P.R. Smith, *Appl.Phys. Lett* **45** , 284, 1984
[3]   R. Mendis, C. Sydlo, J. Sigmund, M. Feiginov, P. Meissner, H. L. Hartnagel, *Solid State Electronics* **58**, 2041, 200
[4]   R. Srnanek, I. Novotny, I. Hotovy, M. El. Gomati, *Mat .Sci. Eng.* **B47,** 127,1997
[5]   M. Herms, G. Irmer. G. Goerigk, E. Bedel, *Mat .Sci. Eng. B* **91-92,** 466, 2002
[6]   T. A. Gant, J. Shen, J. R. Flemish, L. Fotiadis, M. Dutta, *Appl.Phys. Lett* **60**, 1453, 1992

RF plasma deposition of thin amorphous silicon carbide films using a combination of silan and methane

J. Huran[1*], I. Hotový[2], J. Pezoltd[3], N. I. Balalykin[4] and A.P. Kobzev[4]

[1]Institute of Electrical Engineering, Slovak Academy of Sciences, Dúbravská cesta 9, Bratislava, 841 04, Slovakia

[2]Faculty of Electrical Engineering and Information Technology, Slovak University of Technology, Ilkovičova 3, Bratislava, 812 19, Slovakia

[3]Zentrum für Mikro und Nanotechnologien, TU Ilmenau, Germany

[4]Joint Institute for Nuclear Research, Dubna, 141980, Moscow region, Russia

e-mail: huran@savba.sk

A capacitivelly coupled plasma reactor was used for PECVD technology, where both silan and methane were introduced into the plasma reactor through the shower head. The concentration of species in the SiC films was determined by RBS and ERD. Chemical compositions were analyzed by IR spectroscopy. Film morphology was assessed by AFM. The RBS results showed the main concentrations of Si and C in the films. The concentration of hydrogen was approximately 20 at.%. The films contain a small amount of oxygen and nitrogen. IR results showed the presence of Si-C, Si-O, Si-N, Si-H, N-H, C-H, C-N specific bonds. The AFM micrographs revealed the film surface smooth and compact.

1. Introduction

The next generation of electronic devices may benefit from the development of alternatives to silicon and gallium arsenide based devices. The demands of higher power, radiation hardness, operating frequency, temperature and speed are potentially met by silicon carbide with additional attractive properties such as wide band gap, high breakdown field, high electron saturation velocity and physical strength [1]. In addition to high-temperature applications, SiC has potential for use in high-power and high-frequency [2]. However, SiC has several advantages over other wide-band gap semiconductors at the present time including commercial availability of substrates, known device processing techniques, and the ability to grow a thermal oxide for use as masks in processing, device pasivation layers, and gate dielectrics. Furthermore, SiC can also be used as a thin buffer layer for the growth of diamond films on silicon substrates [3]. For example, a-$Si_{1-x}C_x$:H was used as a wide window material to enhance the conversion efficiency of amorphous solar cell. The significance of this material follows from the fact that its electrical and optical properties can be controlled by varying the carbon, silicon and hydrogen composition of the film. PECVD technique offers an attractive opportunity to fabricate amorphous hydrogenated N-doped SiC films at intermediate substrate temperatures and it provides high quality films with good adhesion, good coverage of complicated substrate shapes and high deposition rate [4]. Recently, Si-rich a-SiC_x:H films have attracted new attention in the photovoltaic community, since this material has shown excellent electronic surface passivation of c-Si comparable with thermal SiO_2 and low temperature amorphous silicon nitride (a-SiN_x) passivation [5].

In this contribution the attention has been focused on the properties of a-$Si_{1-x}C_x$:H films prepared by the plasma enhanced chemical vapour deposition (PECVD) of silan SiH_4 and methane CH_4 as a function of the deposition temperature. The properties were investigated by RBS, ERD, IR and AFM measurement techniques.

1-4244-0396-0/06/$25.00 ©2006 IEEE

2. Experimental

The silan and methane were introduced into capacitively coupled plasma reactor through the shower head, which is also an upper electrode with 12 cm diameter. Both gases were flown vertically toward the substrate on bottom electrode. A n-type silicon wafer with resistivity 2-7 Ωcm and (111) orientation was used as the substrate for the a-SiC:H films. Prior to deposition, standard cleaning was used to remove impurities from the silicon surface, and the 5% hydrofluoric acid was used to remove the native oxide on the wafer surface. The wafer was then rinsed in deionized water and dried in nitrogen ambient. The flow rates of SiH_4 and CH_4 gases were 10 sccm and 40 sccm, respectively. The deposition temperatures were 250, 350 and 450°C respectively. The concentration of species in the SiC films was determined by Rutherford backscattering spectrometry (RBS). Chemical compositions were analyzed by infrared spectroscopy. The IR spectra were measured from 4000 to 400 cm^{-1}. The hydrogen concentration was determined by the elastic recoil detection (ERD) method. For this purpose the $^4He^+$ ion beam from a Van de Graaff accelerator at JINR Dubna was applied. The energy of 2.4 MeV was chosen. The target was tilted at an angle of 15° with respect to the beam direction and the recoiled protons were measured in forward direction at an angle of 30°. Film morphology was assessed by Atomic Force Microscopy.

3. Results and discusion

The IR spectra in Fig.1 revealed the main absorption region between 400 and 2000 cm^{-1}. From the spectra we could determine the following vibration frequencies:520 to 530 and 1090

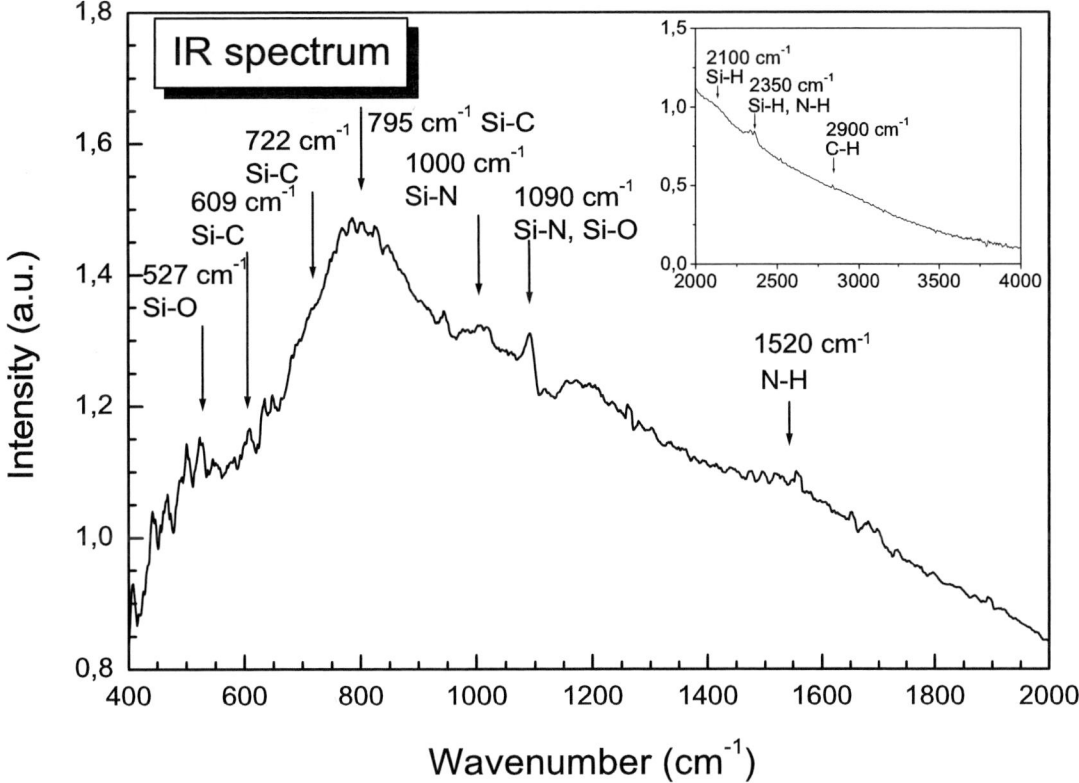

Fig. 1. The IR spectra - the main absorption region between 400 and 2000 cm^{-1} for all samples A1, A2 and A3.

to 1093 can be related to SiO_2 the higher wavenumber can be also related to Si-N bonds close to Si_3N_4 bonds; 944 to 950 and 1000 to 1005: they can be roughly related to Si-N bonds from Si_3N_4; 1260 to 1265: they can be related to C-N bonds; 606 to 610: it can be related to Si-C localised vibration normally found in Si due to C in single crystalline Si. The main phonon or vibration frequency is related to SiC and have the following characteristics determined from the reflection spectra: sample A1: center position 795 cm^{-1}, width 178 cm^{-1}; sample A2: center position 795 cm^{-1}, width 45 cm^{-1}; sample A3: center position 804 cm^{-1}, width 42 cm^{-1}. The non stressed phonon position of cubic SiC is 796 cm^{-1}. In amorphous material a shift to higher values indicate on recrystallisation or nucleation of small crystallites. The IR spectra in right upper Fig.1 show range 2000 to 4000 cm^{-1} were are interesting band approximately 2100 cm^{-1} assigned to Si-H stretching vibrations and the C-H stretching band approximately 2900 cm^{-1}. The band approximately 2300-2400 cm^{-1} assigned to Si-H, N-H bonds. Figure 2 show RBS spectra of three samples A1, A2 and A3 with different deposition conditions of the deposited amorphous silicon carbide films. After modelling, we can show from calculated results the presence of small amounts of oxygen and nitrogen while the concentrations of hydrogen in the SiC films are approximately 20 at.%. Both of the elements nitrogen and oxygen represent 5-7 at.% in the SiC film. The SiC films contained also other species which were under the detection limit of RBS method. In the case of samples A1 and A2 the concentration of silicon and carbon are 30 and 37 at.% respectively. The concentrations of Si and C in sample A3 are 32 and 35 at.% respectively. There was no evidence of the substrate

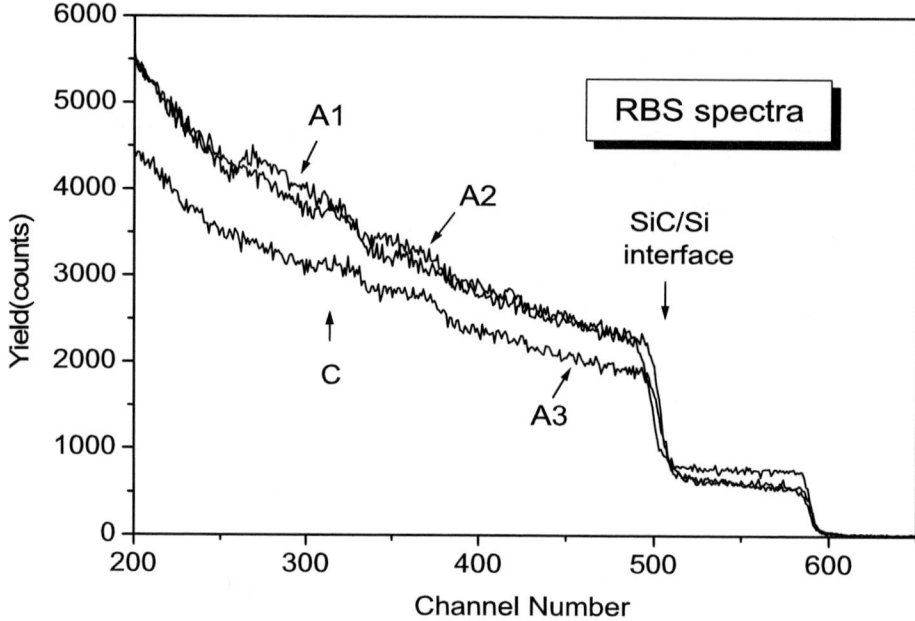

Fig. 2. RBS spectra of SiC films deposited onto a silicon substrate for 2 MeV alfa particles detected at scattering angle of 170°. The spectra are for samples A1 (thickness of film 265 nm), A2 (275 nm) and A3 (314 nm).

mixing into the films. The AFM micrographs (Figure 3.) of the SiC films prepared by PECVD reveal that the film surface is rather smooth and compact. The mean roughness R and the root mean square RMS of the samples, as deduced from the AFM analysis, are similar.

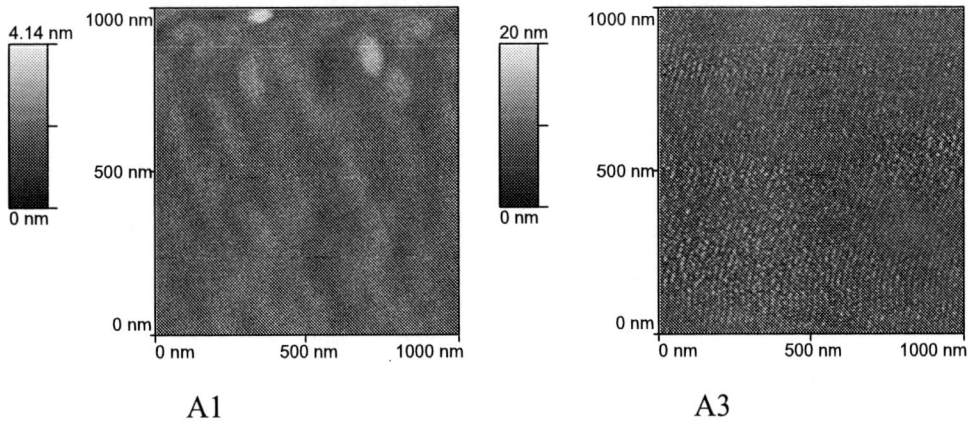

A1   A3

Fig. 3. AFM micrographs showing the surface (1000 x 1000 nm^2) of samples A1 and A3.

4. Conclusion

The experimental results obtained from this work can be summarized as follows. The RBS results showed that the concentration of Si and C in the films dependends a little on the deposition temperature. The concentration of hydrogen was approximately 20 at. %. The films contain a small amount of oxygen and nitrogen. IR results showed the presence of Si-C, Si-O, Si-N, Si-H, N-H, C-H, C-N specific bonds. The AFM micrographs revealed that the film surface is rather smooth and compact.

Acknowledgements

The work was supported by the Scientific Grant Agency of the Ministry of Education of the Slovak Republic and Slovak Academy of Sciences, No. 1/4151/26, 1/3095/26, CENG Slovak Academy of Sciences and by Science and Technology Assistance Agency under the contract No. APVV-99-PO6305.

References

[1] H. Morkoc, S. Strite, G.B. Gao, M.E. Lin, B. Sverdlov and M. Burns, J. Appl. Phys. **76**, 1363, 1994.
[2] R.J. Trew, J.B. Yan, P.M. Mock , Proc. IEEE **79**, 598, 1991.
[3] E.G. Wang, Physica **B185,** 85, 1993.
[4] H. Sachdev, P. Sheid, Diamond Relat. Materials **10**, 1160, 2001.
[5] M. Vetter, C. Voz, R. Ferre, I. Martin, A. Orpella, J. Puigdollers, J. Andreu and R. Alcubila, Thin Solid Films **511-512,** 290, 2006.

Photoluminescence and electrical characterization of transparent Eu and Pd-doped TiO$_2$ thin films

J. Domaradzki, A. Borkowska and D. Kaczmarek

Faculty of Microsystem Electronics and Photonics, Wroclaw University of Technology,
Janiszewskiego 11/17, 50-372 Wroclaw, Poland
e-mail: jaroslaw.domaradzki@pwr.wroc.pl

In this work, optical and electrical characterization of transparent Eu and Pd-doped TiO$_2$ thin films have been presented. Thin films were deposited by low pressure hot target reactive magnetron sputtering form metallic Ti-Pd-Eu mosaic target on silicon. It has been shown that incorporation of Pd and Eu dopants into TiO$_2$ matrix could modify its properties to obtain electrically and optically active oxide-semiconductor with the electron-type (n) of electrical conduction at room temperature. Pd dopant changes the electrical properties of TiO$_2$ from dielectric oxide to conducting oxide. Eu dopant results in enhanced optical activity of Pd-doped TiO$_2$ thin films in ultraviolet range.

1. Introduction

Luminescent properties of lanthanide-doped TiO$_2$ thin films are widely used due to their technological applications (i.a. light emitting devices, flat panel displays) [1, 2]. Now, the possibility of using of rare earth elements photoluminescence for increasing of light conversion in short wavelength range is of great importance.

Connecting oxides with relevant modified properties with conventional microelectronics materials (i.e. silicon) allows fabrication different electronic devices [3-5]. The wide forbidden band gap of TiO$_2$ limits its use in some electrical applications as a transparent conducting electrode and as an active component of transparent heterojunction-based devices.

This work presents photoluminescence and electrical characterization of Eu and Pd-doped TiO$_2$ thin films prepared by magnetron sputtering on silicon. It has been shown that incorporation of Pd and Eu dopants into TiO$_2$ matrix could modify its properties to obtain electrically and optically active oxide-semiconductor with specified type of electrical conduction at room temperature.

2. Experimental

Thin films were deposited by low pressure hot target magnetron sputtering (LP HTRS) [6] from metallic Ti-Pd-Eu mosaic target on conventional silicon wafer.

Optical characterization of Eu and Pd-doped TiO$_2$ thin films was made by means of optical transmission method and photoluminescence spectroscopy. Optical transmission measurements were performed in the spectral range form 200 (nm) to 1000 (nm) using Ocean Optics HR4000 spectrometer. Samples were illuminated with white light from a halogen lamp (100 (W)) at normal incidence. For photoluminescence measurements UV argon laser with the excitation wavelength of 302 nm has been used. The PL signal has been collected by Ocean Optics HR4000 spectrometer, as well.

1-4244-0396-0/06/$25.00 ©2006 IEEE   258

For electrical characterization of thin films of Ti-Pd-Eu oxides two parallel Ag/Ti10W90 metal electrodes were evaporated through the metallic mask into the thin films. Electrical measurements were performed in a temperature range from 300 (K) to 700 (K).

3. Results

Transmission characteristic of TiO_2: (Eu, Pd) thin film has been presented in Fig. 1a. The fundamental absorption edge ($\lambda_{cuttoff}$) was found to be about 450 (nm) (Fig. 1a). Incorporation of Eu and Pd dopants into TiO_2 shifts the position of its $\lambda_{cutt\,off}$ (320 (nm) [6]) toward the longer wavelength region (Fig. 1a). The thickness of prepared thin film, determined from transmission characteristic was about 1200 (nm). The optical band gap energy (E_g^{opt}) for the allowed indirect transition together with the Urbach energy (E_u) have been calculated from the envelope method [7, 8]. The parameters are summarized in Table 1.

Photoluminescence of TiO_2: (Eu, Pd) thin film on silicon has been presented in Fig. 1b. PL spectrum, measured at room temperature, indicates the most intense peaks around 617 (nm) and 700 (nm), due to the standard Eu^{3+} emission (Fig. 1b). Light conversion in short wavelength range can be observed (excitation wavelength 302 (nm)). Dominating transition corresponds to $^5D_0$-$^7F_2$ transition at ~617 (nm) and $^5D_0$-$^7F_4$ transition at ~700 (nm). These wavelengths correspond to working range of typical silicon devices, what opens the possibility of application of TiO_2: (Eu, Pd) thin films to form junction-based devices on silicon substrates.

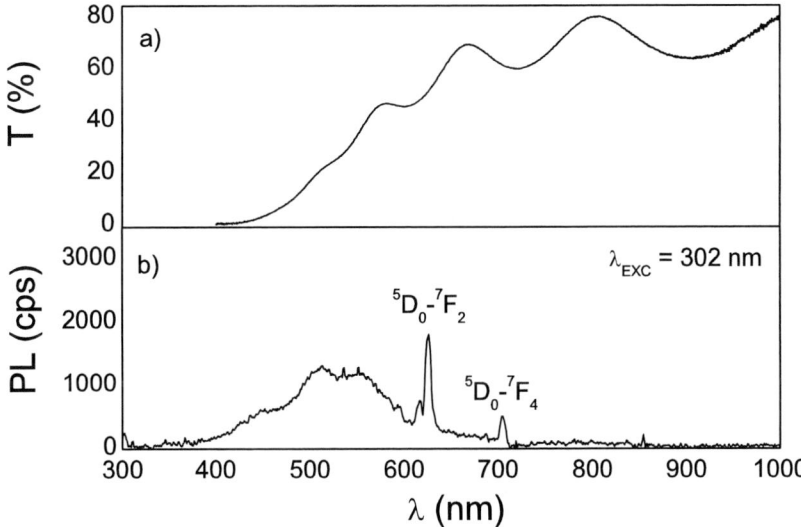

Fig. 1. Transmission (a) and photoluminescence (b) spectra of TiO_2: (Eu, Pd) thin film.

Figure 2 presents dependence of d.c. resistivity (ρ_{dc}) (Fig. 2a) and Seebeck coefficient (S) (Fig. 2b) vs. temperature of TiO_2: (Eu, Pd) thin film. Measurement of d.c. electrical resistivity was done. Experimental data were found to fit straight line and thus the activation energy W_ρ was determined to be of 0.18 (eV) (Fig. 2a).

The temperature dependent characteristic of Seebeck coefficient (Fig. 2b) enabled the determination of the type of electric conduction for examined thin film. Negative sign of Seebeck coefficient indicates the electron-type (n) conduction of Eu and Pd-doped TiO_2 thin film. Our previous studies have shown that doping of TiO_2, as a base oxide, with Pd results in the change of its electrical conductivity from oxide non-conductor to oxide-semiconductor and introduces acceptor levels in the forbidden band gap of this TiO_2: (Pd) oxide-semiconductor [9]. Additional incorporation of Eu results in the change of the type of electrical conduction of prepared thin film from p to n type, which is similar like in the case of vanadium doping [10]. The conduction process could be attributed to the presence of nanocrystalline metallic Pd inclusions randomly distributed in the thin film. They may form discontinuous paths for charge carriers (electrons) excited upon external electric field [10].

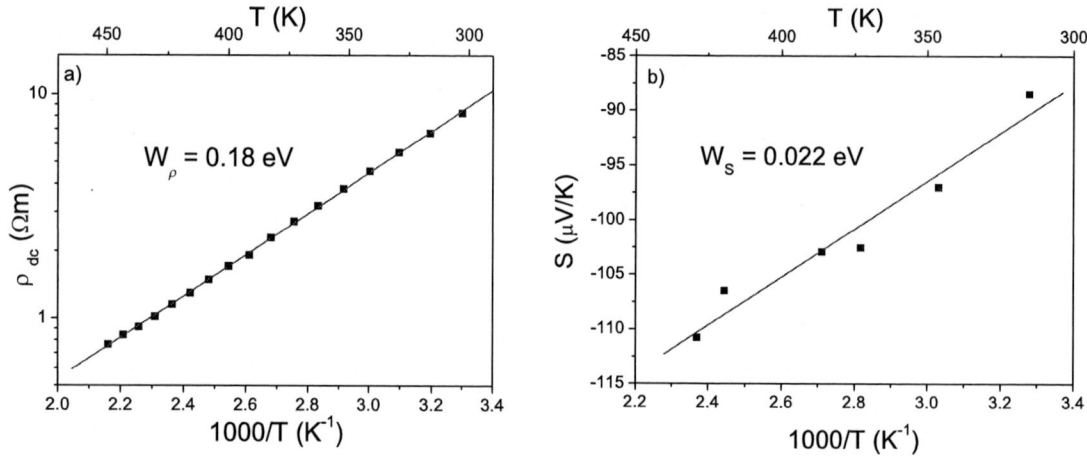

Fig. 2. Dependence of: a) d.c. resistivity (ρ_{dc}) and b) Seebeck coefficient (S) vs. temperature of TiO_2: (Eu, Pd) thin film.

Also, from the slope of S(T) characteristic the thermal activation energy W_S was estimated to be 0.022 (eV).

An order discrepancy between determined values of activation energies ($W_S = 0.022$ (eV) and $W_\rho = 0.18$ (eV)) (Table 1) reflects complex mechanism of electrical conduction in prepared TiO_2: (Eu, Pd) thin film [11]. Estimated values could be affected by all thermally activated processes of charge transport occurring in examined films and thus, the distinction of W_S and W_ρ on the basis of plotted curves (Fig. 2) is difficult.

Table 1. Selected parameters of TiO_2: (Eu, Pd) thin film (1200 (nm) thick).

D (nm)	$\lambda_{cuttoff}$ (nm)	E_g^{opt} (eV)	E_u (eV)	W_ρ (eV)	W_S (eV)
9.8	450	1.71	0.38	0.18	0.022

4. Conclusions

Optical and electrical characterization of Ti-Pd-Eu thin films of oxides prepared by low pressure hot target magnetron sputtering from metallic mosaic target has been presented.

It was shown that the fundamental absorption edge of TiO_2 was shifted through the longer wavelength range due to the Pd and Eu doping. PL spectrum of TiO_2: (Eu, Pd) thin film shows a strong red luminescence from standard Eu^{3+} emission lines. This emission is located in the range, where prepared thin film is transparent (Fig. 1).

From electrical temperature dependent characteristics, the activation energies ($W_\rho = 0.18$ (eV) and $W_S = 0.022$ (eV)) were calculated. Estimated values of W_s and W_ρ reflect complex mechanism of electric conduction in examined thin film. The sign of the Seebeck coefficient indicates n-type electric conduction in TiO_2: (Eu, Pd) thin film.

Semiconducting properties and strong red luminescence, corresponds to working range of typical silicon devices makes possible the application of manufactured Eu and Pd-doped TiO_2 thin films for future wavelength-selective devices based on transparent oxide semiconductors.

Acknowledgement

This work was financed by a grant given by the Polish State Committee for Science Research (KBN) for the years 2005-2007.

Authors would like to thank E.L. Prociow from our Faculty for providing the necessary sample preparation and A. Podhorodecki from Institute of Physics, Wroclaw University of Technology, Poland for providing experimental data from optical measurements.

References

[1] Ch. Jia, E. Xie, A. Peng, R. Jiang, F. Ye, H. Lin, and T. Xu, *Thin Solid Films* **496**, 555, 2006.

[2] Q.G. Zeng, Z.J. Ding, and Z.M. Zhang, *Journal of Luminescence* **118**, 301, 2006.

[3] K. Tonooka, H. Bando, and Y. Aiura, *Thin Solid Films* **445**, 327, 2003.

[4] B. O'Regan, and M. Gratzel, *Nature* **353**, 737, 1991.

[5] P.F. Carcia, and R.S. McLean, *Transparent Oxide Semiconductor Thin Films Transistors*, PCT U.S. Patent WO 2004/034449 A2, April 22, 2004.

[6] J. Domaradzki, D. Kaczmarek, E.L. Prociow, A. Borkowska, D. Schmeisser, and G. Beuckert, *Thin Solid Films* **513** (1-2), 269, 2006.

[7] M. Pal, Y. Tsujigami, A. Yoshikado, and H. Sakata, *Phys. Status Solidi A* **182**, 727, 2000.

[8] J. Tauc, R. Grirorovici, and A. Vancu, *Phys. Status Solidi A*, 627, 1966.

[9] E. Prociow, J. Domaradzki, and D. Kaczmarek, in *Proceedings of the ASDAM'02 Conference*, Smolenice, Slovakia, 2002, p.51.

[10] J. Domaradzki, *Thin Solid Films* **497** (1-2), 245, 2006.

[11] P. Nagels, *Electronic Transport in Amorphous Semiconductors, in: Amorphous semiconductors*, M.H. Brodsky (Ed.), Spinger-Verlag, New York NY, 1979.

Structural, optical and electrical properties of nanocrystalline TiO$_2$ – HfO$_2$ thin films

J. Domaradzki and D. Kaczmarek

Faculty of Microsystem Electronics and Photonics, Wroclaw University of Technology,
Janiszewskiego 11/17, 50-372 Wroclaw, Poland
e-mail: jaroslaw.domaradzki@pwr.wroc.pl

In the present work the properties of mixed Hf$_x$Ti$_{1-x}$O$_2$ solid solution thin films in metal-oxide-semiconductor (MOS) configuration have been studied. Thin films, were grown on monocrystalline silicon substrates using the low pressure hot target reactive sputtering from the Ti:Hf mosaic target. From optical transmission measurements the 3.42 (eV) bandgap of the thin films has been estimated. Electrical characterization of the prepared MOS structures shows classical capacitance- and current-voltage behaviors indicating medium-k gate oxide properties with the low leakage current.

1. Introduction

For fabrication of novel microelectronics devices, stable low leakage high permittivity dielectric thin films are required. Transition metal oxides of IIIB and IVA group elements seem to be a reliable choice to replace SiO$_2$ as a gate oxide in the standard MOS technology. So far, as a gate insulators the typical dioxides (MO$_2$), where M = Ti, Ru, Hf, were commonly applied. On the other hand preparation of dielectrics with large band offset (HfO$_2$) and very high permittivity (TiO$_2$) would give a reasonable solution for fabrication of thin films of an insulator with intermediate properties [1].

In the present work the properties of mixed Hf$_x$Ti$_{1-x}$O$_2$ solid solution thin films have been studied. It have been shown that preparation of mixed oxides, after additional post annealing, gives reasonable solution for preparation of gate oxides with intermediate electrical and optical properties.

2. Experimental

The HfTiO$_4$ thin films, were grown on monocrystalline (100) oriented silicon substrates using the low pressure hot target reactive sputtering [2] from the Ti:Hf mosaic target. The Ti-Hf composition of the thin film was assured by co-sputtering of Hf foil (Aldrich, USA, purity 99.5 %, 0.5 mm thick) placed on the surface of a titanium (purity 99.99 % and 3 mm thick) disc (mosaic target) [3]. Thin films of oxides were deposited in reactive atmosphere using a high purity (99.999 %) oxygen gas. The rate of thin film deposition was kept at about 0.1 nm/s. That assured almost epitaxial conditions of the layer growth. The ratio of components in the whole material coming from the mosaic target, estimated from measuring the deposition rate ratio of the two component oxides was at the amount of Ti:Hf = 46.7 at. %:53.3 at. %. The thickness of the manufactured films measured using standard Fizeau method was 485 nm. After deposition, selected samples were additionally annealed at a temperature of 1000 K for 4 hours in air ambient.

Structure and chemical composition were verified using X-ray diffraction and X-ray Photoelectron Spectroscopy measurements. Additionally, optical transmission measurements

1-4244-0396-0/06/$25.00 ©2006 IEEE

have been performed and band properties (i.e. the width of optical bandgap, the Urbach energy and the width of extended tails) have been estimated in a conventional way. For electrical investigations $Ag/Ti_{10}W_{90}$ circular dots with 3 mm diameter were evaporated through a metallic mask onto the thin film. Electrical properties of prepared metal/oxide/semiconductor structures were examined by means of C-V and I-V measurements. C-V Experiments have been carried out for selected frequency of measuring signal using HP 4192 impedance analyser working in parallel circuit mode. I-V experiments have been done using Keithley 6517A electrometer at room temperature.

3. Results

X-ray diffraction study reveal orthorhombic $HfTiO_4$ phase with the grain size in the range of few tenths of nanometers [3, 4]. Heat treatment enhances crystallinity and causes a stress relief in the thin films.

Chemical composition of the prepared samples was verified by X-ray Photoelectron Spectroscopy (XPS) measurement. The obtained XPS intensities give a composition of x=0.48, y=0.52 and x=0.38, y=0.62 for the as deposited and the annealed $Hf_xTi_yO_4$ thin films, respectively [3].

From optical absorption characteristics the value of the optical band gap (E_g^{opt} = 3.420 (eV) $\pm$ 0.042 (eV) and E_g^{opt} = 3.418 (eV) $\pm$ 0.038 (eV)) for the allowed indirect transition in the as deposited and annealed thin films have been estimated. The values are much lower than for HfO_2 (5.1 (eV) $\div$ 5.5 (eV)) and indicates that resulting bandgap of $HfTiO_4$ is dominated by the oxide with smaller bandgap, i.e. TiO_2 (3.02 (eV)- rutile). The little increase of the bandgap, as compare to pure TiO_2, results from the increasing of the Ti 3d level by combining with Hf 5d metal [5].

The results of structural and optical investigations have been collected in Table 1.

Table 1. Composition, mean grain size determined from XRD data and optical parameters for as deposited and annealed $HfTiO_4$ thin films.

sample	$Hf_xTi_yO_4$ composition		mean grain size (nm)	E_g (eV)	E_u (eV)	n (at 550 nm)
	x	y				
as-deposited	0.48	0.38	8.75	3.420 ±0.042	0.265 ±0.003	2.08
annealed at 1070 K for 4 hours	0.52	0.62	35	3.418 ±0.038	0.217 ±0.005	2.2

Current to voltage characteristics have been presented in semi logarithmic scale in Fig. 1. The run of measured curves is almost symmetrical in both polarization direction. Horizontal line tangent to the experimental curves at the accumulation and inversion current range indicate a very low gate leakage current. No significant influence of additional post annealing could be seen from this I-V measurement.

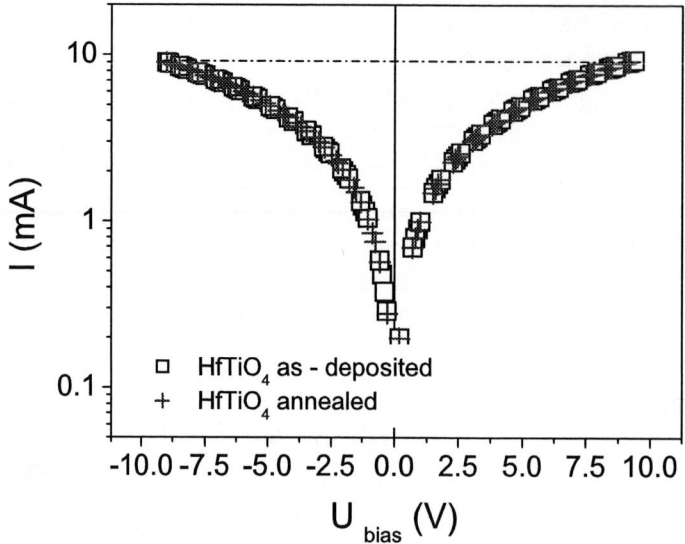

Fig. 1. I-V curves for the MOS structures with thin HfTiO$_4$ thin films. Acceleration voltage 1 (V/s).

Results of capacitance – voltage dependent measurements performed at different frequency of measuring signal for the structures with final annealed thin films have been presented in Fig. 2.

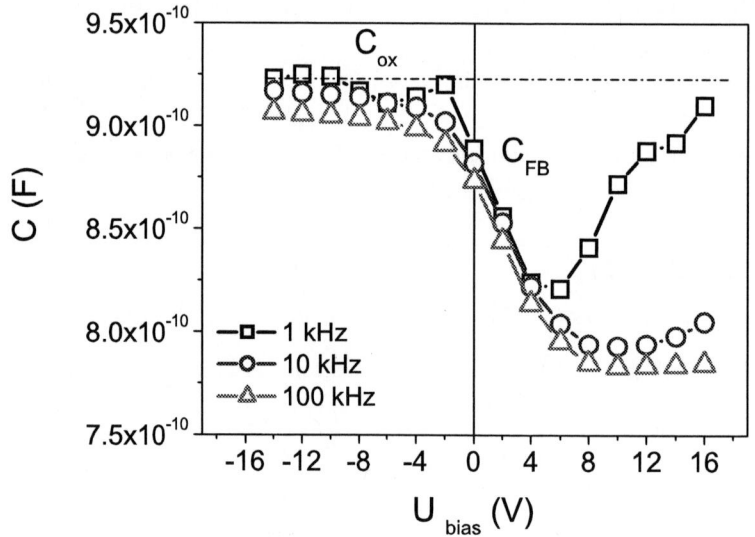

Fig. 2. Capacitance (C) versus bias voltage (U) for the MOS structures with the annealed HfTiO$_4$ thin films.

Experimental characteristics were similar to C-V dependences, observed in the case of classical MOS structures and three typical ranges, i.e. accumulation, depletion and inversion could be easily recognized (Fig. 2). For negatively biased structure, negative electron charge at the gate is balanced by positive hole charge accumulated near the surface of the (p-type) semiconductor. In an ideal MOS system capacitance C measured in this accumulation state is equal to oxide capacitance C$_{ox}$. Switching the bias into positive direction makes semiconductor surface depleted from the holes, so that the additional capacitance of space charge layer C_{sc} is serially connected to C_{ox}. Further positive gate biasing makes that at low frequency conditions inversion of the conduction type occurs. In strong inversion conditions, potential changes at the gate are determined by the presence of electrons at the semiconductor – metal interface. Thus, the total measured capacitance, similar to the accumulation state, approach to the C_{ox} (dotted line in Fig. 2). Taking into account the geometry of the structure and the oxide capacitance (evaluated from the equivalent circuit model [6]) the effective relative permittivity $\varepsilon_r \approx 19$ of thin $HfTiO_4$ oxide has been estimated, as well.

3. Conclusions

In the paper it has been shown that combining of the high permittivity TiO_2 with wide band gap HfO_2 gives reasonable solution for preparation of medium-k gate oxides.

Acknowledgement

This work was financed by a grant given by the Polish State Committee for Science Research (KBN) for the years 2005-2007. The authors would like to thank E.L. Prociow for necessary sample preparation

References

[1] F. Chen, X. Bin, C. Hella, X. Shi, W.L. Gladfelter, and S.A. Campbell, *Microelectron. Eng.* **72**, 263, 2004.

[2] J. Domaradzki, D. Kaczmarek, E.L. Prociow, A. Borkowska, D. Schmeisser, and G. Beuckert, *Thin Solid Films* **513** (1-2), 269, 2006.

[3] J. Domaradzki, D. Kaczmarek, E.L. Prociow, A. Borkowska, R. Kudrawiec, J. Misiewicz, D. Schmeisser, and G. Beuckert, *Surf. Coat. Technol.* **200**, 6283, 2006.

[4] J. Domaradzki, A. Borkowska, D. Kaczmarek, E.L. Prociow, R. Wasielewski, and A. Ciszewski, *Opt. Appl.* **35**, 431, 2005.

[5] G. Lucovsky, J. G. Hong, C. C. Fulton, Y. Zou, R. J. Nemanich and H. Ade, J. Vac. Sci. Technol. B **22**, 2132, 2004.

[6] J. Domaradzki, D. Kaczmarek, A. Borkowska, M. Wolcyrz and B. Paszkiewicz, Phys. Stat. Sol. (a) **9**, 2215, 2006.

Microelectromagnetic matrix for local assembling of magnetic nanoparticles

S. Luby[a], L. Chitu[a], E. Majkova[a], R. Senderak[a], I. Kostic[b], P. Hrkut[b], L. Matay[b], S. Hascik[c], T. Lalinsky[c], I. Capek[d], A. Satka[e]

[a]Institute of Physics, Slovak Acad. Sci., 84511 Bratislava, Slovak Republic
[b]Institute of Informatics, Slovak Acad. Sci., 84507 Bratislava, Slovak Republic
[c]Institute of Electrical Engineering, Slovak Acad. Sci., 84104 Bratislava, Slovak Republic
[d]Polymere Institute, Slovak Acad. Sci., 84236 Bratislava, Slovak Republic
[e]International Laser Center and Faculty of Electrical Engineering and Informatics, Slovak Univ. Technol., 81219 Bratislava, Slovak Republic

Microelectromagnetic matrix fabricated by lithography patterning on Si chip for the manipulation and assembling of nanoparticles (NP) by local magnetic field created by current loaded thin film conductors is described. Co, Fe_3O_4 and $CoFe_2O_4$ NPs are manipulated by matrix elements – meanders, grids and ring traps loaded at the electromigration limit of Ag conductors. The design of matrix is based on calculations of magnetic field over the elements covered by polyimide layer. Amorphous silicon is used as a manipulation surface.

1. Introduction

Nanomagnetism is at the forefront of the emerging nanotechnology era. One of its leading trends is a concept to organize surfactant-mediated nanoparticles (NPs) into regular arrays for futuristic Tbit/in^2 magnetic storage media [1]. Here the first step is a bottom-up formation of NPs by chemical reactions from precursors into a form of colloid solution. Then NPs must be assembled into regular two – dimensional arrays with hexagonal symmetry as a rule. Arrays are created by deposition of NPs onto substrates by drying of drops or by spin coating. Ordering can be improved by deposition in an external magnetic field. Then it results from the magnetic dipole and van der Waals interactions and wetting properties of interface [2]. A disadvantage of the external magnetic field is a lack of its focusing onto small areas. Therefore, in this work attention is paid to design and fabrication of a microelectromagnetic matrix, which will form various local patterns of magnetic field. The basic structure of matrix is an array of current-carrying conductors embedded in the substrate. This way the magnetic field could be directed to the certain areas with the resolution of the respective lithography and the matrix can be used for local assembling of NPs. Various external driving forces have been applied to manipulate NPs up to now. Examples are gradient field of magnetic nanorods [3] and chips with current carrying nonmagnetic conductors [4]. The last approach was used for the manipulation of neutral atoms [5]. The structures described in [4, 5] are studied and applied in this work.

2. Calculations and experiments

We have used: a) spherical Co NPs prepared by thermolysis of $Co_2(CO)_8$ with a mixture of oleic acid and oleyl amine as surfactant. The average radius of NPs is 5.7 ± 0.05 nm, surfactant is 1.8 nm thick [6]; b) spherical Fe_3O_4 and $CoFe_2O_4$ NPs synthesized by high temperature (up to 265°C) solution phase reaction of metal acetylacetonates with 1,2 hexadecanediol, oleic acid and oleyl amine in phenyl ether. The average radius was 3.2 ± 0.3 nm for Fe_3O_4 and 3.8 ± 0.3 nm for $CoFe_2O_4$. Also here the thickness of surfactant was 1.8 nm [7].

Magnetic field of our NPs was computed as follows. We assume that NP is homogeneously magnetized. The magnetic field of sphere can be characterized by a dipole m in its center. In the position A (Fig. 1) its components in the reference system with rotational symmetry are [8] $H_\rho = 3\,mz\rho/r^5$ and $H_z = m(2z^2 - \rho^2)/r^5$ where $r^2 = \rho^2 + z^2$. H_ρ has its maximum $0.86\,m/z3$ at $\rho = z/2$. H_z changes its direction at $2z^2 = \rho^2$. The magnetic dipole $m = M_sV_o$, where $V_o = 775$ nm^3 is the volume of our NP with a = 5.7 nm. For magnetization we take the value of $M_s = 89.4$ emu/g, which was reported for Co NPs with diameter of 12.5 nm [9]. The components of field in the position A, where z = 5.7 nm + 1.8 nm corresponds to the radius of NP covered by surfactant and $\rho = z$, are $H_\rho = 87$ Oe and $H_z = 29$ Oe.

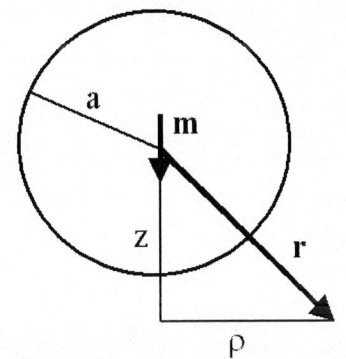

Fig. 1 Configuration for calculation of magnetic field of NP with radius a in position A.

Next, magnetic field of matrix conductor loaded by current I was calculated using the equation $\int \mathbf{H} \cdot d\mathbf{s} = 1$, where s is the closed trajectory embracing the current. To estimate this approach we have computed the field of a published spin valve with the magnetic bias [10]. The current carrying conductor of the spin valve was decomposed into 145 square segments and the fields of the respective segments were numerically summarized at the distance of 0.54 μm from the conductor centre, where it was measured [10]. For the loading current of 130 mA the calculated and measured fields are 13.5 Oe and 22 Oe, respectively. The accuracy of this procedure is about 60 %. The mentioned field was obtained at the current density of 5.3×10^5 A/cm^2. With Cu, Ag or W conductors the current density could be increased by two orders of magnitude. Therefore, the fields of 200 – 2000 Oe could be created in the described configuration. The field linearly decreases with the distance from the conductor. This distance should not be larger than 1 or 2 μm, as can be concluded from our calculations.

4.   Fabrication of matrix

Si wafers (3") and standard photolithography or electron beam lithography (EBL) were used. The lift-off PMMA resist mask was formed using EBL. Alternatively Microposit resist mask was made by photolithography. Both masks were 0.5 μm thick. Onto them Cr (12 nm)/Ag (450 nm)/Au (20 nm) trilayer was e – gun deposited and patterned (Fig. 2). Ag was chosen because of its low resistivity and high electromigration resistance.

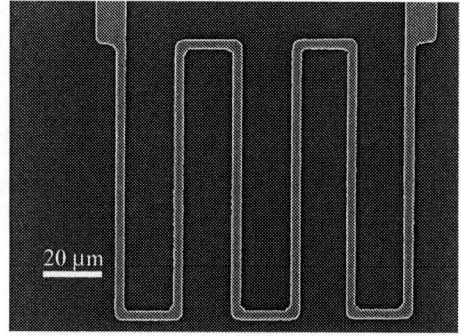 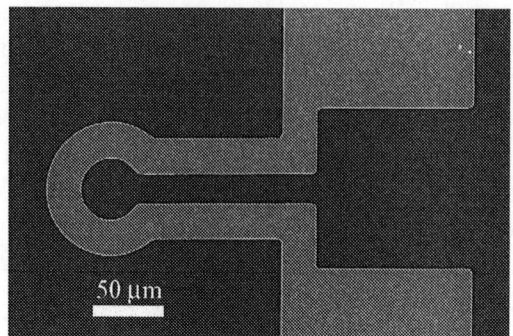

Fig. 2 Details of Cr/Ag/Au meander (left) and trap (right) patterned by lift-off.

The next step was planarization of the matrix surface by 2 or 2.5 μm thick polyimide (PI 2610) layer which was etched by RIE. Degree of planarization was expressed by a ratio DOP = 1 − t_s/t_m, where t_m is the thickness of Cr/Ag/Au line and t_s the height of PI step over the line edge. For ideal planar surface DOP = 1. Our best result was DOP = 0.88 corresponding to the step of 60 nm. Then 40 nm a-Si surface layer was rf sputtered. Bonding windows were opened and Ti/Au (20/200 nm) contact layers were deposited by e-gun. Individual matrix-chips (3.8 x 4.1 mm^2) were Au bonded into ceramic packages. The distance between a-Si surface and the surface of conductor was 640 nm.

Magnetic field was created using meanders, grids and ring traps. The widths of meanders and grid bars were 2, 4 and 6 μm, ring trap conductors were 20 μm wide. On every chip there are 4 meanders, 8 grids and 4 ring traps with internal diameters of 20, 40 or 60 μm. All elements were formed in the interruptions of 120 μm wide peripheral bus bar.

Now we can calculate e.g. the field at the surface of matrix in the middle over the ring trap conductor 20 μm wide at the loading current 100 mA corresponding to the density of 1.04 x 10^6 A/cm^2. The field is 62 Oe. The field could be increased 10 times by applying the current density of 10^7 A/cm^2, which is quite reliable regarding the standard electromigration limit of Ag. These fields should be sufficient for the manipulation of particles in our solution with the viscosity of 1 cP. However, for narrow conductors of meanders and grids (2 μm) we are at the limit even for 100 mA. For this conductors the field at the surface of matrix over the edge of the conductor is 75 Oe. At the distance of 2 μm from the edge it is 33 Oe. Thus, the gradient of the field over the edge is about 20 Oe/μm. The gradient is comparable with that mentioned in [4] (50 G/μm for conductors 10 μm wide). It can be further increased by miniaturization of the conductor size. For nanorods with diameter of 80 nm the gradient at the end of it is more than 10^4 G/μm [3]. Further, it was calculated that the heating of the Si matrix surface with the conductors loaded at 10^7 A/cm^2 is ≤ 10 K. Here the substrate SiO$_2$ layer was 0.1 μm thick. The heating is low because Si is a perfect heat sink.

5. Manipulation and conclusion

Manipulations and assembling of NPs were performed at current loading of 50 or 100 mA. We have shown that NPs agglomerate in the center of ring traps (Fig. 3). Here we have higher field gradient than on the surface of conductor. However, the absolute value of the field is quite low so far, thus the auxiliary effect of the field upon regular assembling is missing (Fig. 3, cf. [7]). Another example is a toroid-like pattern in Fig. 3. Over the edge of the straight conductors of the trap (not shown here) agglomeration of Co NPs is observed because of high local field gradient. Also other configurations due to interactions of particles with field were observed. Thus, the manipulation and assembling of NPs by microelectromagnetic matrix has been demonstrated. Further progress can be achieved by: decreasing the size of conductors [3], taking into account screening and demagnetization effects [11] and short pulse increase of the current densities up to 10^8 A/cm^2 under cooling of the matrix chip [12].

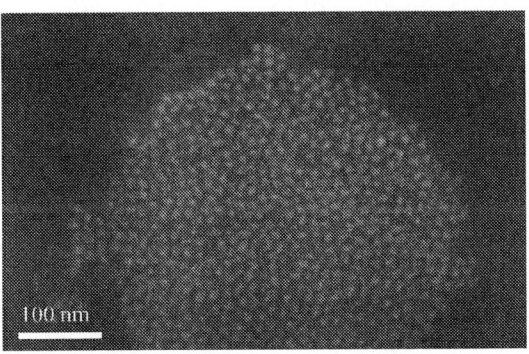 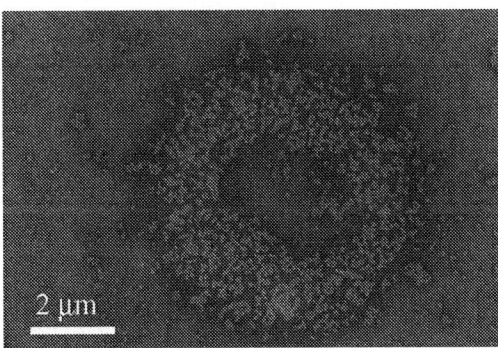

Fig. 3. Agglomeration of Co NPs in the center of the ring trap with internal diameter of 40 μm (left) and toroid-like pattern between straight conductors of the ring trap being 20 μm apart (right) at the loading current of 100 mA.

Acknowledgemet

The support by Science and Technology Assistance Agency (Slovakia) - grant 51-021702, by Center of Excellence SAS -Physics of Information, contract I/2/2006 and by Scientific Grant Agency VEGA, Bratislava, grants 2/4101/26, 2/6030/26 and 2/6184/26, are acknowledged.

References
[1] S. D. Bader, *Rev. Modern Phys.* **78**, 1, 2006.
[2] M. Giersig, and M. Hilgendorf, *J. Phys. D: Appl. Phys.* **32**, L1, 1999.
[3] A. R. Urbach, J. C. Love, M.G. Prentiss, and G. M. Whitesides, *J. Am. Chem. Soc.* **125**, 12704, 2003.
[4] S. C. Lee, H. Lee, and R. M. Westerwelt, *Appl. Phys. Lett.* **79**, 3308, 2001.
[5] S. Groth et al., *Appl. Phys. Lett.* **85**, 2980, 2004.
[6] M. Hilgendorff, B. Tesche and M. Giersig, *Aust. J. Chem.* **54**, 497, 2001.
[7] L. Chitu et al., *EMRS Meeting 2006*, Nice, in press.
[8] M.M. Miller, et al., *J. Magn. Magn. Mater.* **225**, 138, 2001.
[9] X. Sun, A. Gutierrez, M. J. Yacaman, X. Dong, and S. Jin, *Mat. Sci. Engn. A* **286**, 157, 2000.
[10] H. Yamane, J. Mita, and M. Kobayashi, *Jap. J. Appl. Phys.* **36**, L 1591, 1997.
[11] O. Hovorka, B. Yellen, N. Dan, and G. Friedman, *J. Appl. Phys.* **97**, 10Q306, 2005.
[12] N. Drndic, S. K. Johnson, J. H. Thywissen, M. Prentiss, and R. M. Westerwelt, *Appl. Phys. Lett.* **72**, 2906, 1998.

Nickel ohmic contact on silicon carbide

Machac P.[*], Barda B.[*], Sajdl P.[**]

[*] Department of Solid State Engineering
[**] Central Laboratories
Institute of Chemical Technology
Technická 5, 166 28 Prague 6, Czech Republic
e-mail: Petr.Machac@vscht.cz; bardab@vscht.cz; Petr.Sajdl@vscht.cz

We obtained ohmic contact structure with the best contact resistivity of 4.06×10^{-4} $\Omega\ cm^2$ and with excellent thermal stability - the metallization was stable after being tested for 10 hours at 900 °C. The as deposited structure contains oxygen and carbon. Make allowance for the composition of the species in the structure after annealing Ni and Si create Ni_2Si silicides. The reaction release carbon so the structure contains large quantity of vacant carbon.

1. Introduction

Silicon carbide is a semiconductor with useful properties for high temperature, high frequency, high power, and photonic application. Stable ohmic contacts to SiC with low specific contact resistance are important, since parasitic resistances generally limited device operation. Many different metallization for ohmic contacts have been examined up to the present (Al, Ti, W, TiC, Cu, etc. [1-4]), but nickel contacts are commonly used on n-type SiC due to low contact resistivity [5-7]. For fabrication of ohmic contacts on SiC using Ni, annealing processes of Ni/SiC structure are generally performed at temperature around 1000 °C, all resulting in low contact resistivity ($\sim 10^4$-10^{-6} Ωcm^2). The thermal stability of Ni/SiC structures was already demonstrated under operation for several hours at 500 - 600 °C [5, 8, 9], but recent applications (e.g. in automotive) of SiC devices needs higher stability up to 800 °C or more.

The annealing process of Ni/SiC metallization is accompanied by the reaction of Ni with SiC and silicides are created - mainly N_2Si phase. Once a certain region of SiC is decomposed in above reaction, the excess carbon in the contact layer should be considered as well, there is substantial carbon segregation to the top region of contact [10].

The contribution is concentrated on Ni/SiC contact structure. Contact resistivity, thermal stability and the distribution of elements in the metallization after the annealing process are discussed.

2. Sample preparation

N-type 6H-SiC substrate wafers with doping level approximately 1.7×10^{18} cm^{-3} were used in our experiments. The deposition of 50 nm thick nickel contacts was performed using e-beam evaporator at 135 °C in the pressure of 2×10^{-6} mbar. Directly before metal deposition, the wafers were chemically cleaned using by the following process:

10 min in ultrasound bath with acetone,
5 min in ultrasound bath with $NH_4OH:H_2O:H_2O_2$ - 5:1:1,
5 min in HF conc.,
10 min in boiling H_2O,

20 min in DC Ar plasma (the step was produced in situ in evaporator before metal deposition).

Thermal treatment of the contact structures was carried in the evaporator chamber in the same level of vacuum for 45 sec or 10 min up to 1060 °C. The electrical characterisation was performed by measuring the contact resistivity r_C with the four point method. Stability tests were done at the temperature range from 500 to 900 °C in the oven with hydrogen atmosphere for 10 hours.

XPS spectroscopy was used for the structural characterisation of the prepared films prior and after the annealing. We have at disposal ESCAProbeP apparatus produced by Omicron Nanotechnology Ltd. to analyze surface of specimens. The equipment is provided with dual X-ray source, monochromator, two types of ion guns, electron detection with 5 channeltrons, low energy electron gun for charge compensation, UV source for valence band analyses, focused source of electrons, secondary electron detector etc.

3.   Results

Tab. 1 shows the results of the contact resistivity. It is evident that all annealing processes produce nearly the same results; the best is the annealing at 1065 °C for 45 sec – the lowest value of contact resistivity is 4.06×10^{-4} Ωcm^2.

Tab. 1. The contact resistivity of Ni/SiC structures.

Temperature (°C)	961	961	1065	1065
Annealing time (sec)	45	600	45	600
r_C (10^{-4} Ω cm^2)	4.65±0.45	4.16±0.15	4.06±0.16	4.25±0.23

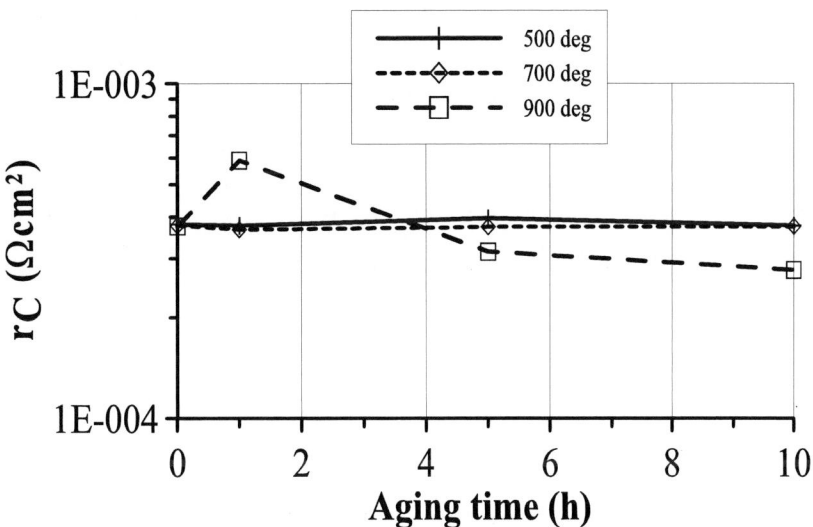

Fig. 1. The ageing test of Ni/SiC contact structures.

The thermal stability of the contact structure was tested at 500, 700, and 900 °C. Fig. 1 shows results of stability tests. The sample was aged at 500 °C at first, because the stability was excellent, we repeated the stability test with the same sample at 700 °C and after that at 900 °C (the duration each test was 10 hours). The value of contact resistivity at zero aging time (500 °C) responds to the value gained by annealing at optimal temperature before the aging process has actually started. In the case of dependencies at 700 and 900 °C it is the final value of previous test. The metallization is very stable at temperature 500 and 700 °C. The structure shoes slightly higher fluctuation at 900 °C, but is still stable.

The reaction of nickel with SiC substrate in the process of contact annealing was studied by XPS measurements. Fig. 2 shows the XPS spectra of the metallization prior the annealing process. Ni layer is strictly separated from the substrate, but it contains relatively large amount of carbon. From the spectra is obvious that structure contains oxygen which is incorporated in the process of the evaporation.

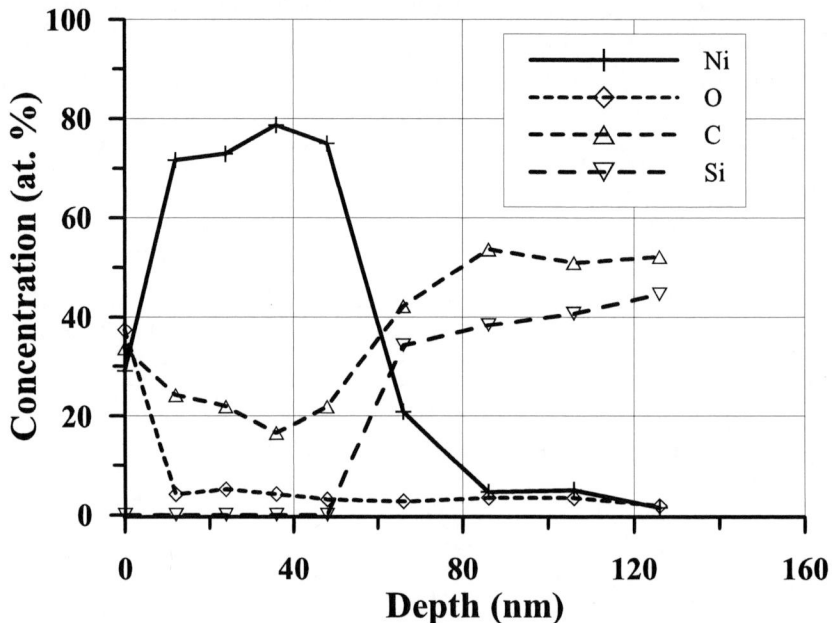

Fig. 2. XPS spectra of Ni/SiC metallization prior the annealing.

Fig. 3 shows the XPS spectra of the metallization after the annealing process at 1065 °C for 10 min. Ni reacts with Si and diffuses relatively deep into the SiC substrate. Accordance to the obtained results it is obvious that the majority phase has composition Ni_2Si. The mentioned reaction consumes silicon from the substrate and so the metallization contains great amount of carbon not only in the volume, but on the surface too.

4. Conclusion

In comparison with published data we obtained ohmic contacts with relatively good contact resistivity ($r_{Cmin} = 4.06 \times 10^{-4}$ Ω cm^2) for SiC with the doping level of 1.7×10^{18} cm^{-3} and with excellent thermal stability (the metallization was stable after being tested for 10 hours at 900 °C).

The structure is contaminated by oxygen that comes from the residual atmosphere in the evaporator and by carbon that is penetrated from the substrate into Ni layer in the deposition process. The annealing process starts reaction of nickel with silicon from SiC substrate. Make allowance for the composition of the species in the structure after annealing Ni and Si create Ni_2Si silicides. The reaction release carbon so the structure contains large quantity of vacant carbon.

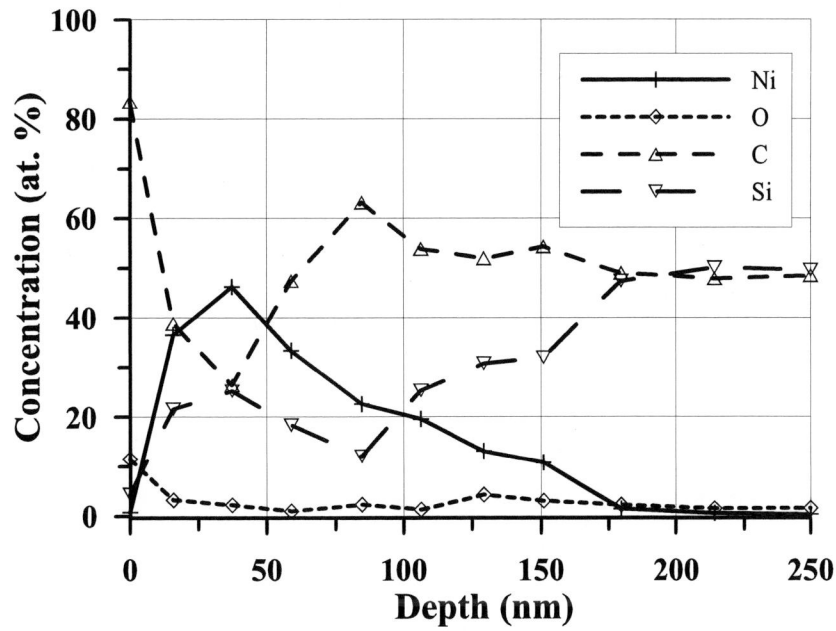

Fig. 3. XPS spectra of Ni/SiC metallization after annealing at 1065 °C.

Acknowledgement

This study was part of research programme MSM 6046137302 (Czech Ministry of Education) – Preparation and research of functional materials and material technologies using micro- and nanoscopic methods.

References:

[1]   Y.S. Park: *SiC materials and device.* Semiconductors and Semimetals **V62**, Academia Press 1998.

[2]   B. Veisz, and B. Pecz: *Appl. Surface Science* **233**, 360, 2004.

[3]   F. La Via, F. Roccaforte, A. Makhtari, V. Raineri, P. Musumeci, L. Calcagno: *Microelectronic Engineering* **60,** 269, 2002.

[4]   R. Getto, J. Freytag, M. Kopnarski, and H. Oechsner: *Materials Science and Engineering* **B61** 270, 1999.

[5]   F. Roccaforte. F. La Via, V. Raineri, L. Calcagno., and Musumeci P.: *Applied Surface Sci.* **184**, 295, 2001.

[6]   S.J. Yang, C.K. Kim, I.H. Noh, S.W. Jang, K.H. Jung, and N.I. Cho.: *Diamond and Related Mat.* **13**, 1149, 2004.

[7]   B.K. Kim, J. Burm, and C. An, in *Proceedings of IEEE*, 2002, p. 97.

[8]   T. Marinova., et al.: *Material Science and Engineering.* **B46**, 223, 1997.

[9]   B.K. Kim, J. Burm, and C. An, in *Proceedings of IEEE*, 2002, p. 97.

[10] B. Pecz: *Applied Surface Science* **184,** 287, 2001.

Two-dimensional electron gas as the THz radiation detector

Michal Horák

Brno University of Technology, Fac. Electr. Eng. Commun., Dept. of Microelectronics
Údolní 53, CZ – 602 00 Brno, Czech Republic, e-mail: horakm@feec.vutbr.cz

The aim of this paper is to find the response of the 2-dim electron gas inside the quantum well to the high-frequency signal within the THz frequency band; that is described by the scalar or vector potential. The electron polarizability as a function signal frequency was calculated and the resonances were found.

1. Introduction

Terahertz frequency band is recently usually considered as the interval 300 GHz – 30 THz that corresponds to the submilimeter wavelength range between 1 mm and 10 μm or to the photon energy within the range 1.25 meV – 125 meV. Below 300 GHz we cross into the milimeter-wave bands, beyond 3 THz we attack the far infrared region. It is clear that many of the characteristic energy scales in nanostructures are of the order of THz radiation energy, especially the heights of potential barriers or the positions of energy levels in quantum wells. The investigation of interaction between THz electromagnetic radiation and electrons gives us not only information on the electronic states and dynamical transport properties of electrons in quantum nanostructures but also information on the THz radiation itself as nanostructures can serve as efficient detectors of this radiation.

The possible application of MOSFET or HEMT as a tunable infrared detector was studied earlier e.g. in [1], [2]. The propagation of plasma waves in the channel of the transistor in [1] was described by equations that are analogous to hydrodynamic equations for shallow water. The perturbative numerical solution of the Schrödinger and Poisson equations was used in [2].

The aim of this paper is to find the response of the 2-dim electron gas inside the quantum well to the high-frequency signal. The signal can be described by the scalar or vector potential. The theoretical approach is based on the numerical solution of the time-dependent one-dimensional Schrödinger equation.

2. Scalar electric modulation

Consider the one-dimensional rectangular well drawn in Fig. 1. If no THz radiation is present, the wave functions of electrons bound in the well and the energy levels can be found in any textbook on quantum physics [4]:

$$|x| \ge d/2 \text{ (outside the well): } \varphi(x) = D\exp(-\kappa|x|) ; \; \hbar\kappa = \sqrt{2m(V_0 - |E|)} \qquad (1)$$

$$|x| \le d/2 \text{ (inside the well): } \varphi(x) = C\begin{Bmatrix} \cos(kx) \\ \sin(kx) \end{Bmatrix} ; \; \hbar k = \sqrt{2m|E|} ; \; \begin{Bmatrix} tg(kd/2) \\ \cot g(kd/2) \end{Bmatrix} = -\frac{\kappa}{k}$$

The last transcendental equation defines the energy level positions inside the quantum well and in general it should be solved numerically.

The triangular quantum well in Fig. 1 can be treated in a similar way: the essential physics is the same, only the mathematics is more complicated as the electron wave function inside the well is a linear superposition of the Airy functions Ai and Bi.

1-4244-0396-0/06/$25.00 ©2006 IEEE

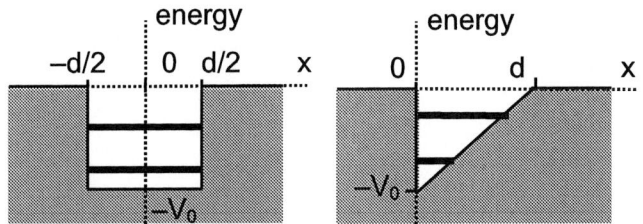

Fig. 1. The rectangular and the triangular quantum well. The triangular well could be a good approximation of the MOSFET inversion layer.

If the depth of the well is modulated by the scalar high frequency signal $V(x,t)$, the situation is described by the time-dependent Schrödinger equation:

$$i\hbar\frac{\partial\psi(x,t)}{\partial t}=-\frac{\hbar^2}{2m}\frac{\partial^2\psi(x,t)}{\partial x^2}+V_0(x)\psi(x,t)+V(x,t)\psi(x,t) \qquad (2)$$

$$|x|\geq d/2:V(x,t)=0;\ |x|\leq d/2:\ V(x,t)=eU\cos(\omega t)$$

Its solution inside the well can be found by means of the Floquet theorem (see e.g. part 2.1 of [3]) and it reads:

$$|x|\leq d/2:\psi(x,t)=\exp(-iEt/\hbar)\sum_{n=-\infty}^{+\infty}C_n\begin{Bmatrix}\cos(k_n x)\\\sin(k_n x)\end{Bmatrix}\exp(-in\omega t)\times\exp[-i\frac{eU}{\hbar\omega}\sin(\omega t)] \qquad (3)$$

$$\hbar k_n=\sqrt{2m(|E|+n\hbar\omega)}$$

Higher order harmonics $\exp(-in\omega t)$ that are generated inside the well should penetrate also outside the well, thus the wave function in this region is:

$$|x|\geq d/2:\psi(x,t)=\exp(-iEt/\hbar)\sum_{n=-\infty}^{+\infty}D_n\exp(-\kappa_n|x|)\exp(-in\omega t) \qquad (4)$$

$$\hbar\kappa_n=\sqrt{2m(V_0-|E|-n\hbar\omega)}$$

The wave function $\psi(x,t)$ and the corresponding single electron current density should be continuous at $x=\pm d/2$; these conditions give at first a transcendental equation similar to that in Eq. (1) that ensures the existence of the solution and they also enable to calculate the unknown coefficients C_n,D_n. However, as it is rather difficult to find analytical formulae, the solution was carried out numerically with the aid of *Matlab*.

If we restrict the infinite summation in the above formulae to $n=0,\pm1,\pm2$, the wave function can be written as the sum of three terms $\psi(x,t)=\psi^{(0)}(x)+\psi^{(\pm1)}(x,t)+\psi^{(\pm2)}(x,t)$ corresponding to the steady state (without signal), to the linear response $\exp(\pm i\omega t)$ and to the nonlinear response $\exp(\pm2i\omega t)$. Similarly the electron concentration can be written as a sum o three terms $n=n^{(0)}+n^{(\pm1)}+n^{(\pm2)}$ that are related to the wave function components by the standard relation $n^{(k)}\propto|\psi^{(k)}|^2$. The effect of the high frequency signal on the electron gas inside the well can be characterized by the polarizability; the linear and the nonlinear second order polarizabilities are defined as [4]

$$\alpha(\pm\omega)\propto\int x\,n^{(\pm1)}(x)\,dx,\ \ \beta(\pm2\omega)\propto\int x\,n^{(\pm2)}(x)\,dx \qquad (5)$$

275

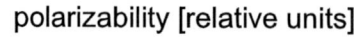

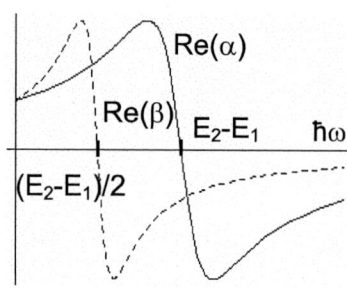
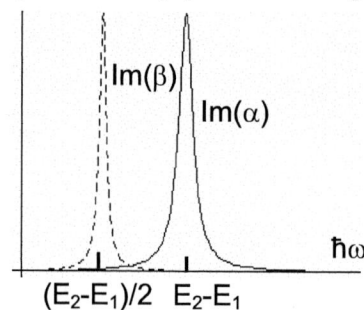

Fig. 2: The real and the imaginary parts of the linear and nonlinear polarizabilities for the rectangular quantum well.

The numerical evaluation was carried out for a rectangular quantum well of depth $V_0 = 55\,\text{meV}$ with two energy levels in positions $E_1 = -51.2\,\text{meV}$, $E_2 = -40.2\,\text{meV}$. As expected the first order linear polarizability $\alpha(\pm\omega)$ exhibits resonance behaviour when $\hbar\omega = E_2 - E_1$, the second order nonlinear polarizability $\beta(\pm 2\omega)$ has resonance character at frequency $\hbar\omega = (E_2 - E_1)/2$, see Fig. 2. These resonances correspond to the signal frequency $\nu = \omega/2\pi = 26.6\,\text{THz}$.

3. Vector electromagnetic modulation

The investigation in the previous part was focused on the high-frequency response of the electron gas in the quantum well on the to time-varying *scalar* electrostatic potential. As a next step we generalize the investigation to the coupling of quantum well to electromagnetic field described by the *vector* potential.

The time-dependent Schrödinger equation now reads

$$i\hbar\frac{\partial}{\partial t}\psi(\vec{r},t) = \left[\frac{1}{2m}(\hat{p} - e\vec{A})^2 + V(\vec{r})\right]\psi(\vec{r},t) \tag{6}$$

where $\vec{A}(\vec{r})$ is the vector potential of the electromagnetic field and $V(\vec{r})$ is the scalar electrostatic potential energy that describes the profile of the quantum well. It is known that the so called Kramers-Henneberger unitary transformation [5] can be applied to Eq. (6) that transforms Eq. (6) into a different form:

$$i\hbar\frac{\partial}{\partial t}\varphi(\vec{r},t) = \left[\frac{1}{2m}\hat{p}^2 + V(\vec{r} + \vec{\alpha}(t))\right]\varphi(\vec{r},t),\ \text{where}\ \frac{d^2}{dt^2}\vec{\alpha}(t) = -\frac{e}{m}\frac{d}{dt}\vec{A}(t) = \frac{e}{m}\vec{E}(t) \tag{7}$$

In the high-frequency limit of the signal that correspond to the THz-frequency band the electron is subjected to the time-averaged potential $\langle V(\vec{r} + \vec{\alpha}(t))\rangle$. This means that only the zero-order term in the Fourier expansion of $V(\vec{r} + \vec{\alpha}(t))$ determines the electron state; thus the time-independent averaged potential $\langle V(\vec{r} + \vec{\alpha}(t))\rangle$ can be used in Eq. (7). The result is that the profile of the quantum well is slightly modified and the energy levels E_1, E_2 inside the well are slightly shifted, as it is demonstrated in Fig. 3 for the rectangular quantum well. The Schrödinger equation (7) is in this way converted into the equation with time independent and spatially varying potential. The profile of the modified quantum well and the position of the shifted energy levels should be calculated numerically.

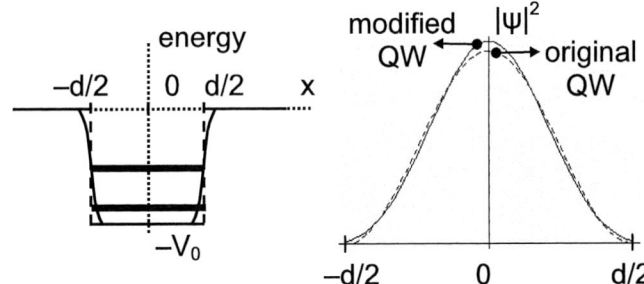

Fig. 3. The original (dashed) and the modified (full line) quantum well according to Eqs. (6), (7) and $|\psi|^2$ for the lowest energy level.

The variation of the quantum well profile and the energy level shift depend on the intensity (or on the amplitude) of the incident electromagnetic signal and especially the energy level shift is negligible for small signals. If the profile of the well is modified, the electron wave functions should be also slightly modified, see Fig. 3, and the same should be true also for the electron charge density. The polarizability of the electron gas is now given as the difference between electron concentrations in the original well (without radiation) and in the disturbed well (with profile affected by the radiation). However, our numerical calculation does not offer convincing results for the polarizability, especially for small signal.

4.   Conclusions

The response of the 2-dim electron gas inside the quantum well to the high-frequency signal was investigated by two different methods. If the signal is described by the scalar potential, the Floquet theorem was applied to the Schrödinger equation and the linear and nonlinear polarizabilities of the electron gas were calculated. The polarizability resonances occurs at energy that corresponds to the difference between energy levels inside the quantum well. If the signal is described by the electromagnetic vector potential, the method of unitary transformation was applied; however, no convincing results for the polarizability were achieved.

Acknowledgment

This research has been supported by the Czech Ministry of Education in the frame of Research Plan MSM 0021630503 *MIKROSYN New trends in microelectronic systems and nanotechnologies.*

References

[1]   Shur S. M., Dyakonov M.: Two-dimensional electrons in field effect transistors. In: Quantum-based electronic devices and systems (edited by Dutta M., Stroscio M. A.), pp. 65-99. World Scientific, 1998.

[2]   Mahan G. D.: Gusmao M. S.: Field-effect transistors as tunable infrared detectors. *J. Appl. Phys.* **79** (1996), pp. 2752-2754.

[3]   Grifoni M., Hanggi P.: Driven quantum tunnelling. *Physics Reports* **304** (1998), pp. 229-354.

[4]   Bransden B. H., Loachain C. J.: Introduction to Quantum Mechanics. Addison-Wesley Longman, 1989.

[5]   Hennenberger, W. C.: Perturbation method for atoms in intense light field. Phys. Rev. Lett. **21**, 1968, pp. 838-841.

Influence of mechanical strain on essential characteristics of GMR structures

V. Áč[a], B. Anwarzai[a], S. Luby[b], and E. Majkova[b]

[a] A. Dubček University of Trenčín, Študentská 1, 91151 Trenčín, Slovak Republic
[b] Institute of Physics, Slovak Academy of Sciences, 84511 Bratislava,
Slovak Republic
e-mail: vac@tnuni.sk and luby@savba.sk

Giant magnetoresistance (GMR) of Co and Fe-Co based structures with Cu and Au spacers were e-beam evaporated onto Si wafers. The thickness of layers was obtained from the simulation of X-ray reflectivity spectra. The GMR ratio was between 3.3 and 5.6 %. The effect of strain upon samples was studied in a bending configuration. The different dependences of coercivity (H_c) vs. strain were found. E.g. for sample with Co(5)/Au(2.2)/Co(2) core structure (where numbers denote thickness in nm) H_c increases with increasing compressive stress, whereas for sample with Co(0.5)/Cu(3)/Co(5) core structure it increases with tensile stress. The relative change of GMR ratio vs. loading in the strain interval $\pm 280 \times 10^{-6}$ is 1-2 % near to the point of inflection of GMR vs. H curve (3.6 kA/m) for the second sample. The structures can be further optimized and applied in sensors of mechanical quantities.

1. Introduction

The giant magnetoresistance effect (GMR) was discovered nearly 20 years ago [1, 2]. Nowadays GMR-based products are commercially available and research moved into diverse areas, e.g. combining standard GMR with magnetostriction [3–5]. This way combined sensors of mechanical quantities, like deformation, pressure or displacement are produced. In these sensors the magnetization orientation in magnetostrictive material is influenced by stress (strain), i.e. inverse magnetostriction is effective [3]. The measured quantity is detected due to the change of GMR in external magnetic field. The effects of stress (strain) on MR were studied also in other papers. GMR sensor response onto magnetoelastic stress is described in [6], strain induced GMR changes in CoFe/Cu structures are presented in [7]. Also tunnel MR structures with strain gauge factor 300 – 600 were studied [8].

The early GMR structures relied on a very thin spacer layers in multilayers where the magnetic coupling provides antiferromagnetic alignment of magnetic layers at zero field. A simpler arrangement is the spin-valve (SV) [9], where two magnetic layers are separated by a non-magnetic spacer and magnetization of one layer is pinned so that it cannot be rotated in a moderate field. In [3-5] SV strain sensors with one layer pinned by antiferromagnet layer were studied.

Pseudo SV structures with pinning of one layer due to different coercivities were also reported [10]. Sometimes they don't show the true asymmetric SV behaviour. This type of strain sensors based on Co layers is studied in this work. The idea is to make the structure more simple, avoiding generally used magnetic alloys. Magnetostriction in Co layers is well understood, being $- 6 \times 10^{-5}$ for pure element [11] (constriction in magnetic field), whereas in alloys it is mostly positive (elongation).

1-4244-0396-0/06/$25.00 ©2006 IEEE

2. Experimental

Si (100) wafers (0.45 mm) with native oxide were used as substrates. Ten different SV structures were deposited by e-beam evaporation at RT in an UHV (10^{-7} Pa). The growth rates were 0.1 nm/s [12]. The Co/Au/Co and Co/Cu/Co core structures were deposited onto adhesion/buffer underlayer and they were covered by a top layer to prevent oxidation. Sometimes Fe-Co bilayer was used instead of simple Co. Thickness was measured in situ by quartz monitor. In this paper results for three representative samples will be shown (Table 1).

The samples were analysed by specular X-ray reflectivity (XRR) and X-ray diffraction (XRD) using CuKα radiation. GMR vs. strain measurements were done in bending configuration using 15 x 3 mm^2 Si strips. Samples were seated on two supports l = 12 mm apart. Force F was exerted by the mass of the block of 50, 100, 150, 200, 250, 300 or 350 g. The stress in the spin-valve is tensile or compressive for the stack on bottom or top surface of the strip, respectively. GMR was measured by four-point probe along the distance of 6 mm at RT, current and field in the plane of layered structure being mutually perpendicular. The bowing depth y was 40 μm/50 g increasing linearly with the mass. Using well known formula $y = F.l^3/48\ E.J$, where $E = 1.3 \times 10^4$ N/mm^2 is the Young's modulus of Si(100) and J is the moment of inertia of Si strip cross section vs. neutral axis ($J = 0.03$ mm^4 in our case), the calculated value is $y = 45$ μm, which seems reasonable.

Table 1. Studied pseudo SV structures (3s – example of XRR thickness simulation)

#	Underlayer [nm]	Core structure [nm]	Top layer [nm]	GMR [%]
1	Cr(3)	Fe(3)Co(0.5)/Cu(3)/Co(5)	Cu(2)/Cr(2)	5.2
2	Cr(5)/Au(2.5)	Co(5)/Au(2.2)/Co(2)	Au(1)	5.6
3	Cr(5)/Cu(5)	Co(2)/Cu(2.2)/Co(5)	---	3.3
3s	Cr(6.45)/Cu(7.39)	Co(1.99)/Cu(1.92)/Co(5.12)	---	3.3

3. Results and discussion

The X-ray diffraction revealed a partial crystallinity of the samples. By simulation of XRR spectra the thickness data were obtained which show small differences with those obtained in situ (Table 1). In the GMR structures an oscillatory exchange coupling between ferromagnetic layers through non-magnetic spacer is a fairly general behaviour. It depends on the nature of spacer and also on the growth direction [13]. The coupling energy (field) correlates with GMR. At large negative maxima of coupling GMR is high and sensitivity of sensor is increased [14]. In Co/Au/Co structure the maxima of GMR were reported for spacer thickness 1.35 nm, 2.4 – 2.5 nm and 3.9 nm [13, 15]. For Co/Cu/Co structure the negative maxima of coupling are at 2.1 and 3 nm [14]. The design of our structures respects these facts. Further, the change of the spacer thickness due to the deformation of sensor will not shift the GMR considerably on the oscillatory GMR vs. spacer thickness curve. Assuming that the period of oscillations, as shown above, is about 1 nm, the change of spacer thickness of 0.25 nm would shift GMR to the point of inflection of the oscillatory dependence and this could influence the output considerably (cf. [16]). However, from our calculations it follows that at the bowing depth of 40 μm the elongation/constriction of SV structure is only $\pm$ 6 μm and corresponding change of the whole SV thickness is $\pm$ 0.01 nm, which can be omitted. The same conclusion might be done in respect of the change of roughness which could also influence the GMR signal [16].

The results of our strained samples from Table1 are shown in Figs. 1, 2, 3 (samples # 1, 2, 3). The figures contain the relative magnetoresistance $MR=R(H)/R_o$ vs. intensity of magnetic field H characteristics, details of these characteristics and strain dependences of coercivity field H_c and field in the point of inflection of MR characteristics H_{ip}. R_o is the resistance of the strip in demagnetized state. H_c is the field at $dMR/dH=0$ and H_{ip} at $d^2MR/dH^2=0$ on full characteristics. Virgin characteristics are displayed as well.

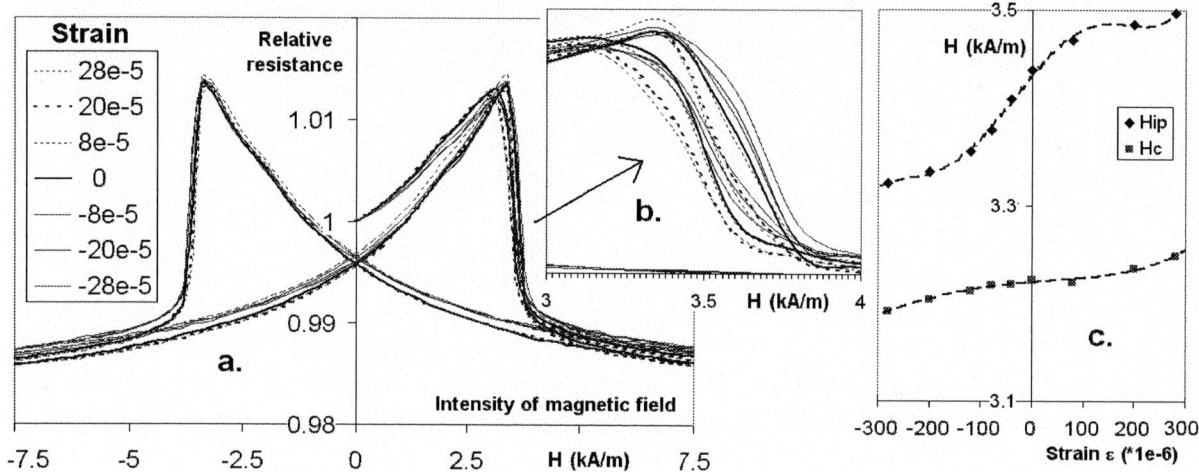

Fig. 1 Magnetoresistance and H_c, H_{ip} strain dependences of sample # 1 (Table 1) with the Fe(3)Co(0.5)/Cu(3)/Co(5) core structure.

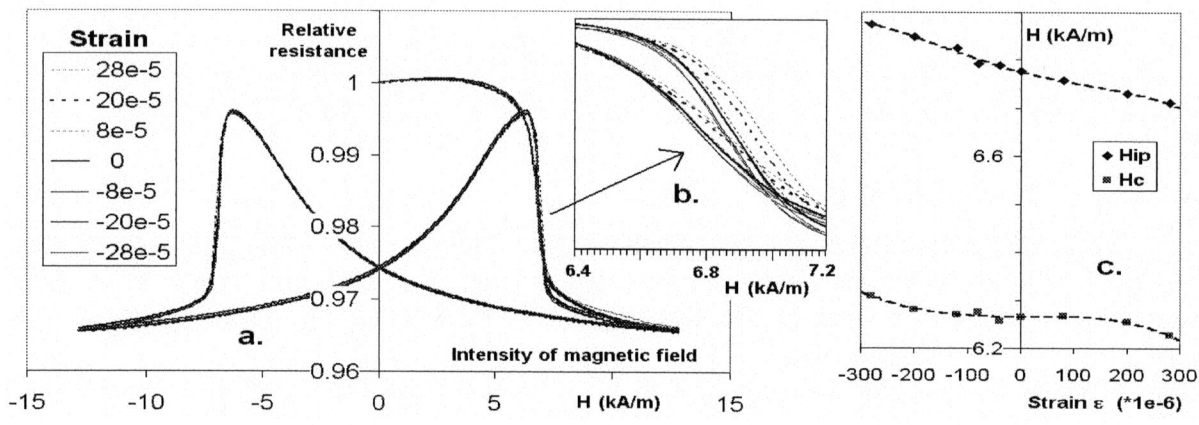

Fig. 2 Magnetoresistance and H_c, H_{ip} strain dependences of sample # 2 (Table 1) with the Co(5)/Au(2.2)/Co(2) core structure.

H_c and H_{ip} are displayed to make an impression of MR vs. H curve shift under bending. If the sensor working point is chosen near to point of inflection of the MR curve, the change of resistance is 1 - 2% and it can be simply detected. Nevertheless, the sensitivity of sensor should be further increased by more appropriate design of structure. The dependences of H_c and H_{ip} vs. strain differ, showing increase (# 1), decrease (#2) or mixed behaviour. Remarkable is the difference between samples # 1 and # 2, which could be explained by the Fe-Co layer in the sample # 1, which is mixed, in fact, because 0.5 nm thick Co is discontinuous and therefore we have Fe-Co with a positive magnetostriction. Further measurements on different structures will give more coherent insight into the studied effects.

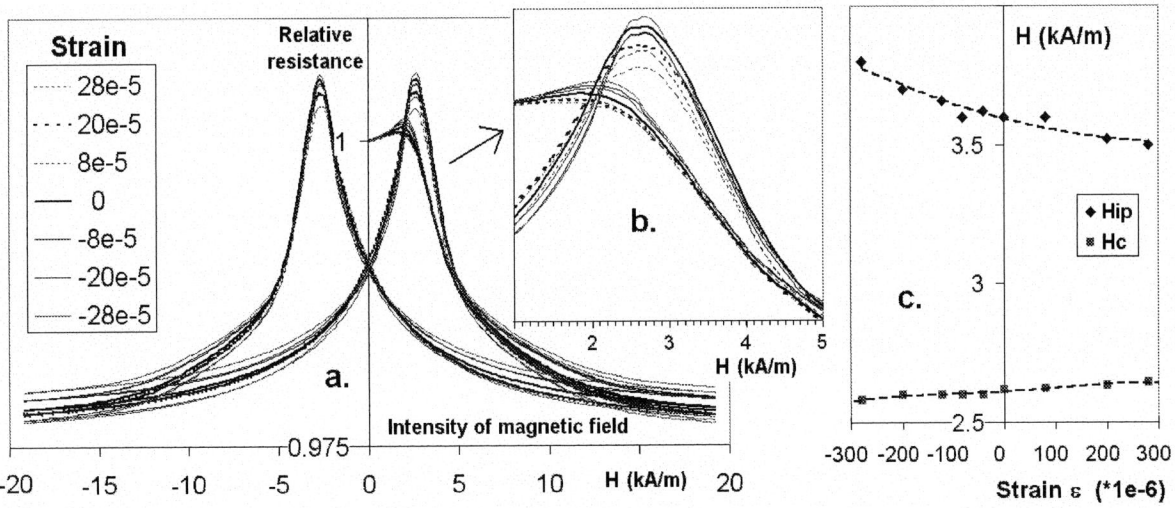

Fig. 3 Magnetoresistance and H_c, H_{ip} strain dependences of sample # 3 (Table 1) with the Co(2)/Cu(2.2)/Co(5) core structure.

4. Conclusion

The different behaviour of coercivity vs. strain was found in SV structures with Co/Cu/Co and Co/Au/Co core structures. They seem to be related to the positive or negative magnetostriction and thickness of layers. The relative change of MR in the strain interval $\pm 280 \times 10^{-6}$ is 1-2 %. Structures can be used as sensors of mechanical quantities.

Acknowledgement

Work was supported by Scientific Grant Agency VEGA, Bratislava, grants 2/4101/24 and 2/6030/26 and by Centre of Excellence SAS – Physics of Information, contract I/2/2006.

References

[1] N. N. Baibich, et al., *Phys. Rev. Lett.* **71**, 2472, 1988.
[2] G. Binasch, P. Grünberg, F. Saurenback, aqnd W. Zinn, *Phys. Rev. B* **39**, 4828, 1989.
[3] H. J. Mamin, B. A. Gurney, D. R. Wilhoit, and V. S. Speriosu, *Appl. Phys. Lett.* **72**, 3220, 1998.
[4] L. Baryl, B. Gurney, D. Wilhoit, and V. Speriosu, *J. Appl. Phys.* **85**, 5139, 1999.
[5] T. Duenas, et al., *J. Magn. Magn. Mater.* **242-245**, 1132, 2002.
[6] W. Ricken, J. Liu, and W. J. Becker, *Sensors and Actuators A* **91**, 42, 2001.
[7] S. H. Florenz, and R. D. Gomez, *IEEE Trans. on Magnetics* **39**, 3411, 2003.
[8] M. Löhndorf, et al., Appl. *Phys. Lett.* **81**, 313, 2002.
[9] B. Dieny, et al., *Phys. Rev B* **43**, 1297, 1991.
[10] J.-Q. Wang, et al., *Mater. Sci. Engn. B* **76**, 1, 2000.
[11] L. Louail, D. Maouche, and A. Cheriat, *Mater. Lett.* **4369**, 1, 2003.
[12] M. Jergel, E. Majkova, M. Ozvold, and R. Senderak, *acta physica slov.* **56**, 145, 2006.
[13] V. Grolier, et al., *Phys. Rev. Lett.* **71**, 3023, 1993.
[14] V. S. Speriosu, B. Dieny, P. Humbert, B. A. Gurney, and H. Lefakis, *Phys. Rev. B.* **44**, 5358, 1991.
[15] C. Christides, R. Lopusnik, J. Mistrik, S. Stavroyiannis, and S. Visnovsky, *J. Magn. Magn. Mater.* **198-199**, 36, 1999.
[16] K. Suenaga, et al., *IEEE Trans. Magn.* **42**, 1499, 2006.

Energy band diagram and charge distribution in AlGaN/GaN heterostructure studied by classical approach

J. Osvald

Institute of Electrical Engineering , Slovak Academy of Sciences,
Dúbravská cesta 9, 841 04 Bratislava, Slovakia
e-mail: elekosva@savba.sk

We used a classical approach to calculate energy band diagrams of AlGaN/GaN heterostructures. This approach enables to calculate the band diagram and carrier concentrations also when the external bias is applied on the structure. Also the potential on the Schottky barrier is more rigorously defined as in a self-consistent solution of Poisson and Schrödinger equations. Dependence of the band profile and the carrier concentration of the two-dimensional gas on the piezoelectric charge can also be calculated by this approach.

1. Introduction

AlGaN/GaN material systems are well known candidates for making high power and high frequency transistors. On the basis of spontaneous and piezoelectric polarization it is possible to obtain electron concentration in the channel of the transistor in the order of 10^{13} cm^{-2} or even higher in undoped or unintentional doped structures. The polarization fields caused by the spontaneous polarization in AlGaN and the piezoelectric polarization made by the strain on the AlGaN/GaN interface are in order of 10^6 V/cm. Investigation of the energy band diagram and the charge distribution in such a structure is of primary importance.

Currently the energy profile in AlGaN/GaN structure is studied by the self-consistent solution of Poisson and Schrödinger equations [1-3]. By the solution of this system of equations we obtain eigenenergy states in the subband structure, their population, the sheet electron concentration in the channel region and finally the potential distribution in the system. However, by this approach it is a problem to calculate the channel concentration of electrons and potential curve when the structure is biased by external potential on the Schottky diode made on the AlGaN surface. For the two-dimensional electron gas the Fermi level is a part of the overall solution of the equations and is not known as an input of the equations. That is why it is not possible to state the boundary potential at the Schottky contact side and thus the Schottky barrier is defined only approximately as a potential difference between the Fermi level and the conduction band minimum CBM. On the other hand the present approach enables to calculate carrier concentrations also when the current is flowing through the structure and the thermodynamic equilibrium is disturbed.

2. Formalism

There are two types of polarization charge in heterostructures as AlGaN/GaN. The spontaneous polarization is a function of a composition parameter x and can be approximately expressed as [4]

$$P^{sp}_{Al_xGa_{1-x}N} = -0.090x - 0.034(1-x) + 0.019x(1-x).$$

1-4244-0396-0/06/$25.00 ©2006 IEEE

Vegard's law is approved to be valid for the calculation of piezoelectric polarization and the dependence can be expressed as a function of the strain-dependent bulk piezoelectric polarizations of the relevant binary compounds

$$P^{pz}_{Al_xGa_{1-x}N}(x) = xP^{pz}_{AlN}[\varepsilon(x)] + (1-x)P^{pz}_{GaN}[\varepsilon(x)].$$

These polarizations are

$$P^{pz}_{AlN} = -1.808\varepsilon + 5.624\varepsilon^2, \qquad \text{for } \varepsilon < 0,$$

$$P^{zp}_{AlN} = 1.808\varepsilon - 7.888\varepsilon^2, \qquad \text{for } \varepsilon > 0,$$

$$P^{pz}_{GaN} = -0.918\varepsilon + 9.541\varepsilon^2.$$

The basal strain $\varepsilon(x)$ is given as

$$\varepsilon(x) = [a_{subs} - a(x)]/a(x),$$

with a_{subs} and $a(x)$ being the lattice constants of the unstrained alloy and of the substrate.

The total polarization and simultaneously the polarization charge can be calculated according to the above equations.

The orientation of the piezoelectric polarization field depends on the surface orientation of GaN at the heterointerface [5]. When the GaN layer at the heterostructure is terminated by Ga atoms the piezoelectric field reduces the two-dimensional electron density. If there are N atoms the surface is N-oriented and the piezoelectric field enhances the two-dimensional electron density. The conditions for two-dimensional gas formation at the AlGaN/GaN interface are therefore fulfilled only for N-oriented GaN crystals [1].

We used a classical approach to study such systems. It is based on the simultaneous solution of Poisson and continuity equations

$$\Delta\varphi = -(q/\varepsilon_s)(p - n + N_d + N_a)$$

$$\frac{\nabla \cdot \mathbf{J}_n}{q} = U, \qquad \mathbf{J}_n = q(-\mu_n n \nabla\varphi + D_n \nabla n)$$

$$\frac{\nabla \cdot \mathbf{J}_p}{q} = -U, \qquad \mathbf{J}_p = q(-\mu_p p \nabla\varphi - D_p \nabla p),$$

where φ is the electrostatic potential, q is the elementary charge, ε_s is the dielectric constant of semiconductor, n and p are the electron and hole densities, respectively, μ_n, μ_p, D_n, D_p and $\mathbf{J}_n$, $\mathbf{J}_p$ are respectively the mobilities, diffusion coefficients and the currents of electrons and holes, and U is the net recombination rate. The approach enables to simulate energy bands and carrier concentrations also under applied external bias. It was shown [6] that the polarization charge at the AlGaN/GaN interface may be simulated by insertion of very thin δ-doped layer with a sheet carrier concentration determined by the polarization effects. The δ-doped layer with the same sheet carrier concentration but with opposite charge must be put at the AlGaN surface layer. The layer at the heterointerface creates for N-oriented GaN surface the potential well which attracts the electrons. On the other hand the negative sheet charge placed at the AlGaN surface creates the potential peak which enhances the Schottky barrier height. This last effect was not taken into account in the previous work [6].

3. Results

The thickness of the δ-doped layer was chosen in our calculations to be in the range of one monolayer thickness – 0.25 nm and the Schottky barrier height assumed was 1 V. AlGaN layer was 30 nm thick. The doping concentration of AlGaN layer was assumed to be 2×10^{18} cm^{-3} and of GaN layer it was 1×10^{16} cm^{-3}. In Fig. 1 the potential of the CBM for three external voltages is shown. The potential step in the conduction band at the AlGaN/GaN heterojunction was taken to be 0.343 V. The peak in the potential near the Schottky contact is a result of the negative polarization sheet charge. This charge effectively enhances the

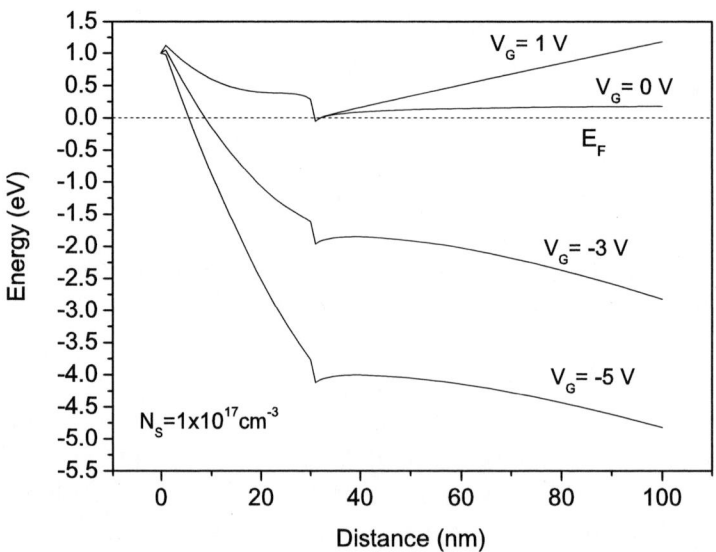

Fig. 1. Potential of the conduction band minimum of the AlGaN/GaN heterostructure for different applied external voltages.

Schottky barrier height of the contact [7]. It is seen that the lowest states in the potential well are situated under the Fermi level at equilibrium and at temperature 300 K. In Fig. 2 the concentration of electrons near the heterointerface region is shown as a function of external

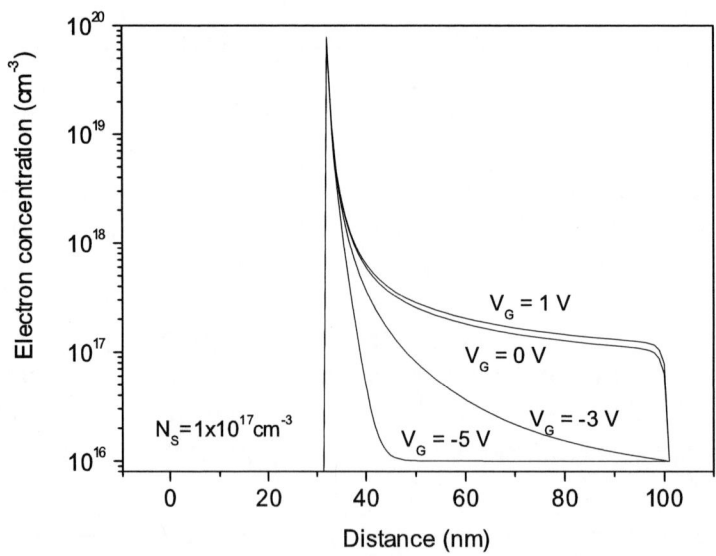

Fig. 2. Electron concentration near AlGaN/GaN interface for different applied voltages.

bias. Analysis of this plot leads to the conclusion that the electron concentration directly at the AlGaN/GaN interface is not seriously influenced by external bias. As a reaction to the applied voltage which simulates pinching-off the channel layer only the concentration outside the δ-doped layer is significantly influenced and diminished. The electrons move deeper into the semiconductor by decreasing the reverse voltage. The electrons at the boundary are in the lowest subband and their binding is stronger than that of the electrons in higher energy subbands. It is also seen that enlarging the gate voltage from 1 to -5 V changes the electron concentration preferably in the bulk of GaN. The concentration changes proceed with increasing reverse voltage from the bulk of GaN towards the AlGaN/GaN interface.

4. Conclusion

We have shown that the basic semiconductor equations - Poisson and the continuity equations can be used to calculate energy bands profile and the free electron concentration near the AlGaN/GaN interface. It was also shown how the electron concentration in the channel region in GaN changes with external bias.

Acknowledgment

The author gratefully acknowledges the financial support of Slovak Grant Agency for Science under Contract No. 2/6097/26.

References

[1]   Y. Zhang and J. Singh, *J. Appl. Phys.* **85**, 587, 1999.
[2]   R. M. Chu, Y. D. Zheng, Y. G. Zhou, P. Han, B. Shen, S. L. Gu, *Appl. Phys. A* **75**, 387, 2002.
[3]   Y. C. Kong, Y. D. Zheng, S. I. U, R. Zhang, P. Han, Y. Shi, R. I. Jiang, *Appl. Phys. A* **84**, 95, 2006.
[4]   V. Fiorentini, F. Bernardini and O. Ambacher, *Appl. Phys. Lett.,* **80**, 1204, 2002.
[5]   M. Stutzmann, O. Ambacher, M. Eickhoff, U. Karrer, A. Lima Pimenta, R. Neuberger, J. Schalwig, R. Dimitrov, P. J. Schuck,, and R. D. Grober, *phys. stat. sol. (b)* **228**, 505, 2001.
[6]   N. Li, D. G. Zhao, H. Yang, *Solid State Commun.* **132**, 701, 2004.
[7]   U. Karrer, O. Ambacher, and M. Stutzmann, *Appl. Phys. Lett.* **77**, 2012, 2000.

Spectroscopic ellipsometric study of LPCVD-deposited Si nanocrystals in SiN_x and Si_3N_4

P. Basa, P. Petrik, M. Fried

Hungarian Academy of Sciences,
Research Institute for Technical Physics and Materials Science,
Budapest 114, P.O.Box 49, H-1525 Hungary
e-mail: basa@mfa.kfki.hu

Low-pressure chemical vapour deposited and annealed SiN_x/nc-Si/SiN_x and Si_3N_4/nc-Si/Si_3N_4 layers prepared on Si substrates were characterized by spectroscopic ellipsometry. The effective medium approximation model was used to obtain the thickness, the composition and homogeneity of the layers. It was obtained that the deposited nc-Si layer thickness depended on the stoichiometry of the underlying silicon nitride layer.

1. Introduction

Dielectric layers with embedded semiconductor nanocrystals have been widely studied recently, in order to overcome difficulties of non-volatile memory devices connected with technology scale-down, and to develop Si-based light emitting diodes (LED's) [1-5]. It is reported in the literature, that silicon nanocrystal formation takes place during the direct low pressure chemical vapour deposition (LPCVD) of a thin silicon layer [6-7]. Varying the deposition parameters allows the nanocrystal size to be tuned.

Spectroscopic ellipsometry proved to be a powerful optical characterization technique for studying thin film structures [8,9]. In this work spectroscopic ellipsometry is applied to study the effect of deposition time on the composition, homogeneity, microstructure, and thickness of SiN_x/nc-Si/SiN_x and Si_3N_4/nc-Si/Si_3N_4 layers. Cross-sectional transmisison electron microscopy (XTEM) and electron energy loss spectroscopy (EELS) was used to obtain direct thickness and composition information on the layers.

2. Sample preparation

The multistructured layers were deposited by LPCVD at 810^OC and 30 mPa using SiH_2Cl_2 and NH_3 in three steps. During the deposition, the flow rates for the top and bottom layers were 100/25 sccm for the non-stoichiometric layers, and they were 21/90 sccm for the stoichiometric layers (SiH_2Cl_2/NH_3, respectively). For the middle layers they were 100/0 sccm in case of all samples. Five different structures were prepared on Si wafers depending on the deposition times of the layers as it is shown in Table 1 and 2.

Deposition time for…	Sample A1	Sample A2	Sample A3
…top SiN_x layer	5 min	5 min	3 min
…middle nc-Si layer	5 min	2 min	2 min
…bottom SiN_x layer	5 min	5 min	3 min

Table 1. Deposition times for the examined SiN_x/nc-Si/SiN_x structures

Deposition time for...	Sample B1	Sample B2
...top Si₃N₄ layer	10 min	10 min
...middle nc-Si layer	5 min	2 min
...bottom Si₃N₄ layer	5 min	5 min

Table 2. Deposition times for the examined Si₃N₄/nc-Si/Si₃N₄ structures

For the ellipsometric measurements, Woollam M88 and Woollam M2000 ellipsometers were used in the photon energy range 1.55–4.42 eV.

3. Ellipsometric modeling

For parametrizing the non-stoichiometric nitrides and the surface roughness layer between the nc-Si layer and the top nitride layer, the Bruggeman effective medium approximation (EMA) model was applied. This model was successfully used previously for modeling of such layers [8,10]. It mixes the dielectric functions of constituent materials by varying their volume fractions in an isotropic matrix.

As for the non-stoichiometric silicon nitride (SiN_x) layers, their composition was modeled by using stoichiometric silicon nitride (Si_3N_4), and fine-grained polycrystalline silicon (fp-Si, [11]) as constituent materials. The ratio of these components was determined by evaluating single-layer SiN_x reference samples deposited with the same SiH_2Cl_2/NH_3 input gas ratio. The silicon excess in addition to the Si_3N_4 was found to be 7.6% in this layer (it means a $SiN_{x=1.22}$ layer).

The surface roughness layer between the middle nc-Si layer and the top silicon nitride layer was modeled by the EMA mixture of the top and the middle layer with the fixed 50% ratio of them.

The middle nc-Si layer was modeled by the fp-Si material taken from the literature.

4. Results and discussion

The best-fit results for the samples are shown in Table 3. The thicknesses are in good agreement with the values obtained from XTEM images. All samples were laterally homogeneous within the error range. For Sample A2 and A3, the deposition time was the same for the middle layer, and the fit shows that the middle layer thicknesses are the same within the error range.

	Sample A1	Sample A2	Sample A3	Sample B1	Sample B2
Top nitride layer thickness (nm)	17.4 ± 0.6	12.3 ± 0.2	6.6 ± 0.2	30.7 ± 0.1	30.1 ± 0.1
Surface roughness layer thickness (nm)	0 ± 0.9	5.1 ± 1.0	5.0 ± 1.0	4.3 ± 0.2	9.1 ± 1.0
Middle nc-Si layer thickness (nm)	17.2 ± 0.3	2.5 ± 0.3	2.5 ± 0.2	16.4 ± 0.2	3.5 ± 0.2
Bottom nitride layer thickness (nm)	16.6 ± 0.1	14.2 ± 0.6	8.9 ± 0.6	14.3 ± 0.2	13.9 ± 0.6
MSE	6.45	5.22	5.00	1.26	1.45

Table 3. Fitted layer thickness and the mean-squared error (MSE) for the examined samples

Systematic dependences have been found in the middle and bottom layer thickness as a function of deposition time (see Table 3). The thickness of the bottom SiN_x layers for 5 minutes of deposition was 14-17 nm, while for 3 minutes it was about 9 nm. The thickness of the middle nc-Si layers for 5 minutes of deposition on SiN_x bottom layer was 17 nm, while for 2 minutes of deposition it was about 2.5 nm. The thickness of the surface roughness layer was the same for Sample A2 and A3. Both the thickness of the middle layer and the thickness of the surface roughness layer increased in case of Sample B2 in respect to Sample A2 and A3. The explanation could be the different nucleation mechanisms depending on the stoichiometry of the bottom silicon nitride layer.

5. Conclusions

In this work low-pressure chemical vapour deposited SiN_x/nc-Si/SiN_x and Si_3N_4/nc-Si/Si_3N_4 heterostructures on Si wafers were investigated as a function of deposition time. A multilayer optical model based on the EMA model was developed to determine the layer thickness, compositions and homogeneity. Systematic dependence have been found in the middle layer thickness as a function of deposition time. It was obtained that the deposited nc-Si layer thickness depended on the stoichiometry of the underlying silicon nitride layer.

Acknowledgements

This work was partially supported by the European Commission through project SEMINANO (Contract NMP4-CT-2004-505285) and by the (Hungarian) National Scientific Research Fund (OTKA) under Grant No. T048696, T047011 and K61725.

References

[1]   J. U. Schmidt, B. Schmidt, *Materials Science and Engineering B,* vol. 101, pp. 28-33, 2003.

[2]   D. Pacifici, A. Irrera, G. Franzo, M. Miritello, F. Iacona, F. Priolo, *Physica E,* vol. 16, pp. 331-340, 2003.

[3]   A. Irrera, M. Miritello, D. Pacifici, G. Franzo, F. Iacona, F. Priolo, D. Sanfilippo, G. Di Stefano, P. G. Fallica, *Nuclear Instruments and Methods in Physics Research B,* vol. 216, pp. 222–227, 2004.

[4]   P. Basa, Zs. J. Horváth, T. Jászi, A. E. Pap, L. Dobos, B. Pécz, L. Tóth, and P. Szöllősi, *Physica E,* accepted, 2006.

[5]   P. Basa, P. Petrik, M. Fried, L. Dobos, B. Pécz, L. Tóth, *Physica E,* accepted, 2006.

[6]   T. Baron, F. Mazen, J.M. Hartmann, P. Mur, R.A. Puglisi, S. Lombardo, G. Ammendola, C. Gerardi, *Solid-State Electronics,* vol. 48, pp. 1503-1509, 2004.

[7]   R.A. Rao, R.F. Steimle, M. Sadd, C.T. Swift, B. Hradsky, S. Straub, T. Merchant, M. Stoker, S.G.H. Anderson, M. Rossow, J. Yater, B. Acred, K. Harber, E.J. Prinz, B.E. White Jur., R. Muralidhar, *Solid-State Electronics,* vol. 48, pp. 1463-1473, 2004.

[8]   T. Lohner, N. Q. Khánh, and Zs. Zolnai, "Spectroellipsometric characterization of ion implanted semiconductors and porous silicon", *Acta Physica Slovaca,* vol. 48, pp. 441–450, 1998.

[9]   M. Serényi, M. Rácz, T. Lohner, "Refractive index of sputtered silicon oxynitride layers for antireflection coating", *Vacuum,* vol. 61, 245–249, 2001.

[10] P. Petrik, T. Lohner, M. Fried, L. P. Biro, N. Q. Khanh, J. Gyulai, W. Lehnert and C. Schneider, H. Ryssel, "Ellipsometric study of polycrystalline silicon films prepared by low-pressure chemical vapor deposition", *Journal of Applied Physics,* vol. 87, no. 4, pp. 1734–1742, 2000.

[11] G. E. Jellison, Jr., M. F. Chrisholm, S. M. Gorbatkin, *Applied Physics Letters,* vol. 62, p. 3348, 1993.

Annealing behaviour of low temperature grown GaAs investigated by SIMS

A. Vincze[1], J. Kováč[1,2], R. Srnánek[2]

[1]International Laser Centre, Ilkovicova 3, 812 19 Bratislava, Slovak Republic
[2]Department of Microelectronics, Slovak University of Technology Bratislava, Ilkovicova 3,
812 19 Bratislava, Slovak Republic
e-mail: vincze@ilc.sk

Epitaxial growth of LT GaAs at low temperatures of around 200°C with an excess of a group V element leads to a non-stoichiometric layer properties. The electrical and optical parameters of such a layer are significantly changed and strongly depend on the post growth annealing conditions. To find the optimal annealing temperature of LT GaAs layers subsequent Secondary Ion Mass Spectroscopy (SIMS) depth profile was investigated. The contaminants in a form of C and O are determined, which alter due to temperature treatments the SIMS depth profile. Segregations of C and O near surface regions as a function of annealing of the LT GaAs structures were observed. The heat treatment causes C concentration out-diffusion in the dependence of annealing temperatures. It was confirmed that these two elements belong to the dominant dopants in the LT GaAs layers.

1. Introduction

The GaAs grown at low temperatures and its unique properties has proven to be a useful material, both as insulation layer and as active layer in the field of fast photoconductive switches. Low Temperature grown GaAs (LT GaAs) at high electrical bias is used in optical heterodyne conversion or photo mixing that generates continuous wave radiation at THz frequencies. [1-2]. The highest continuous wave output power measured in any photo mixer has been approximately 1μW in the THz region [3]. Growth at low temperatures of around 200°C with an excess of a group V element leads to a non-stoichiometric layer, which alters the electrical and optical characteristics significantly. Moreover the characteristics depend on the post growth annealing conditions. The layer resistivity of LT GaAs and the photo generated carrier lifetime typically 150-350 fs were measured by different methods [4]. These parameters depend also upon annealing process after the growth of LT GaAs layers. Studies of LT GaAs layers indicate the presence of numerous point defects in the material, such as antisite defects, gallium vacancies, and As interstitials [5]. It has been estimated that LT GaAs contains as much as 1-2% of arsenic-antisite defects (As_{Ga}) [5]. Carriers generated by absorption of optical phonons in the LT GaAs exist for a period of time before recombining, typically around 300 fs. These carriers contribute to a photocurrent in the presence of electric field [3]. For LT GaAs, it is known that the dominant conduction mechanism is a hopping process between deep-level defect states (As_{Ga}), which have a concentration of up to 10^{20} cm^{-3} in the unannealed condition. The main carrier transport at the metal/semiconductor interface is therefore also via electrons in deep-level defects, rather than conduction band electrons. The resulting barrier at the interface for the former electrons is low, allowing a thermionic emission at room temperature and leading to non-alloyed ohmic contacts. For nonstoichiometric LT GaAs, it is known that during annealing at sufficient high temperatures, metallic As-precipitates are formed, which reduces the number of As_{Ga} point defects resulting in a higher resistance. Additionally, the formation of metallic precipitates results in depletion regions surrounding the clusters, which will further support a high resistance. For LT GaAs, the resistance is increased for a higher annealing temperature [6].

1-4244-0396-0/06/$25.00 ©2006 IEEE

Several characterization methods were employed to investigate LT GaAs as a basic material for the Auston switches [7]. Secondary Ion Mass Spectroscopy (SIMS) is widely used to study the surface and depth profile from the chemical point of view of the solids and semiconductors. Another approach is the post annealing investigation of the LT GaAs surface of the grown layers. In this paper the influence of the annealing conditions on the LT GaAs properties in accordance with SIMS analysis of the structures is reported.

2. Experimental

The structures were grown on semi-insulating (001) GaAs substrates in a Riber 32 MBE system equipped with a valve cracker cell for As, which enables on-demand generation of either As_2 or As_4 molecules at an extrapolated substrate temperature with a V/III beam equivalent pressure of 2.5, calibrated by a RHEED pattern [6].

Different LT GaAs samples were investigated after annealing under the same conditions in one run for one temperature. For comparing the results one as grown sample was proven. Parts of the 2 inch wafers were *ex situ* annealed at 550°C, 600°C and 650°C for 10 min in a nitrogen atmosphere using rapid thermal annealing (RTA). During the annealing process, a sample sandwich is used to minimize group V out-diffusion from the surface. Due to this sample geometry, the wafer surface remained mirror-like after annealing [6].

The structure of the sample A consists of 600 nm thick epitaxial LT GaAs layer (T_s= 185°C) grown on 400 nm thick AlGaAs and 40 nm GaAs buffer layer at normal growth temperature. The structure of sample B is the same, except the lower growth temperature of LT GaAs T_s= 135°C. Sample C1 differ from the previous two samples in its structure. The sample doesn't contain AlGaAs layer and the LT GaAs layer is 1350 nm thick grown at T_s= 135°C.

After the electrical and optical characterisation of LT GaAs layers in dependence of the annealing temperature [8] SIMS analyse was performed. Time of flight secondary ion mass spectrometry using Ion-TOF SIMS instrument with high energy Au^+ primary source was employed for LT GaAs structures composition profile determination. High energy pulsed primary source (25 keV) was combined with low energy sputter gun at 1 keV (Cs^+) in 45° to sample surface because of low erosion rate of analysing gun. Sputtering ion beam was raster over 300x300 μm^2 area while the primary beam acquired signal within 80x80 μm^2 area in centre of the sputtered area.

SIMS profiling was done using Cs^+ ion sputtering, while using of Cs^+ ions enhance the ionisation probability of negative charged secondary ions. The SIMS profile of the structure was prepared for the chemical composition investigations. From the profile the contamination of C and O elements in the layers were examined. The sputtered crater depth was measured by using Tallystep (Taylor Hobson). The results are comparable with AFM image measurements and growth data. Similar measurements were accomplished via Raman spectroscopy also [8, 9].

3. Results

The SIMS profile of main matrix signals Ga, Al, As, C and O of sample A after the annealing at 600°C is shown in Fig.1. The Al containing layers are bonding the C and O, which is preferably building into the AlGaAs layer and the concentration in the LT GaAs is lowering. The AlGaAs layer in the interface contains higher concentration of C and O. From the comparison of SIMS profiles for sample set A the concentration of C and O decreases with annealing (not shown here). All other concentrations of for Al, Ga a As concentrations remains in the same level. After the annealing process the concentration of the C and O elements shows the redistribution and decreases in the concentration. The recombination time, measured by optical time domain reflectometry for this sample has been determined at ~350 fs. This value of carrier lifetime proposed to use this sample as a basic material for THz sources fabrication.

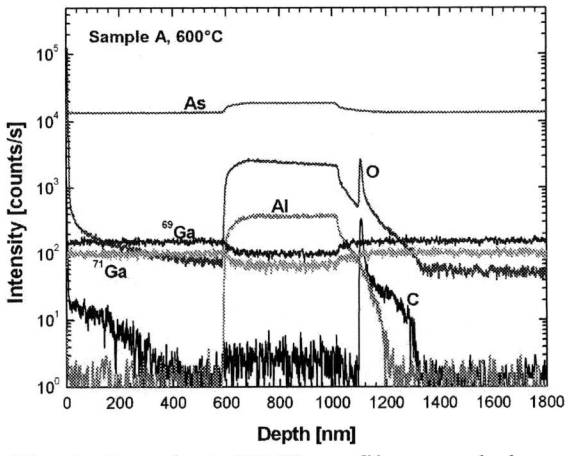

Fig. 1. Sample A SIMS profile annealed at 600°C

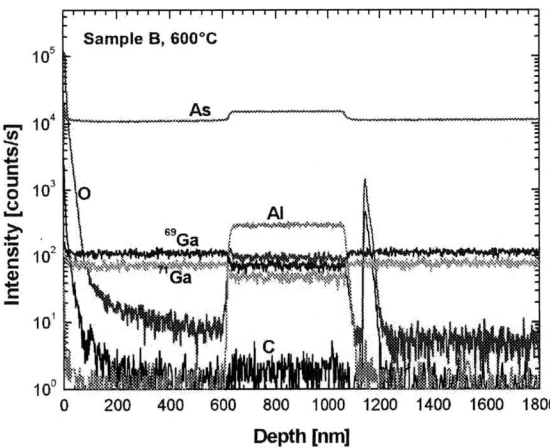

Fig. 2. SIMS profile of the structure B after the annealing at 600°C

In the as grown SIMS profile of sample B the interfaces between GaAs buffer/AlGaAs and AlGaAs/LT GaAs due to low growth temperature were not so expressive and the concentration of C and O at the surface region shows increased accumulation. For the annealed sample at 550°C the concentration of C and O under the surface shows only small changes. On the other hand the annealing at 600°C, Fig. 2 results in a sharp interface AlGaAs/GaAs due to out diffusion of C and O impurities. The higher annealing temperature at 650°C does not result is concentration lowering for C and O elements.

The SIMS profiles of sample C1 revealed the influence of long growth process to absorption of C and O elements in LT GaAs layers. Due to annealing process the adsorbed C and O diffuse out toward the sample surface. Comparison of C and O concentrations for the sample C1 as a function of annealing temperatures is showed in Fig. 3 and 4. The lowest depth of C concentration 400 nm was found for the annealed sample at 600°C and 700 nm for O element concentration. These values changed due to annealing process.

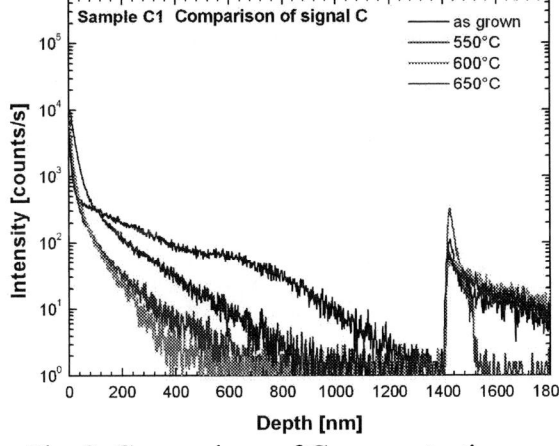

Fig. 3. Comparison of C concentration in the sample C1 as a function of annealing temperatures

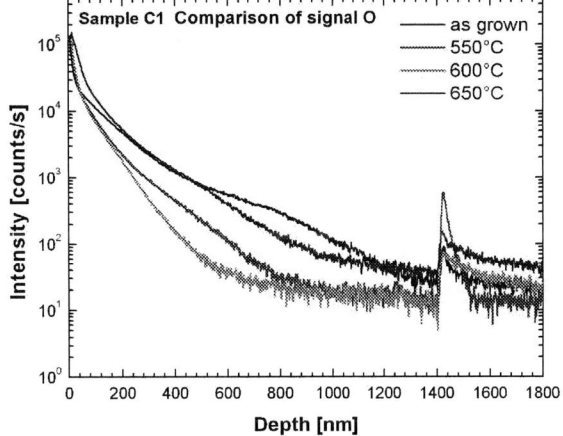

Fig. 4. Comparison of O concentration in the sample C1 as a function of annealing temperatures

The change of the resistivity before and after the annealing process is summarized in Table 1. The measurements of the sheet resistance R_s for sample A shows the lowest value for annealing at 600°C. The sheet resistance dependence shows similar behaviour as the results from Raman spectroscopy and roughness characterisation by using AFM [10, 11] with minimum value at 600°C.

	Sample A	Sample C1
Annealed at [°C]	Sheet resistance R_s [Ω]	Sheet resistance R_s [Ω]
NA	$8.53 \ 10^9$	$5.83 \ 10^8$
550	$6.48 \ 10^9$	$3.04 \ 10^9$
600	$2.32 \ 10^8$	$1.33 \ 10^8$
650	$8.46 \ 10^9$	$1.5 \ 10^{10}$

Table 1. Sheet resistance R_s for Sample A and C1

4. Conclusion

The structural and chemical properties were evaluated in the terms of impurities C and O for LT GaAs samples after annealing. By SIMS surface analysis the segregation of As and Ga elements as well as the concentration distribution of C and O in the structures was revealed. The results show segregation of C and O near surface regions and their out diffusion due to annealing. It was confirmed that this two elements belong to the dominating dopants in the LT GaAs layers. The annealing of the LT GaAs structures causes C concentration out-diffusion profile in the dependence of the annealing temperatures where a lowest depth of 400 nm was revealed for 600°C annealing temperature of the structure C1. From the carrier lifetime point of view the sample A shows the shortest response ~350 fs and was proposed to use as a basic material for THz sources fabrication

Acknowledgements

The authors would like to thank J. Sigmund and H. L. Hartnagel, TU Darmstadt, for kindly providing the samples. This work was partially supported by the Scherer foundation TU Darmstadt, Deutsche Forschungsgemeinschaft (DFG HA 1132/36-4), Germany and by the Slovak Grant Agency (VEGA) Grants No. 1/3076/06, 1/3108/06 and 1/3111/06.

References

[1] E. R. Brown, F. W. Smith and K.A. McIntosh, Journal of Applied Physics, Vol 73, pp. 1480, 1993
[2] K. A. McIntosh, E. R. Brown, K. B. Nicols and O. B. McMahon, Applied Physics Letters, Vol 67, pp.3844, 1995
[3] R. Mendis, C. Sydlo, J. Sigmund, M. Feiginov, P. Meissner, H. L. Hartnagel, Solid state electronics 58, pp.2041-2045, 2004
[4] H. Abe, H. Harima *et al.*, Japan Journal of Applied Physics 35, part 1, No. 12A, pp. 5955-5563, 1996
[5] D. C. Look, Thin Solid Films 231, pp. 61-73, 1993
[6] J. Sigmund, H. L. Hartnagel, Journal of Crystal growth 278, pp. 209-213, 2005
[7] D. H. Auston, K. P. Cheung and P.R. Smith, Applied Physics Letters, 45 (3), pp.284, 1984
[8] A. Vincze, R. Srnanek, J. Sigmund, J. Kovac, H. L Hartnagel, G. Irmer, in *Proceedings of the ASDAM`04 Conference*, Smolenice, Slovakia, pp.115-118, ISBN 0-7803-8535-7
[9] A. Vincze, R. Srnanek, J. Sigmund, J. Kovac, H.L. Hartnagel, in *Proceedings of the WOCSDICE`04 Conference*, pp.141-142, ISBN 80-227-2050-X
[10] A. Vincze, M. Michalka, J. Podskocova, J. Sigmund, J. Kovac, H. L. Hartnagel, in *Proceedings of the APCOM`04 Conference*, pp. 282-286, ISBN 80-227-2073-9
[11] A. Vincze, Epitaxial growth and characterisation of GaAs and LT GaAs layers prepared by MBE, Dissertation thesis, 2006

Evaluation of parameters of Schottky junctions with large excess currents

Zs. J. Horváth

(The contribution was not been delivered in time)

Deposition of AZ5214-E Layers
on Non-planar Substrates with a "Draping" Technique

P. Eliáš [1], D. Gregušová [1], P. Štrichovanec [1], I. Kostič [2], and J. Novák [1]

[1] Institute of Electrical Engineering, Slovak Academy of Sciences,
Dúbravská cesta 9, 841 04 Bratislava, Slovak Republic
[2] Institute of Informatics, Slovak Academy of Sciences,
Dúbravská cesta 9, 845 07 Bratislava, Slovak Republic
e-mail: elekelia@savba.sk

A draping technique was tested to deposit AZ5214-E resist on non-planar (100)-oriented III-V substrates that contained various three-dimensional topographies. In each draping experiment, an AZ5214-E sheet was: (1) formed floating on the water surface, (2) lowered onto a non-planar substrate, and (3) draped over it during drying. Self-sustained and conformal AZ5214-E layers were formed over the non-planar substrates depending on drying temperature. Interactions between water and AZ5214-E can result in the depression of the glass transition temperature T_g of AZ5214-E material during drying. Hence, an AZ5214-E sheet that was formed glassy can become rubbery. At room temperature $T < T_g$, the sheet is glassy, and it can form a self-sustained or bridging layer over a 3D topography. By contrast, at $T \approx$ or $> T_g$, the sheet becomes rubbery and mouldable by adhesion and capillary forces. As the result, it can contour a 3D topography.

1. Introduction

A comprehensive approach to 3D device processing necessitates the coating of various 3D topographies with uniform resist layers that are either conformal or planarizing or bridging. However, resist deposition on 3D surfaces is a challenge. Spin-coating techniques can coat such surfaces uniformly and conformally only with difficulty due to liquid phenomena combined with centrifugal forces [1]. Ultrasonic spray-coating is more promising, as resist is deposited from aerosols onto non-planar substrates spun at low speed [2]. Other techniques are also used, such as meniscus coating or extrusion coating, electro-deposition [3], and plasma deposition.

Zhou et al. used a "float-coating" method to deposit polymethyl methacrylate resist from the water surface on non-planar substrates to realize a variety of sensors on AFM tip apexes [4, 5]. We used their approach to deposit positive-tone AZ5214-E photo-resist onto non-planar III-V semiconductor substrates ("draping" technique) to transfer device topologies into various facets of micromachined objects in (100) InP [6] and to process vector magnetic field sensors based on (100) GaAs [7]. A preliminary study of the draping technique was presented recently [8].

This contribution shows the draping technique can be used to deposit conformal and bridging AZ5214-E layers onto non-planar III-V semiconductor substrates.

2. Experiment

The experiments were carried out using positive-tone AZ5214-E resist that is a mixture of solid (28.3%) [9] and liquid substances. The solid constituents are a phenol-formaldehyde polymer resin (novolac) and a diazonaphthoquinone photo-active compound (DNQ) (Fig. 1). DNQ, blended into the novolac polymer matrix, enables or inhibits the dissolution of the matrix in aqueous base solutions if irradiated or not irradiated by ultraviolet light, respectively [10]. The main liquid constituent is the PGMEA solvent, i.e. propylene glycol monomethyl ether-1,2-acetate. AZ5214-E also contains nominally 0.5 % of water [9].

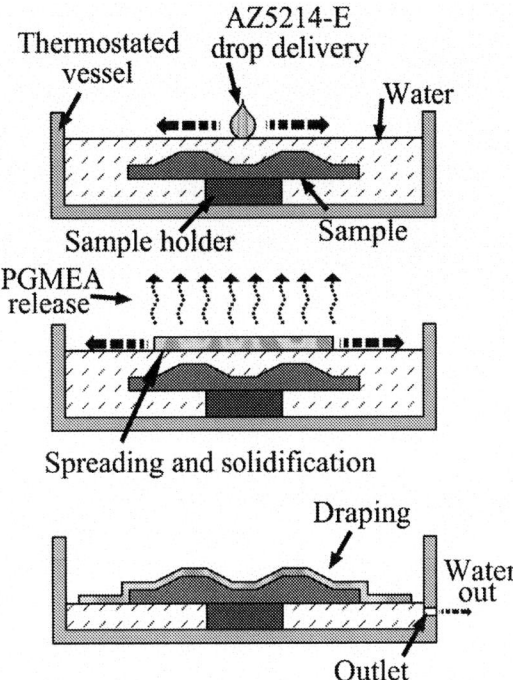

Fig. 1 Components of AZ5214-E.

AZ5214-E was deposited onto non-planar semi-insulating (100) GaAs and InP substrates prepared by etching via Ti mask in xH$_3$PO$_4$:yH$_2$O$_2$:zH$_2$O and via InGaAs mask in 3HCl:1H$_3$PO$_4$, respectively.

In each experiment, an AZ5214-E layer was: (1) formed floating on water, (2) lowered onto a non-planar substrate, and (3) draped over it during drying. Fig. 2 shows a sketch of the set-up used to realize steps (1) and (2). A sample was placed on a holder submerged in distilled water in a temperature-controlled vessel at 20 ± 0.2 °C. A plastic frame (34.5 × 34.5 mm²) was used to define the working area on the water surface. The sample was positioned 2 mm below the working area near one of its corners. An AZ5214-E drop was delivered using a Hamilton syringe with a 0.1 µl precision onto the water surface within the working area. Once the drop contacted the water surface, it detached via adhesive forces from the needle, spilled onto the water surface, and formed a thin, floating layer that assumed the shape of a quarter-circle centred about the spot at which the drop contacted the water surface. Within the quarter-circle areas, the thickness of the AZ5214-E layers followed a linear dependence versus drop volume [8]. When a sheet was lowered onto a substrate during step (2), the substrate was positioned under the quarter-circle area. Using this approach, AZ5214-E layers of controllable thickness and acceptable thickness uniformity can be formed over an area of several square centimetres [8]. The layer was then immediately lowered onto the sample at ~ 0.5 mm.s^{-1} by letting the water out from the vessel. Within one minute, the sample was transferred into an oven to be dried.

Fig. 2 Draping technique: steps (1) and (2).

3. Results and discussion

Fig. 3 shows a SEM image of a cleavage of a ~ 21-μm-high mesa confined at the sides to facets tilted to (100) at 56° at the top and at 36° at the bottom. It was revealed in (100) GaAs in $1H_3PO_4:1H_2O_2:4H_2O$ at ~ 2 °C during 20 min. The mesa was coated conformally in a ~ 0.8-μm-thick AZ5214-E layer formed from ~ 4-μl-sized drop and soft-baked at 92 ± 2 °C for 5 min. Fig. 4 (a), (b) shows SEM images of ~ 16.4-μm-high V-grooves formed in (100) InP by etching in $3HCl:1H_3PO_4$ at 16 ± 0.1 °C. The V-grooves were confined at the sides to ~ 35°-tilted {211}A-related facets. The thinner layer (Fig. 4 (a)) was more conformal to the V-grooved substrate compared with the thicker one (Fig. 4 (b). One can expect the conformal resist coverage of such V-grooves is possible if the thickness of the resist layer is adjusted to their geometry. Fig. 5 (a)-(d) shows SEM images of the InP substrate (Fig. 4) coated with a ~ 3.5-μm-thick AZ5214-E layer formed from ~ 6-μl-sized drop. The layer was dried in the air at room temperature (Fig. 5 (a), (b)), soft-baked at 92 ± 2 °C during 60 s (Fig. 5 (c)) and during 120 s (Fig. 5 (d)). The layer attached partly and fully to the grooves when soft-baked for 60 s and 120 s, respectively. If it was dried at room temperature, it attached itself less to the substrate (Fig. 5 (b)) or formed bridges over the grooves (Fig. 5 (a)).

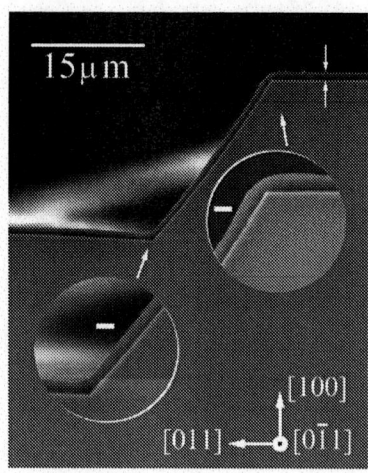

Fig. 3 ~ 0.8-μm-thick AZ5214-E layer on ~ 21-μm-high GaAs mesa.

The drying of AZ5214-E sheets deposited on non-planar substrates from the water surface crucially influenced the draping process. A sheet was more likely to attach itself fully to a non-planar substrate, copying or even planarizing its 3D topography, the higher the temperature and the longer the duration of the soft-bake process. Conversely, the lower the temperature and the shorter the duration of the soft-bake, the sheet was more likely to form shortcuts, bridges, or even a planar sheet over the topography.

The technique is based on the formation of glassy AZ5214-E sheets on the water surface and on the depression of their glass transition temperature T_g at elevated temperatures during drying. In step (1), a floating AZ5214-E sheet is formed on the water surface from a drop of AZ5214-E because of surface tension and evaporation phenomena. Upon contact with the water, the drop spills immediately onto the surface because the surface tension of water is larger than that of AZ5214-E. The spreading is mediated by interactions between PGMEA and water molecules because PGMEA can form hydrogen bonds to water via its ester endings [8]. PGMEA is concurrently evaporating from the spreading sheet, which increases the internal viscosity of AZ5214-E so much that the sheet stops spreading, assumes visco-elastic properties, and becomes glassy. However, the glassy sheet can be made rubbery via the depression of its glass transition temperature T_g, which occurs at elevated temperatures during

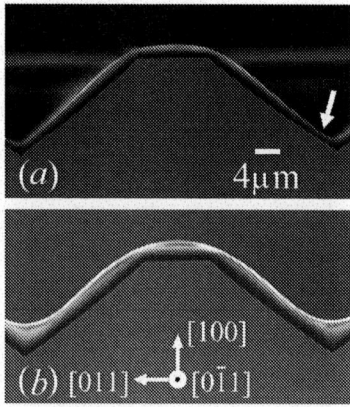

Fig. 4 ~ 16.4-μm-high V-grooves in (100) InP coated with AZ5214-E layers of different thickness.

the drying stage. The transition is a temperature-activated process induced by water molecules permeating via the sheet during drying. While the glassy sheet is stiff enough to form bridges over non-planar substrates, the rubbery sheet can conform to such substrates under the action of capillary and adhesion forces. The T_g of DNQ/novolac photo-resists lies between 110 and 125 °C [10]. This experiment suggests the glassy-to-rubbery transition in thin AZ5214-E sheets permeated by water molecules occurred at lower temperatures.

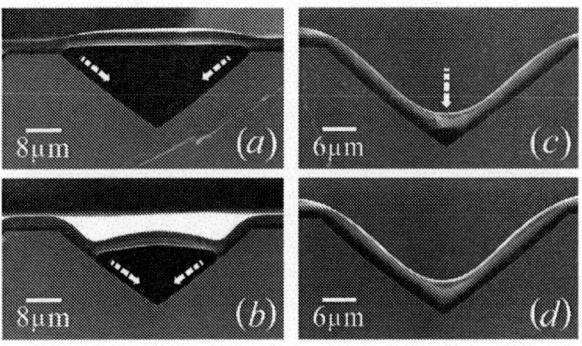

Fig. 5 ~ 16.4-μm-high V-grooves in (100) InP coated with a ~ 3.5-μm-thick AZ5214-E layer dried at room temperature (*a,b*) and at 92 ± 2 °C during 60 s (*c*) and 120 s (*d*).

4. Conclusion

A draping technique was used to deposit AZ5214-E sheets from the water surface onto non-planar III-V substrates. Depending on drying temperature, the sheets attached themselves to the substrates nearly conformally or formed bridging, self-sustained layers over the substrates. The technique is based on the formation of glassy AZ5214-E sheets on the water surface that can become rubbery as their glass transition temperature becomes lowered at elevated temperatures. The draping technique is very promising for three-dimensional device processing.

Acknowledgement

This research was sponsored under projects APVV-51-045705, VEGA Agency 2/6096/26, and VEGA Agency 2/6184/26.

References

[1] A. Ramam and S. Chua , *J. Electrochem. Soc.* **141**, 576, 1994.
[2] N. P. Pham, J. N. Burghartz, P. M. Sarro, *J. Micromech. Microeng.* **15**, 691, 2005.
[3] R. Schnupp, R. Baumgärtner, R. Künhold, H. Ryssel, *Sens. Actuators* A **85**, 310, 2000.
[4] H. Zhou, G. Mills, B. K. Chong, A. Midha, L. Donaldson and J. M. R. Weaver, *J. Vac. Sci. Technol.* A **17**, 2233, 1999.
[5] H. Zhou, B. K. Chong, P. Stopford, G. Mills, A. Midha, L. Donaldson and J. M. R. Weaver, *J. Vac. Sci. Technol.* B **18**, 3594, 2000.
[6] P. Eliáš, Š. Haščík, J. Martaus, I. Kostič, J. Šoltýs, I. Hotový, *Electrochem. Sol.-St. Lett.* **9**, G27, 2006.
[7] D. Gregušová, P. Eliáš, Z. Oszi, R. Kúdela, J. Šoltýs, J. Fedor, V. Cambel, I. Kostič, *Microel. J*, in press.
[8] P. Eliáš, D. Gregušová, J. Martaus, and I. Kostič, *J. Micromech. Microeng.* **16**, 191, 2006.
[9] AZ Product Data Sheets on AZ5214-E.
[10] R. Dammel, *Diazonaphthoguinone-based resists* **TT 11**, SPIE Optical Engineering Press Bellingham, Washington, 1993.

Author Index

A

Absil, P.	7
Ác, V.	278
Anwarzai, B.	278
Aoulaiche, M.	60
Atanassova, E.	170
atka, A. -	56
Augendre, E.	7

B

B., Barda	270
Balalykin, N. I.	254
Ballo, P.	238
Basa, P.	118, 286
Bedyk, W.	204
Benkovic, M.	238
Benkovská, J.	106
Beno, Peter	180
Biesemans, S.	7
Blanco, F.	98
Boeykens, S.	246
Borkowska, A.	258
Boura, A.	90
Boura, Adam	86
Bruncko, J.	162
Búc, D.	106
Buca, D.M.	48

C

Cambel, V.	11, 174
Cané, C.	98, 220
Cané, Carles	188
Capek, I.	266
Caymax, M.	48
Chiarella, T.	7
Chitu, L.	266
Chovan, J.	38
Chvala, Ales	180
Cico, K.	110, 166
Correig, X.	98

D

De Keersgieter, A.	7
Degraeve, R.	60
Dobrocka, E.	126, 242
Domaradzki, J.	258, 262
Donoval, D.	56, 162
Donoval, Daniel	180
Dubecky, F.	52, 74, 130, 250
Dubecky, Frantisek	134

E

Eliái, P.	294
Espinos, J. P.	216

F

Faes, A.	204
Falepin, A.	7
Ferraria, C.	52
Figueras, E.	98, 220
Figueras, Eduard	188
Flickyngerova, S.	82
Florovic, M.	38
Foit, J.	234
Fonseca, L.	98, 220
Fonseca, Luís	188
Fox, A.	15, 146, 158
Franchi, S.	154
Franta, M.	242
Fried, M.	286
Frigeri, P.	154
Frollo, I.	130
Fröhlich, K.	110, 126, 216, 242

G

García, Isabel	188
Gendt, S. De	60
Georgakilas, A.	110
Gombia, E.	154
Goryll, M.	48
Gràcia, I.	98, 220
Gregu, J.	78
Greguiová, D.	11, 110, 166, 174, 294
Grobelny, D.	78
Groeseneken, G.	60
Grym, J.	70

H

Hardy, V.	118
Harmatha, L.	82, 106, 238
Hascik, S.	266
Hascik,	65
Hashizume, Tamotsu	138
Hašík, Š.	200
Hasko, D.	162
Hazdra, Pavel	114
Heidelberger, G.	15, 146, 158
Heigl, A.	30
Hlinka, B.	106
Hoffmann, T.	7
Horák, Michal	274
Horváth, Zs. J.	69, 118, 293
Hotový, I.	200, 254
Houssa, M.	60
Hrkut, P.	266
Hubik, P.	246
Huieková, K.	126, 216
Huran, J.	254
Husak, M.	90
Husák, Miroslav	86, 102

Author Index

I

I., Vávra 150
Irmer, G. 42, 56, 250
Ivanov, P. 98, 220

J

J., Novák 150
Jakabovic, J. 38, 90, 162, 208, 238
Jakubek, J. 130
Janicek, V. 234
Janos, L. 162
Jászi, T. 118
John, J. 52, 250
Juras, J. 130
Jurczak, M. 7

K

Kaczmarek, D. 258, 262
Kaiser, A. B. 19
Kaluza, N. 158
Kaneko, Masamitsu 138
Kapels, H. 184
Kauerauf, T. 60
Kerner, C. 7
Kim, Insung- 122
Kindl, D. 246
Kobzev, A.P. 254
Komarnitskyy, Volodymyr 114
Kordoi, P. 15, 26, 146, 158, 166
Kostic, I. 11, 65, 266, 294
Kotani, Junji 138
Kovác, J. 38, 56, 162, 208, 228, 289
Krc, J. 224
Kristofik, J. 246
Kubicek, S. 7
Kudela, P. 162
Kudela, R. 11, 174, 250
Kudelab, R. 52
Kuzmik, J. 110
Kytka, M. 208

L

Ladziansky, Milan 134
Lalinsky, T. 65, 78, 110, 200, 266
Lengyel, O. 208
Leys, M. R. 246
Liday, J. 65
Loo, R. 48
Luby, S. 266
Luby, S. 278
Lupták, R. 216
Lüth, H. 146, 158

M

Majer, L. 196
Majkova, E. 266, 278

M

Mantl, S. 48
Mare, J. J. 246
Marek, Juraj 180
Margesin, B. 204
Marman, P. 196
Marso, M. 15, 26, 146, 158
Martaus, J. 174
Matay, L. 266
Matsuo, Kazushi 138
Michael, E. A. 26
Michalka, M. 162
Mikulics, M. 26
Min, Bokki- 122
Mir, S. 78
Miroslav, Husak 192
Moers, J. 48
Mosca, R. 154
Motta, A. 154
Mozolova, Z. 65, 78
Mudron, J. 130
Mullerova, J. 82

N

Nagy, K. 118
Narducci, Margarita S. 188
Nasi, L. 154
Necas, Vladimir 134
Necasc, V. 74
Niessner, M. 204
Nigrovicova, M. 82, 94
Noda, T. 7
Novái, J. 166
Novak, J. 234, 294
Novotny, I. 82, 162, 228, 250

O

Osvald, J. 282
Oszi, Zs. 78

P

P., Machac 270
P., Sajdl 270
P., Strichovanec 150
Pantisano, L. 60
Pap, A. E. 118
Paskaleva, A. 170
Pawlak, B. 7
Permthammasin, K. 184
Peternai, L. 208
Petrik, P. 286
Pezoltd, J. 254
Písecný, P. 238
Podör, B. 22, 69
Pogany, D. 110
Prezioso, M. 154
Pribil, J. 130
Procházková, O. 70, 74

Author Index

R

R., Kúdela 150
Radziewicz, D. 38
Rakovics, V. 69
Reháková, A. 200
Reményi, G. 22
Roeckerath, M. 158
Romanus, H. 200
Roozeboom, F. 126
Rosová, A. 242
Roth, S. 19
Roth, Siegmar 1
Rubio, R. 220
Rubio, R. 98
Rufer, L. 78

S

Sabaté, N. 98, 220
Sagatova-Perdochova, Andrea 134
Santander, J. 98, 220
Satka, A. 11, 266
Schmitt, M. 184
Schrag, G. 204
Schubert, J. 158
Sciana, B. 38
Senderak, R. 266
Seravalli, L. 154
Severi, S. 7
Shtereva, K. 228
Siu, G. G. 106
Skákalová, V. 19
Skakalova, Viera 1
Skriniarova, J. 162
Soltys, J. 11, 174
Song, Jaesung- 122
Spiess, L. 200
Srnánek, R. 38, 56, 150, 228, 250, 289
Srnanekd, R. 52
Stancek, S. 26
Stefaniak, M. 158
Steins, R. 158
Stoklas, R. 166
Stopjaková, V. 196
Strichovanec, P. 294
Stuchlíková, L. 106, 238
Suchánek, Pavel 102
Sutta, P. 82, 228
sVincze, A. 162
Szöllosi, P. 118

T

Tapajna, M. 126, 216, 238, 242
Tengeri, D. 200
Tlaczala, M. 38
Trevisi, G. 154
Tvarožek, V. 82, 94, 196, 228

U

Uherek, F. 65

V

Vanko, G. 65, 78
Vavrinský, E. 196
Vavrinsky, E. 82
Veháek, V. 200
Víglaský, R. 94
Vilanova, X. 98
Vincze, A. 56, 65, 250, 289
Vladimir, Janicek 192
Vogrincic, P. 65
Woo, Y.-S. 19
Voves, J. 34
Vrbicky, Andrej 180
Vyborny, Z. 246

V

Wachutka, G. 30, 184, 204
Wachutka, Gerhard 178
Wang, Jiangling 1
Weis, M. 94, 196

Z

Záluský, R. 250
Zatko, B. 130
Zatko, Bohumir 134
Zatkoa, B. 74
Zavadil, J. 70
Zdánský, K. 70
Zeman, M. 224

9781424403967